AF333274

BRAIDS

Introductory Lectures on Braids, Configurations
and Their Applications

Lecture Notes Series, Institute for Mathematical Sciences, National University of Singapore Vol. 19

BRAIDS

Introductory Lectures on Braids, Configurations and Their Applications

Editors

A. Jon Berrick
National University of Singapore, Singapore

Frederick R. Cohen
University of Rochester, USA

Elizabeth Hanbury
National University of Singapore, Singapore and Durham University, UK

Yan-Loi Wong
Jie Wu
National University of Singapore, Singapore

World Scientific

NEW JERSEY · LONDON · SINGAPORE · BEIJING · SHANGHAI · HONG KONG · TAIPEI · CHENNAI

Published by

World Scientific Publishing Co. Pte. Ltd.

5 Toh Tuck Link, Singapore 596224

USA office: 27 Warren Street, Suite 401-402, Hackensack, NJ 07601

UK office: 57 Shelton Street, Covent Garden, London WC2H 9HE

British Library Cataloguing-in-Publication Data
A catalogue record for this book is available from the British Library.

Lecture Notes Series, Institute for Mathematical Sciences, National University of Singapore — Vol. 19
BRAIDS
Introductory Lectures on Braids, Configurations and Their Applications

ISBN-13 978-981-4291-40-8
ISBN-10 981-4291-40-4

Printed in Singapore by Mainland Press Pte Ltd

CONTENTS

FOREWORD

The Institute for Mathematical Sciences at the National University of Singapore was established on 1 July 2000. Its mission is to foster mathematical research, both fundamental and multidisciplinary, particularly research that links mathematics to other disciplines, to nurture the growth of mathematical expertise among research scientists, to train talent for research in the mathematical sciences, and to serve as a platform for research interaction between the scientific community in Singapore and the wider international community.

The Institute organizes thematic programs which last from one month to six months. The theme or themes of a program will be chosen from areas at the forefront of current research in the mathematical sciences and their applications.

Generally, for each program there will be tutorial lectures followed by workshops at research level. Notes on these lectures are usually made available to the participants for their immediate benefit during the program. The main objective of the Institute's Lecture Notes Series is to bring these lectures to a wider audience. Occasionally, the Series may also include the proceedings of workshops and expository lectures organized by the Institute.

The World Scientific Publishing Company has kindly agreed to publish the Lecture Notes Series. This Volume, "Braids: Introductory Lectures on Braids, Configurations and Their Applications", is the nineteenth of this Series. We hope that through the regular publication of these lecture notes the Institute will achieve, in part, its objective of promoting research in the mathematical sciences and their applications.

June 2009

Louis H. Y. Chen
Ser Peow Tan
Series Editors

PREFACE

These lecture notes arose from the program *"Braids"* which was held from 14 May to 13 July 2007 at the National University of Singapore's Institute for Mathematical Sciences (IMS). The program's highlights included an international conference on braids, from 25 to 29 June 2007, and a summer school for three weeks (4 to 22 June) leading up to the conference. In addition, there were numerous informal seminars and a public lecture by Robert Ghrist.

The main theme of the program was the mathematical structure of the braid group, together with applications arising from this structure both within mathematics, and outside mathematics such as magnetohydrodynamics, robotics, cryptography and molecular biology. These diverse aspects of braids were reflected in the wide range of talks at the conference. At the same time, we were also delighted to have the opportunity to celebrate the 80th birthday of one of the great pioneers of the subject, Joan Birman.

The purpose of the summer school (the first of PRIMA, the Pacific Rim Institutes of Mathematics Association) was to introduce graduate students and others not necessarily familiar with braids to the topic, as preparation for more technical material in the conference itself. More than 30 students from Canada, China, Germany, Japan, Korea, Malaysia, Mexico, UK, USA and Singapore attended the PRIMA summer school. We were very fortunate to have some gifted expositors to provide a relatively gentle introduction to the topic. Their lectures form the basis of this book. Their work was complemented by talks from a number of graduate students, among whom we should mention those of E-Jay Ng, who gave a review of the algebraic topology needed for later lectures.

Dale Rolfsen's introduction set up notational conventions and showed how braids interweave geometry, topology and algebra, leading to group theory associated with the question of ordering the elements of the braid group. From there, there were two strands to the theoretical discussion. Jie Wu introduced the simplicial approach to algebraic topology, and showed how fruitful it can be in studying the topology of braid groups. On the other hand, Fred Cohen focused on braids as paths in configuration spaces, an idea that is rich in applications. One source of such applications is robotics, which was explored by Robert Ghrist in a way that showed that questions arising from applications can lead to interesting theoretical research. Mitch Berger's lectures covered a lot of territory — literally, as they demonstrated remarkable braiding phenomena in astrophysics. In a completely different vein, David Garber explored to what extent difficult problems in the group theory of braids can be harnessed to create new techniques in cryptography.

These were generally agreed to be excellent courses, as we are sure will be apparent from a reading of the chapters within. We thank them most warmly for the effort that they have put into the preparation of the chapters, and the patience with which they have dealt with all our editorial queries.

As well as the summer school lecturers, we would like to thank Joan Birman and Toshitake Kohno for help with the program's organization, and their part in ensuring such a high quality of participation. The conference organizers also thank IMS and its staff for providing such an idyllic setting for mathematical discussions and collaborations. Funding support from IMS and its director, Louis Chen, and from NUS Research Grant R-146-000-097-112, was vital to the success of the program. The cooperation of the NUS Department of Mathematics and the Singapore Mathematical Society is also gratefully acknowledged.

February 2009

A. Jon Berrick
National University of Singapore, Singapore

Frederick R. Cohen
University of Rochester, USA

Elizabeth Hanbury
National University of Singapore, Singapore and
Durham University, UK

Yan-Loi Wong
National University of Singapore, Singapore

Jie Wu
National University of Singapore, Singapore
Editors

TUTORIAL ON THE BRAID GROUPS

Dale Rolfsen

Department of Mathematics
University of British Columbia
Vancouver, BC, V6T 1Z2, Canada
E-mail: rolfsen@math.ubc.ca

This is an introduction to the theory of braids, one of the most beautiful examples of algebra and topology working hand-in-hand. The emphasis is on the variety of points of view from which one can view braids, and the information one gains from each aspect. As preparation for further lectures in a summer school and conference on the braid groups, these talks (and therefore this chapter) are intended for students with some basic background in topology and the theory of groups.

1. Introduction

The braid groups B_n were introduced by E. Artin eighty years ago [1], although their significance to mathematics was possibly realized a century earlier by Gauss, as evidenced by sketches of braids in his notebooks, and later in the nineteenth century by Hurwitz. The braid groups provide a very attractive blending of geometry and algebra, and have applications in a wide variety of areas of mathematics, physics, and recently in polymer chemistry and molecular biology.

Despite their ripe old age, and an enormous amount of attention paid to them by mathematicians and physicists, these groups still provide us with big surprises. The subject was given a major boost when V. Jones discovered, in the mid 1980's, new representations of the braid groups, which led to the famous Jones polynomial of knot theory. In this chapter, we will touch on that discovery, and more recent ones, regarding this wonderful family of groups, and raise a few questions about them which are still open. Along the way, we will also discuss some properties of these groups which deserve to be better known. Among our personal favorites is the orderability of braid groups, which implies that they have some special algebraic properties.

This discussion will not include many interesting aspects of braid theory, such as the conjugacy problem as solved by Garside, dynamics of braids, applications to cryptography and connections with homotopy groups of spheres. Many proofs will be omitted in this presentation. Some proofs will be left as exercises for the students, often with hints or references.

The emphasis here is that there are many ways of looking at braids, and each point of view provides new information about these groups. Though much of the motivation for studying braids comes from knot theory, we will be stressing the algebraic aspects of braid groups.

1.1. *Prerequisites*

Students will be assumed to have background in basic topology and group theory. No other specialized expertise is necessary.

1.2. *Suggested reading*

At the time of this writing, an excellent book on braid theory is in press; we predict it will become the "bible" of the theory:

- C. Kassell and V. Turaev, *Braids*, Springer, 2008.
  Following are some other useful introductions to the subject.
- Joan Birman and Tara Brendle, Braids, A Survey, in *Handbook of Knot Theory* (eds. W. Menasco and M. Thistlethwaite), Elsevier, 2005.
  Available electronically at:
  http://www.math.columbia.edu/~jb/Handbook-21.pdf.
- Joan Birman, *Braids, Links and Mapping Class Groups*, Annals of Mathematics Studies, Princeton University Press, 1974.
  This classic is still a useful reference.
- Gerhard Burde and Heiner Zieschang, *Knots*, De Gruyter Studies in Mathematics 5, second edition, 2003.
  This book on knots contains a very readable chapter on braid groups.
- Patrick Dehornoy, Ivan Dynnikov, Dale Rolfsen, and Bert Wiest, *Ordering Braids*, Mathematical Surveys and Monographs, Amer. Math. Soc. (2008).
  More than you ever wanted to know about braid ordering appears in this monograph.

- Vaughan F. R. Jones, *Subfactors and Knots*, CBMS Regional Conference series in Mathematics, Amer. Math. Soc., 1991.
  This gem focuses on the connection between operator algebras and knot theory, via the braid groups.
- Vassily Manturov, *Knot Theory*, Chapman & Hall, 2004.
- Kunio Murasugi and Bohdan Kurpita, *A Study of Braids*, Mathematics and Its Applications, Springer, 1999.
  Also see research papers and other books in the reference section.

2. Braids as Strings, Dances or Symbols

One of the interesting things about the braid groups B_n is they can be defined in so many ways, each providing a unique insight. We will briefly describe some of these definitions, punctuated by facts about B_n which are revealed from the various points of view. You can check the excellent references [1, 2, 5, 7, 11, 17] for more information.

2.1. *Definition 1: Braids as strings in 3-D*

This is the usual and visually appealing picture. An n-braid is a collection of n strings in (x, y, t)-space, which are disjoint and monotone in (say) the t direction.

We require that the endpoints of the strings are at fixed points, say the points $(k, 0, 0)$ and $(k, 0, 1)$, $k = 1, \ldots, n$. We also regard two braids to be equivalent (informally, we will say they are equal) if one can be deformed to the other through the family of braids, with endpoints fixed throughout the deformation.

Although most draw braids vertically, we prefer to view them horizontally with the t-axis running from left to right. The reason is that we do algebra (as in writing) from left to right, and the multiplication of braids is accomplished by concatenation (which is made more explicit in the second definition below), which we take to be in the same order as the product. In the vertical point of view, some view the product $\alpha\beta$ of two braids α and β with α above β, while others would put β on top of α. The horizontal convention eliminates this ambiguity.

There is, of course, a strong connection between braids and knots. A braid β defines a knot or link $\hat{\beta}$, its closure, by connecting the endpoints in a standard way, without introducing further interaction between the strings.

Equivalent braids give rise to equivalent links, but different braids may give rise to the same knot or link. We will discuss how to deal with this

Fig. 1.   The closure of a braid.

ambiguity later. B_n will denote the group formed by n-braids with concatenation as the product, which we will soon verify has the properties of a group, namely, an associative multiplication, with identity element and inverses.

2.2. *Definition 2: Braids as particle dances*

If we take t as a time variable, then a braid can be considered to be the time history of particles moving in the (x, y) plane, or if you prefer, the complex plane, with the usual notation $x + y\sqrt{-1}$.

This gives the view of a braid as a dance of noncolliding particles in $\mathbb{C}$, beginning and ending at the integer points $\{1, \ldots, n\}$. We think of the particles moving in trajectories

$$\beta(t) = (\beta_1(t), \ldots, \beta_n(t)), \quad \beta_i(t) \in \mathbb{C},$$

where t runs from 0 to 1, and $\beta_j(t) \neq \beta_k(t)$ when $j \neq k$. A *braid* is then such a time history, or dance, of noncolliding particles in the plane which end at the spots they began, but possibly permuted. Equivalence of braids roughly reflects the fact that choreography does not specify precise positions of the dancers, but rather their relative positions. However, notice that if particle j and $j + 1$ interchange places, rotating clockwise, and then do the same move, but counterclockwise, then their dance is equivalent to just standing still!

The product of braids can be regarded as one dance following the other, both at double speed. Formally, if α and β are braids, we define their product $\alpha\beta$ (in the notation of this definition) to be the braid which is $\alpha(2t)$ for $0 \le t \le 1/2$ and $\beta(2t-1)$ for $1/2 \le t \le 1$.

Exercise 2.1. Verify that the product is associative, up to equivalence. The identity braid is to stand still, and each dance has an inverse; doing the dance in reverse. More formally, if β is a braid, as defined in this section, define $\bar{\beta}(t) = \beta(1-t)$. Write a formula for the product $\gamma = \beta\bar{\beta}$ and verify that there is a continuous deformation from γ to the identity braid. [Hint: Let s be the deformation parameter, and consider the dance which is to perform β up till time $s/2$, then stand still until time $1 - s/2$, then do the dance $\bar{\beta}$ in the remaining time.]

A braid β defines a permutation $i \to \beta_i(1)$ which is a well-defined element of the permutation group Σ_n. This is a homomorphism $B_n \to \Sigma_n$ with kernel, by definition, the subgroup $P_n < B_n$ of *pure* braids. P_n is sometimes called the *colored* braid group, as the particles can be regarded as having identities, or colors. P_n is of course normal in B_n, of index $n!$, and there is an exact sequence

$$1 \to P_n \to B_n \to \Sigma_n \to 1.$$

Exercise 2.2. Show that any braid is equivalent to a piecewise-linear braid. Moreover, one may assume that under the projection $p : \mathbb{R}^3 \to \mathbb{R}^2$ given by $p(x, y, t) = (y, t)$, there are only a finite number of singularities of p restricted to the (image of the) given braid. One may assume these are all double points, where a pair of strings intersect transversely, and at different t-values.

Exercise 2.3. Viewing the projection in the (y, t)-plane, with the t-axis as horizontal, for any fixed value of t at which a crossing does not occur, label the strings $1, 2, \ldots, n$ counting from the bottom to the top in the y direction. Let σ_i for $i = 1, \ldots, n-1$ denote the braid with all strings horizontal, except that the string $i+1$ crosses over the i string, resulting in the permutation $i \leftrightarrow i+1$ in the labels of the strings. (Think of it as a right-hand screw motion of these two strings. Argue that any braid is equivalent to a product of these "generators" of the braid group.) [In Fig. 1, for example, the braid could be written either as $\sigma_2^{-1}\sigma_1^{-1}\sigma_3\sigma_2\sigma_1^{-1}$ or $\sigma_2^{-1}\sigma_3\sigma_1^{-1}\sigma_2\sigma_1^{-1}$ (or indeed many other ways).]

Finally, note that the inverse of σ_i is the braid with all strings horizontal, except that the string i crosses over the $i+1$ string. The first sketch in Fig. 2 shows that these are indeed inverse to each other.

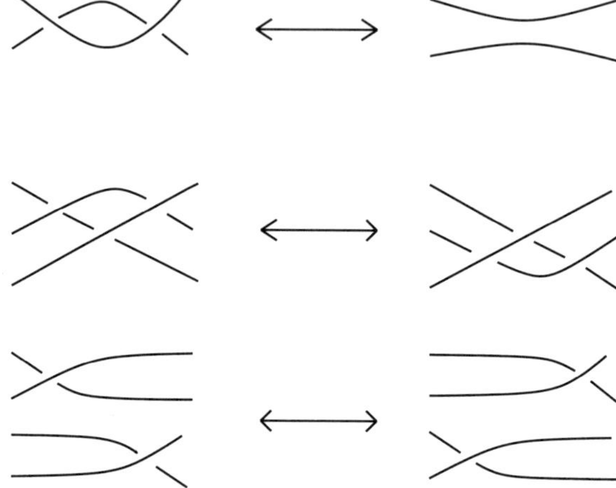

Fig. 2.   Equivalent braids.

2.3. *Definition 3: The algebraic braid group*

B_n can be regarded algebraically as the group presented with generators $\sigma_1, \ldots, \sigma_{n-1}$, where σ_i is the braid described in the previous exercise. These generators are subject to the relations (as suggested in Fig. 2):

$$\sigma_i \sigma_j = \sigma_j \sigma_i, \quad |i - j| > 1,$$

$$\sigma_i \sigma_j \sigma_i = \sigma_j \sigma_i \sigma_j, \quad |i - j| = 1.$$

It was proved by Artin that these are a complete set of relations to abstractly define B_n. This means that any relation among the σ_j can be deduced from the above relations.

Exercise 2.4. Verify that, in B_n, all the generators σ_j are conjugate to each other. Show the total degree $deg(\beta)$ which is the sum of the exponents which appear in an expression of β in the σ_j, is well-defined, i.e. independent of the expression. Verify that the abelianization of B_n, for $n \geq 2$, is the infinite cyclic group $\mathbb{Z}$, and $deg : B_n \to \mathbb{Z}$ is equivalent to the abelianization homomorphism. Thus the commutator subgroup of B_n consists of all words of total degree zero.

Exercise 2.5. In B_3, consider the special element $\Delta = \Delta_3 = \sigma_1 \sigma_2 \sigma_1$, which consists of a clockwise "half-twist" of the three strings. Verify that $\Delta \sigma_1 = \sigma_2 \Delta$ and $\Delta \sigma_2 = \sigma_1 \Delta$. Conclude that Δ^2 commutes with both generators of B_3 and therefore is in the center of B_3. (It actually generates the center, but we will not prove this.)

Show that in B_3 there are two braids α and β such that $\alpha \neq \beta$ but $\alpha^3 = \beta^3$. What about squares?

Exercise 2.6. For $n \geq 3$, define $\gamma = \gamma_n = (\sigma_1 \sigma_2 \cdots \sigma_{n-1})^n$. Draw this braid and convince yourself that it is a "full-twist" and that γ is central in B_n.

In fact γ_n generates the (infinite cyclic) center of B_n.

We can take a whole countable set of generators $\sigma_1, \sigma_2, \ldots$ subject to the above relations, which defines the infinite braid group B_∞. If we consider the (non-normal) subgroup generated by $\sigma_1, \ldots, \sigma_{n-1}$, these algebraically define B_n. Notice that this convention gives "natural" inclusions $B_n \subset B_{n+1}$ and $P_n \subset P_{n+1}$.

Exercise 2.7. Does B_∞ have a nontrivial center?

Going the other way, if one forgets the last string of an $(n+1)$-braid the result is an n-braid. But strictly speaking, this is only a well-defined homomorphism for pure braids, (or at best for the subgroup of braids in which the string beginning at the point $n+1$ also ends there). Later, we will have a use for this forgetful map

$$f : P_{n+1} \to P_n \,.$$

It is easy to see that f is a left inverse of the inclusion, or in other words a retraction in the category of groups.

2.4. *Artin combing*

We now have the ingredients for the combing technique, which gives a solution to the word problem for pure braid groups, and therefore for the full braid groups.

Solving the word problem in B_n reduces to P_n: Given a word in the σ_j, first work out its corresponding permutation. If the permutation is non-trivial, so is the group element it represents. If trivial, the braid is a pure braid, and we have reduced the problem to the word problem for P_n.

Let β be a pure n-braid and $f(\beta)$ the pure $n-1$ braid obtained by forgetting the last string, but then by inclusion, regard $f(\beta)$ in P_n. Then β and $f(\beta)$ can be visualized as the same braid, except the last string has been changed in $f(\beta)$ so as to have no interaction with the other strings.

Let K be the kernel of f. Then it is easy to verify that $f(\beta)^{-1}\beta \in K$ and the map

$$\beta \to (f(\beta), f(\beta)^{-1}\beta)$$

maps P_n bijectively onto the cartesian product $P_{n-1} \times K$. However, the multiplicative structure is that of a *semidirect* product, as happens whenever we have a split exact sequence of groups, in this case

$$1 \to K \to P_n \to P_{n-1} \to 1.$$

Also notice that every element of K can be represented by a braid in which the first $n-1$ strands go straight across. In this way we identify K with the fundamental group of the complement of the points $\{1, \ldots, n-1\}$ in the plane, which is a free group: $K \cong F_{n-1}$.

Exercise 2.8. Draw the braid $\alpha_i = \sigma_{n-1}\sigma_{n-2}\cdots\sigma_{i+1}\sigma_i^2\sigma_{i+1}^{-1}\cdots\sigma_{n-2}^{-1}\sigma_{n-1}^{-1}$ and verify that $\alpha_1, \ldots, \alpha_{n-1}$ represent a free basis for the group K above.

To solve the word problem in P_n: let $\beta \in P_n$ be a braid expressed in the standard generators σ_i; the goal is to decide whether β is really the identity. Consider its image $f(\beta)$, which lies in P_{n-1}. We may assume inductively that the word problem is solved in P_{n-1}. If $f(\beta) \neq 1$ we are done, knowing that $\beta \neq 1$.

On the other hand, if $f(\beta) = 1$, that means that β lies in the kernel K, which is a free group. Express β as a product of the generators α_i in K, and then we can decide if β is the identity, as K is freely generated by the α_i.

The semidirect product decomposition process can be iterated on P_{n-1} to obtain the Artin normal form: P_n is an iterated semidirect product of free groups $F_1, F_2, \ldots, F_{n-1}$. Thus every $\beta \in P_n$ has a unique expression

$$\beta = \beta_1\beta_2\cdots\beta_{n-1}$$

with $\beta_i \in F_i$.

For later reference, we will call the vector

$$(\beta_1, \beta_2, \ldots, \beta_{n-1})$$

the "Artin coordinates" of β.

In the above discussion, the subgroup F_j of P_{j+1} (with $j < n$) consists of the pure braids in the subgroup P_{j+1} of P_n, which become trivial if the strand labeled $j+1$ is removed. A set of generators for F_j is given by the braids α_{ij}, $i < j$, which link the i and j strands over the others:

$$\alpha_{ij} = \sigma_{j-1}\sigma_{j-2}\cdots\sigma_{i+1}\sigma_i^2\sigma_{i+1}^{-1}\cdots\sigma_{j-2}^{-1}\sigma_{j-1}^{-1}\,.$$

Note that α_{in} is just the braid we called α_i above.

The semidirect product decomposition permits us to write a finite presentation for P_n. An example is the following:

Generators are α_{ij} with $1 \le i < j \le n$.

Relations are:

$\alpha_{ij}\alpha_{ik}\alpha_{jk} = \alpha_{ik}\alpha_{jk}\alpha_{ij} = \alpha_{jk}\alpha_{ij}\alpha_{ik}$ whenever $1 \le i < j < k \le n$.

$\alpha_{ij}\alpha_{kl} = \alpha_{kl}\alpha_{ij}$ and $\alpha_{il}\alpha_{jk} = \alpha_{jk}\alpha_{il}$ whenever $1 \le i < j < k < l \le n$.

$\alpha_{ik}\alpha_{jk}\alpha_{jl}\alpha_{jk}^{-1} = \alpha_{jk}\alpha_{jl}\alpha_{jk}^{-1}\alpha_{ik}$ whenever $1 \le i < j < k < l \le n$.

Exercise 2.9. Verify these relations for specific values of i, j, k and l and convince yourself that they are "obvious".

Exercise 2.10. Show that the abelianization of P_n is free abelian of rank $\binom{n}{2}$. Conclude that the set of generators for P_n given above is minimal: no smaller set of generators is possible.

Exercise 2.11. By contrast, show that the full braid group B_n can be generated by just two generators. Hint: Recall that the generators σ_i are all conjugates in B_n.

It may seem strange that even though P_n is a subgroup of B_n, its abelianization is, in general, considerably larger. An even more striking example of this phenomenon can be obtained from the free group on two generators, together with its commutator subgroup, which is an infinitely generated free group. After abelianization they become, respectively, $\mathbb{Z}^2$ and $\mathbb{Z}^\infty$.

We mention an important theorem of W. Thurston. An infinite group is called *automatic* if is well-modeled (in a well-defined technical sense which we will not elaborate here) by a finite-state automaton. The standard reference is [10].

Theorem 2.12 (Thurston). *B_n is automatic.*

This implies, for example, that the word problem can be solved by an algorithm which is quadratic in the length of the input.

Exercise 2.13. What is the complexity of the Artin combing algorithm?

Exercise 2.14. Show that K is not normal in B_n, but is normalized by the subgroup B_{n-1}. Moreover, the action of B_{n-1} on K by conjugation is essentially the same as the Artin presentation, given in the next section.

3. Mapping Class Groups and Braids

3.1. *Definition 4: B_n as a mapping class group*

Going back to the second definition, imagine the particles are in a sort of planar jello and pull their surroundings with them as they dance about. Topologically speaking, the motion of the particles extends to a continuous family of homeomorphisms of the plane (or of a disk, fixed on the boundary). This describes an equivalence between B_n and the mapping class of D_n, the disk D with n punctures (marked points). That is, B_n can be considered as the group of homeomorphisms of D_n fixing ∂D and permuting the punctures, modulo isotopy fixing $\partial D \cup \{1, \ldots, n\}$.

3.2. *Definition 5: B_n as a group of automorphisms*

A mapping class $[h]$, where $h : D_n \to D_n$, gives rise to an automorphism $h_* : F_n \to F_n$, where F_n is the fundamental group of the punctured disk, a free group of rank n. Using the interpretation of braids as mapping classes, this defines a homomorphism

$$B_n \to Aut(F_n),$$

which Artin showed to be faithful, i.e. injective.

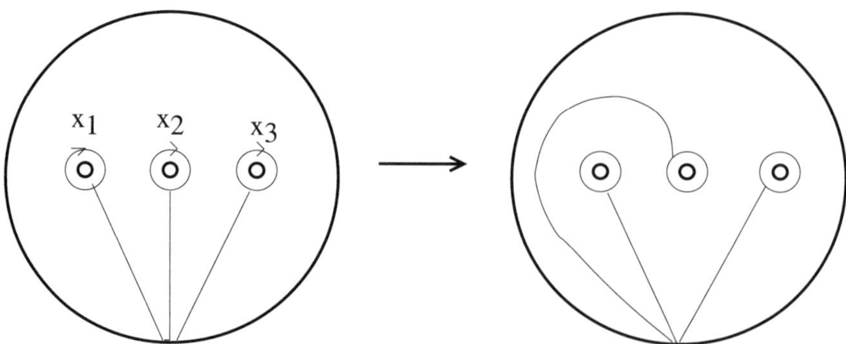

Fig. 3.   The action of σ_1 on D_3.

The generator σ_i acts as

$$x_i \to x_i x_{i+1} x_i^{-1}$$
$$x_{i+1} \to x_i$$
$$x_j \to x_j, j \neq i, i+1. \tag{3.1}$$

This is shown in the figure above, for the case $n = 3$, and the action of σ_1.

Theorem 3.1 (Artin). *Under the identification described above, B_n is the set of automorphisms $h \in Aut(F_n)$ of the form*

$$h(x_i) = w_j^{-1} x_j w_j,$$

where w_j are words in F_n, and satisfying $h(x_1 \cdots x_n) = x_1 \cdots x_n$.

The last equation of Artin's theorem reflects the fact that h comes from a map that is fixed on the boundary.

This point of view gives further insight into the group-theoretic properties of braid groups. A group G is *residually finite* if for every nonidentity element $g \in G$ there is a homomorphism $h : G \to F$ onto a finite group F such that $h(g)$ is not the identity. In other words, any element other than the identity can be proved nontrivial by looking at some homomorphism of G to a finite group.

Exercise 3.2. Show that subgroups of residually finite groups are residually finite.

Exercise 3.3. Show that the automorphism group of any finitely generated residually finite group is itself residually finite. [This is a result of Baumslag. You may want to consult [21] for a proof, and further discussion on the subject.]

Exercise 3.4. Show that finitely-generated abelian groups are residually finite. The matrices $\begin{bmatrix} 1 & 2 \\ 0 & 1 \end{bmatrix}$ and $\begin{bmatrix} 1 & 0 \\ 2 & 1 \end{bmatrix}$ generate a free group in the group $SL(2, \mathbb{Z})$, two-by-two integral matrices with determinant 1, which is a group of automorphisms of $\mathbb{Z}^2$. Thus $SL(2, \mathbb{Z})$ is residually finite and so is the rank 2 free group. Conclude that a free group of any finite or countable rank is residually finite.

From these observations, we see that $Aut(F_n)$ is residually finite and conclude the following.

Theorem 3.5. B_n *is residually finite.*

A group G is said to be Hopfian if it is not isomorphic with any nontrivial quotient group. In other words, if $G \to H$ is any surjective homomorphism with nontrivial kernel, then $H \not\cong G$.

Exercise 3.6. A finitely-generated residually finite group is Hopfian.

It follows immediately that

Theorem 3.7. B_n *is Hopfian.*

3.3. *The word problem*

If a group G is given by generators, say $\{g_i\}$, and relations, it may be difficult to determine whether two words in the generators actually represent the same element of G. This is called the *word problem*. It easily reduces to the question: given a word in the g_i, does it represent the identity element of G?

If G is residually finite, then in principle a given word which does *not* represent the identity, can be proven so by finding a homomorphism of G onto a finite group in which the word is sent to a nontrivial element, where verifying this is a finite problem. Of course, finding such a homomorphism may well be terribly difficult. But for particular groups there may be more algorithmic methods for solving the word problem. Giving such an algorithm, for a particular group, or class of groups, is what is meant by a *solution* to the word problem.

Such solutions exist for free groups and for the braid groups. First, consider a free group, with free generators $\{x_i\}$. If w is a given word in these generators, simply perform all free cancellations which are possible: that is, remove any two consecutive letters which happen to be $x_i x_i^{-1}$ or $x_i^{-1} x_i$. Continue doing this until no such cancellations are possible. If the result is the empty word, w represents the identity; otherwise it does not.

Artin's theorem provides a means for solving the word problem algorithmically in B_n by reducing it to several word problems in a free group. Suppose a word w in the Artin generators σ_i is given. Then we may consider w to be an automorphism of the free group F_n. Using equations (3.1), we can explicitly calculate the values $w(x_1), \ldots, w(x_n)$ as words in the x_i. We can then solve the word problem n times in F_n to decide whether $w(x_i) = x_i$ for each $i = 1, \ldots, n$. If this is the case, w represents the identity of B_n; otherwise, it does not.

Another solution to the word problem in braid groups, again reducing it to free groups, will be discussed later, using the technique called "Artin combing". Yet another, due to Garside, also solves the so-called *conjugacy* problem, which is to decide whether two given words in the generators are conjugates in the group. We will not discuss Garside's method here, as it is covered quite thoroughly in the notes on cryptography in this workshop.

3.4. *The modular group*

There is an interesting connection between the braid group B_3 and two-by-two integral matrices.

Consider the matrices

$$S = \begin{bmatrix} 0 & -1 \\ 1 & 0 \end{bmatrix}, \qquad T = \begin{bmatrix} 0 & -1 \\ 1 & 1 \end{bmatrix}.$$

One can easily check that $S^2 = -I = T^3$. These matrices generate cyclic subgroups of $SL(2, \mathbb{Z})$, $\mathbb{Z}_4$ and $\mathbb{Z}_6$, respectively (we use the abbrviation $\mathbb{Z}_n = \mathbb{Z}/(n\mathbb{Z})$). It is well-known that, using the generators S and T, there is an amalgamated free product structure

$$SL(2, \mathbb{Z}) \cong \mathbb{Z}_4 *_{\mathbb{Z}_2} \mathbb{Z}_6.$$

The *modular group* $PSL(2, \mathbb{Z})$ is the quotient of $SL(2, \mathbb{Z})$ by its center $\{\pm I\}$. One can also regard $PSL(2, \mathbb{Z})$ as the group of Möbius transformations of the complex plane

$$z \mapsto \frac{az + b}{cz + d} \quad \text{corresponding to} \quad \begin{bmatrix} a & b \\ c & d \end{bmatrix} \in SL(2, \mathbb{Z}).$$

The modular group has the structure of a free product $\mathbb{Z}_2 * \mathbb{Z}_3$, with (the cosets of) S and T generating the respective factors.

Exercise 3.8. Show that the braid group $B_3 \cong \langle \sigma_1, \sigma_2 | \sigma_1 \sigma_2 \sigma_1 = \sigma_2 \sigma_1 \sigma_2 \rangle$ also has the presentation $\langle x, y | x^2 = y^3 \rangle$, by finding appropriate expressions of x, y in terms of σ_1, σ_2, and vice versa, so that the transformations respect the relations and are mutual inverses.

Verify that the element $x^2 (= y^3)$ is central in B_3 and in fact generates the center of B_3.

Show that B_3 modulo its center is isomorphic with $PSL(2, \mathbb{Z})$.

3.5. *Definition 6: B_n as a fundamental group*

In complex n-space $\mathbb{C}^n$ consider the big diagonal

$$\Delta = \{(z_1, \ldots, z_n); z_i = z_j, \text{ some } i < j\} \subset \mathbb{C}^n.$$

14   *D. Rolfsen*

Using the basepoint $(1, 2, \ldots, n)$, we see that

$$P_n = \pi_1(\mathbb{C}^n \backslash \Delta).$$

In other words, pure braid groups are fundamental groups of complements of a special sort of complex *hyperplane arrangement*, itself a deep and beautiful subject.

To get the full braid group we need to take the fundamental group of the *configuration space*, of orbits of the obvious action of Σ_n upon $\mathbb{C}^n \backslash \Delta$. Thus

$$B_n = \pi_1((\mathbb{C}^n \backslash \Delta)/\Sigma_n).$$

Notice that since the singularities have been removed, the projection

$$\mathbb{C}^n \backslash \Delta \longrightarrow (\mathbb{C}^n \backslash \Delta)/\Sigma_n$$

is actually a covering map. As is well-known, covering maps induce injective homomorphisms at the π_1 level, so this is another way to think of the inclusion $P_n \subset B_n$.

It was observed in [13] that $\mathbb{C}^n \backslash \Delta$ has trivial homotopy groups in dimension greater than one. That is, it is an Eilenberg-Maclane space, also known as a $K(P_n, 1)$. Therefore its cohomology groups coincide with the group cohomology of P_n. By covering theory, the quotient space $(\mathbb{C}^n \backslash \Delta)/\Sigma_n$ also has trivial higher homotopy, so it is a $K(B_n, 1)$. Since these spaces have real dimension $2n$, this view of braid groups gives us the following observation.

Theorem 3.9. *The groups B_n and P_n have finite cohomological dimension.*

If a group contains an element of finite order, standard homological algebra implies that the cohomological dimension of the group must be infinite. Thus there are no braids of finite order.

Corollary 3.10. *The braid groups are torsion-free.*

Finally, we note that the space $(\mathbb{C}^n \backslash \Delta)/\Sigma_n$ can be identified with the space of all complex polynomials of degree n which are monic and have n distinct roots

$$p(z) = (z - r_1) \cdots (z - r_n).$$

This is one way in which the braid groups play a role in classical algebraic geometry, as fundamental groups of such spaces of polynomials.

3.6. Definition 7: The fundamental group of a space of polynomials

B_n is the fundamental group of the space of all monic polynomials of degree n with complex coefficients and only simple roots.

3.7. Surface braids

Let Σ denote a surface, with or without boundary. One can define a braid on Σ in exactly the same way as on a disk — namely, a collection of disjoint paths $\beta_1(t), \ldots, \beta_n(t)$ in $\Sigma \times I$, with each $\beta_i(t) \in \Sigma \times \{t\}$ so that they begin and end at the same points of Σ, possibly permuted. They form a group, which is usually denoted $B_n(\Sigma)$.

The subgroup of braids for which the permutation of the points is the identity is the *pure* surface braid group $P_n(\Sigma)$. Thus $B_n(D^2)$ and $P_n(D^2)$ are the classical Artin braid groups which we have discussed before. Much of the above discussion holds for surface braid groups; for example, they may be interpreted as fundamental groups of configuration spaces.

On the other hand, there are differences. For example the braid groups of the sphere S^2 and the projective plane $\mathbb{R}P^2$ have elements of finite order. Van Buskirk showed that these are the only closed surfaces to have torsion in their braid groups, however. There are finite presentations for surface braid groups in the literature, and these groups are currently an active area of research. One reason for their interest is that they play an important part in certain topological quantum field theories (TQFT's).

4. Knot Theory, Braids and the Jones Polynomial

4.1. Knots and Reidemeister moves

One often pictures a knot or link by drawing a projection onto the plane, with only double points, and indicating which string goes under by putting a small gap in it, as in Fig. 1. If one deforms the plane by a planar isotopy, an equivalent projection results. One can also make local changes to a knot diagram. The first two vignettes in Fig. 2 are sometimes called Reidemeister moves of types 2 and 3 (respectively). One can remember the numbering, as type 2 involves two strands and type 3 involves three. There is also a type 1 Reidemeister move.

Theorem 4.1 (Reidemeister). *Two knot diagrams present equivalent knots if and only if one can be transformed to the other by a finite sequence of Reidemeister moves (plus an isotopy of the plane).*

Fig. 4.   Reidemeister move of type 1.

4.2. *Markov moves*

It has already been mentioned that different braids can, upon forming the closure, give rise to the same knot or link. You can easily convince yourself that if α and β are n-string braids, then the closure of $\alpha^{-1}\beta\alpha$ is equivalent to the closure of β. Because of the trivial strings added in forming the closure, α and its inverse can annihilate each other! Therefore we see that conjugate braids close to the same link.

Another example of this phenomenon is to consider the n-braid β as an element of B_{n+1} by adding a trivial string at the top, and compare the original closure $\hat{\beta}$ with the closure of $\beta \times \sigma_n$, the latter braid taken in B_{n+1}. If you sketch this, you can easily convince yourself that these result in the same link (up to isotopy). The first move (conjugation) and the second move just described constitute the two Markov moves. They are the key to understanding the connection between braid theory and knot theory. The following theorem is due to J. W. Alexander (first part) and A. A. Markov. A full proof appeared first, to our knowledge, in [5].

Theorem 4.2. *Every knot or link is the closure of some braid. Two braids close to equivalent knots or links if and only if they are related by a finite sequence of the two Markov moves.*

Exercise 4.3. If you want to learn a nice, elementary proof of the above theorem, read the paper [24].

Exercise 4.4. Consider the 3-braid $\beta = \sigma_1\sigma_2$ and identify the knot or link, simplifying, if possible by using Reidemeister moves: $\hat{\beta}$, $\hat{\beta^2}$, $\hat{\beta^3}$. For which integers k is the closure of β^k a knot?

4.3. *Kauffman's bracket and Jones polynomial*

The original construction of the Jones polynomial involved a family of representations

$$B_n \to \mathcal{A}$$

of the braid groups into an algebra $\mathcal{A}$ involving a parameter t. This algebra has a linear trace function into the ring of (Laurent) polynomials. The composite of the representation and the trace, together with some correction terms to account for the Markov moves, defined the original Jones polynomial.

We will present this in a sort of reverse order, as there is a very simple derivation of the Jones polynomial discovered a few years later by L. Kauffman. From this, we can define an algebra, called the Temperley-Lieb algebra, and reconstruct what is essentially Jones' representation.

Consider a planar diagram D of a link L, which has only simple transverse crossings. Kauffman's bracket $\langle D \rangle$ is (at first) a polynomial in three commuting variables, a, b and d defined by the equations:

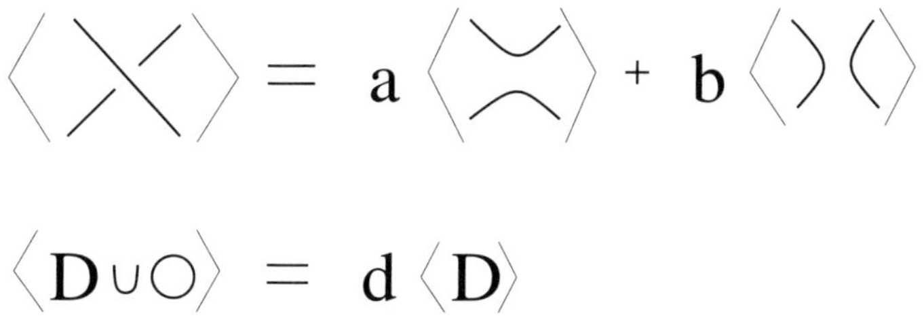

$$\left\langle \times \right\rangle = a \left\langle \asymp \right\rangle + b \left\langle)(\right\rangle$$

$$\langle D \cup \bigcirc \rangle = d \langle D \rangle$$

Fig. 5. Equations defining Kauffman's bracket polynomial.

Some explanation is in order. The vignettes in the brackets in the first equation stand for complete link diagrams, which differ only near the crossing in question. Those on the right side represent diagrams with fewer crossings. In the second equation, one can introduce a closed curve which has no intersections with the remainder of the diagram D, resulting in a diagram whose bracket polynomial is d times the bracket of D.

Exercise 4.5. Verify that the bracket polynomial is well-defined, if we decree that the bracket of a single curve with no crossings is equal to 1.

Exercise 4.6. Show that the bracket will be invariant under the type 2 Reidemeister move, if we have the following relations among the variables:

$$a^2 + b^2 + abd = 0 \quad \text{and} \quad ab = 1.$$

Thus we make the substitutions $b = a^{-1}$ and $d = -a^2 - a^{-2}$ and now consider the bracket to be a Laurent polynomial in the single variable a.

Exercise 4.7. Show that the invariance under the type 2 Reidemeister moves implies the bracket is invariant under the type 3 move, too.

Exercise 4.8. Calculate the bracket of the two trefoil knot diagrams with 3 crossings. Show that they are the same, except for reversal of the sign of the exponents.

Exercise 4.9. Investigate the effect of Reidemeister move 1 on the bracket of a diagram, and show that it changes the bracket by a factor of $-a^{\pm 3}$, the sign of the exponent depending on the sense of the curl removed.

Because of this, one can define a polynomial invariant under all three Reidemeister moves, by counting the number of positive minus the number of negative crossings, and modifying the bracket polynomial by an appropriate factor. A positive crossing corresponds to a (positive) braid generator, if both strings are oriented from left to right. This gives, up to change of variable, the Jones polynomial of the knot. Specifically, we define the writhe of an *oriented* diagram $\vec{D}$ for an oriented knot (or link) $\vec{K}$ to be

$$w(D) = \sum_c \epsilon_c$$

where the sum is over all crossings and $\epsilon_c = 1$ if the crossing c is positive, and -1 if negative. Then we define

$$f_{\vec{K}}(a) = (-a^3)^{-w(\vec{D})}\langle D \rangle.$$

This is an invariant of the oriented link $\vec{K}$. If K happens to be a knot, it is independent of the orientation, as reorienting both strands of a crossing does not change its sign. It is related to the Jones polynomial $V_K(t)$ by a simple change of variables:

$$V_{\vec{K}}(t) = f_{\vec{K}}(t^{-1/4}).$$

Exercise 4.10. Show that all exponents of a in $\langle D \rangle$ are divisible by 4 if D is a diagram of a knot (or link with an odd number of components). In other cases, they are congruent to 2 mod 4. Thus the Jones polynomial is truly a (Laurent) polynomial in t for knots and odd component links, but a polynomial in $\sqrt{t}$ in the other cases.

5. Representations

This is a very big subject, which we will just touch upon. By a representation of a group we will mean a homomorphism of the group into a group of matrices, or more generally into some other group, or ring or algebra. Often, but not always, we want the target to be finite-dimensional. We've already encountered the Artin representation $B_n \rightarrow Aut(F_n)$, which is faithful. Here the target group is far from being "finite-dimensional".

5.1. *Jones representation and the Temperley-Lieb algebra*

Another very important representation is the one defined by Jones [15] which gave rise to his famous knot polynomial, and the subsequent revolution in knot theory. The version we will discuss is more thoroughly described in [17]; it is based on the Kauffman bracket, an elementary combinatorial approach to the Jones polynomial. First we need to describe the Temperley-Lieb Algebras $\mathcal{T}_n$, in their geometric form. The elements of $\mathcal{T}_n$ are something like braids: we consider strings in a box, visualized as a square in the plane, endpoints being exactly n specified points on each of the left and right sides. The strings are not required to be monotone, or even to run across from one side to the other. There also may be closed components. Really what we are looking at are "tangle" diagrams. Two tangle diagrams are considered equal if there is a planar isotopy, fixed on the boundary of the square, taking one to the other.

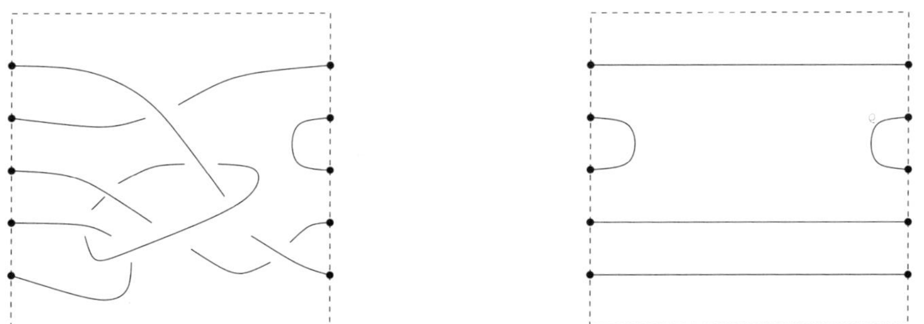

Fig. 6.   A typical element in $\mathcal{T}_5$ and the generator e_3.

Now we let A be a fixed complex number (regarded as a parameter), and formally define $\mathcal{T}_n$ to be the complex vector space with basis the set of all tangles, as described above, but modulo the following relations, which correspond to similar relations used to define Kauffman's bracket version of the Jones polynomial (we have promoted the variable a to upper case).

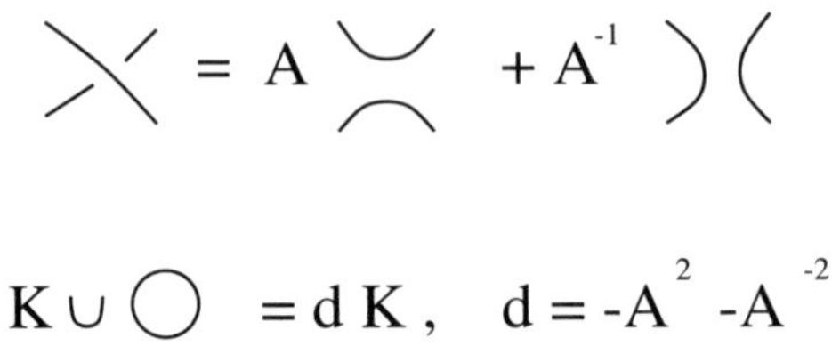

$$\times = A \;\smile\!\frown\; + A^{-1} \;)\,($$

$$K \cup \bigcirc = d\,K, \quad d = -A^2 - A^{-2}$$

Fig. 7.   Relations in $\mathcal{T}_n$.

The first relation means that we can replace a tangle with a crossing by a linear combination of two tangles with that crossing removed in two ways. As usual, the pictures mean that the tangles are identical outside the part pictured. The second relation means that we can remove any closed curve in the diagram, if it does not have any crossings with the rest of the tangle, at the cost of multiplying the tangle by the scalar $-A^2 - A^{-2}$. Using the relations, we see that any element of $\mathcal{T}_n$ can be expressed as a linear combination of tangles which have no crossings and no closed curves — that is, disjoint planar arcs connecting the $2n$ points of the boundary. This gives a finite generating set, which (for generic values of A) can be shown to be a basis for $\mathcal{T}_n$ as a vector space. But there is also a multiplication of $\mathcal{T}_n$, a concatenation of tangles, in exactly the same way braids are multiplied. This enables us to consider $\mathcal{T}_n$ to be generated *as an algebra* by the elements $e_1, \ldots, e_{n-1}$. In e_i all the strings go straight across, except those at level i and $i+1$ which are connected by short caps; the generator e_3 of $\mathcal{T}_5$ is illustrated in Fig. 2. The identity of this algebra is simply the diagram consisting of n horizontal lines (just like the identity braid). $\mathcal{T}_n$ can be described abstractly as the associative algebra with the generators $e_1, \ldots, e_{n-1}$, subject to the relations:

$$e_i e_j = e_j e_i \text{ if } |i - j| > 1 \qquad e_i e_{i\pm1} e_i = e_i \qquad e_i^2 = (-A^2 - A^{-2})e_i .$$

It is an enjoyable exercise to verify these relations from the pictures. Now the Jones representation $J : B_n \to \mathcal{T}_n$ can be described simply by considering a braid diagram as an element of the algebra. In terms of generators, this is just

$$J(\sigma_i) = A + A^{-1} e_i .$$

5.2. *The Burau representation*

One of the classical representations of the braid groups is the Burau representation, which can be described as follows. Consider the definition of

B_n as the mapping class group of the punctured disk D_n (Definition 4). As already noted, the fundamental group of D_n is a free group, with generator x_i represented by a loop, based at a point on the boundary of the disk, which goes once around the ith puncture. Consider the subgroup of $\pi_1(D_n)$ consisting of all words in the x_i whose exponent sum is zero. This is a normal subgroup, and so defines a regular covering space

$$p \colon \bar{D}_n \to D_n.$$

The group of covering translations is infinite cyclic. Therefore, the homology $H_1(\bar{D}_n)$ can be considered as a module over the polynomial ring $\Lambda := \mathbb{Z}[t, t^{-1}]$, where t represents the generator of the covering translation group. Similarly, if $*$ is a basepoint of D_n, the relative homology group $H_1(\bar{D}_n, p^{-1}(*))$ is a Λ-module.

A braid β can be represented as (an isotopy class of) a homeomorphism $\beta \colon D_n \to D_n$ fixing the basepoint. This lifts to a homeomorphism $\bar{\beta} \colon \bar{D}_n \to \bar{D}_n$, which is unique if we insist that it fix some particular lift of the basepoint. The induced homomorphism on homology, $\bar{\beta}_* \colon H_1(\bar{D}_n, p^{-1}(*)) \to H_1(\bar{D}_n, p^{-1}(*))$ is a linear map of these finite-dimensional modules, and so can be represented by a matrix with entries in $\mathbb{Z}[t, t^{-1}]$. The mapping

$$\beta \to \bar{\beta}_*$$

is the Burau representation of B_n.

Let us illustrate this for the case $n = 3$. D_3 is replaced by the wedge of three circles, which is homotopy equivalent to it, to simplify visualization. The covering space $\bar{D}_n$ is shown as an infinite graph. Although as an abelian group $H_1(\bar{D}_3, p^{-1}(*))$ is infinitely generated, as a Λ-module it has three generators $g_i = \tilde{x}_i$, the lifts of the generators x_i of D_3, $i = 1, 2, 3$ beginning at some fixed basepoint in $p^{-1}(*)$. The other elements of $H_1(\bar{D}_3, p^{-1}(*))$ are represented by translates $t^n g_i$.

Now consider the action of σ_1 on D_3, at the fundamental group level. As we saw in Definition 5, σ_{1*} takes x_1 to $x_1 x_2 x_1^{-1}$. Accordingly $\bar{\sigma}_{1*}$ takes g_1 to the lift of $x_1 x_2 x_1^{-1}$, which is pictured in the lower part of the illustration. In terms of homology, this is $g_1 + t g_2 - t g_1$. Therefore

$$\bar{\sigma}_{1*}(g_1) = (1 - t) g_1 + t g_2.$$

Similarly

$$\bar{\sigma}_{1*}(g_2) = g_1, \quad \bar{\sigma}_{1*}(g_3) = g_3.$$

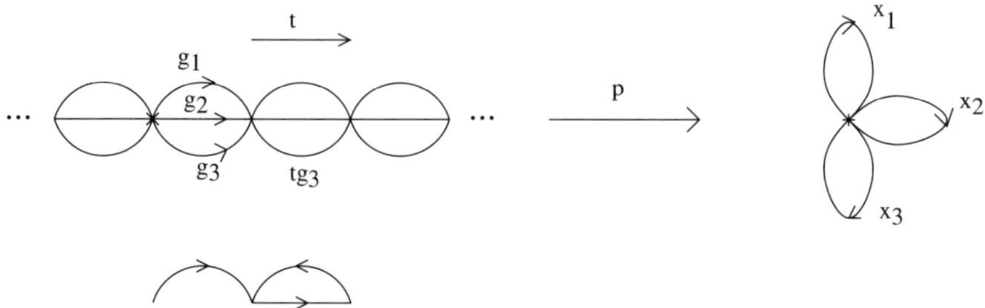

Fig. 8.   Illustrating the Burau representation for $n = 3$.

Exercise 5.1. Show that, with appropriate choice of basis, the Burau representation of B_n sends σ_j to the matrix

$$I_{j-1} \oplus \begin{bmatrix} 1 - t & t \\ 1 & 0 \end{bmatrix} \oplus I_{n-j-1},$$

where I_k denotes the $k \times k$ identity matrix.

A probabilistic interpretation. Vaughan Jones offered the following interpretation of the Burau representation, which we will modify slightly because we have chosen opposite crossing conventions. Picture a braid as a system of trails which cross only at bridges, where one goes over the other. Wherever one trail crosses under another there is a probability t that a person will jump from the lower trail to the upper trail; the probability of staying on the same trail is $1 - t$. The i, j entry of the Burau matrix corresponding to a braid then represents the probability that, if a person starts on the trail at position i, she will finish on the trail at position j.

It has been known for many years that this representation is faithful for $n \leq 3$, and it is only within the last decade that it was found to be unfaithful for any n at all. John Moody [23] showed in 1993 that it is unfaithful for $n \geq 9$. This has since been improved by Long and Paton [20] and more recently by Bigelow to $n \geq 5$. The case $n = 4$ remains open, at the time of this writing.

5.3. *Linearity of the braid groups*

It has long been questioned, whether the braid groups are *linear*, meaning that there is a faithful representation $B_n \to GL(V)$ for some finite-dimensional vector space V. A candidate had been the so-called Burau representation, but as already mentioned it has been known for several years that Burau is unfaithful, in general.

The question was finally settled recently by Daan Krammer and Stephen Bigelow, using equivalent representations, but different methods. They use a representation defined very much like the Burau representation. But instead of a covering of the punctured disk D_n, they use a covering of the *configuration space* of pairs of points of D_n, upon which B_n also acts. This action induces a linear representation in the homology of an appropriate covering, and provides just enough extra information to give a faithful representation!

Theorem 5.2 (Krammer [19], Bigelow [4]). *The braid groups are linear.*

In fact, Bigelow has announced that the BMW representation (Birman, Murakami, Wenzl) [6] is also faithful. Another open question is whether the Jones representation $J : B_n \to \mathcal{T}_n$, discussed earlier, is faithful.

6. Ordering Braid Groups

This is, to me, one of the most exciting of the recent developments in braid theory. Call a group G *right orderable* if its elements can be given a strict total ordering $<$ which is right-invariant:

$$\forall\, x, y, z \in G, \quad x < y \Rightarrow xz < yz.$$

Theorem 6.1 (Dehornoy [8]). B_n *is right-orderable.*

Interestingly, we know of three quite different proofs. The first is Dehornoy's, the second is one that was discovered jointly by myself and four other topologists. We were trying to understand difficult technical aspects of Dehornoy's argument, then came up with quite a different way of looking at exactly the same ordering, but using the view of B_n as the mapping class group of the punctured disk D_n. Yet a third way is due to Thurston, using the fact that the universal cover of D_n embeds in the hyperbolic plane. Here are further details.

6.1. *Dehornoy's approach*

It is routine to verify that a group G is right-orderable if and only if there exists a subset Π (positive cone) of G satisfying:

(1) $\Pi \cdot \Pi = \Pi$.
(2) The identity element does not belong to Π, and for every $g \neq 1$ in G exactly one of $g \in \Pi$ or $g^{-1} \in \Pi$ holds.

One defines the ordering by $g < h$ iff $hg^{-1} \in \Pi$.

Exercise 6.2. Verify that the transitivity law holds, and that the ordering is right-invariant. Show that the ordering defined by this rule is also left-invariant if and only if $g\Pi g^{-1} = \Pi$ for every $g \in G$; that is, Π is "normal".

Dehornoy's idea is to call a braid i-positive if it is expressible as a word in $\sigma_j, j \geq i$ in such a manner that all the exponents of σ_i are positive. Then define the set $\Pi \subset B_n$ to be all braids which are i-positive for some $i = 1, \ldots, n - 1$. To prove (1) above is quite easy, but (2) requires an extremely tricky argument.

6.2. *Ordering braids as mapping classes*

Here is the point of view advocated in [12]. Consider B_n as acting on the complex plane, as described above. Our idea is to consider the image of the real axis $\beta(\mathbb{R})$, under a mapping class $\beta \in B_n$. Of course there are choices here, but there is a unique "canonical form" in which (roughly speaking) $\mathbb{R} \cap \beta(\mathbb{R})$ has the fewest number of components. Now declare a braid β to be positive if (going from left to right) the first departure of the canonical curve $\beta(\mathbb{R})$ from $\mathbb{R}$ itself is into the *upper* half of the complex plane. Amazingly, this simple idea works, and gives exactly the same ordering as Dehornoy's combinatorial definition.

6.3. *Braids acting on the real line*

Finally, Thurston's idea for ordering B_n again uses the mapping class point of view, but a different way at looking at ordering a group. This approach, which has the advantage of defining infinitely many right-orderings of B_n is described by H. Short and B. Wiest in [29]. The Dehornoy ordering (which is discrete) occurs as one of these right-orderings – others constructed in this way are order-dense. A group G acts on a set X (on the right) if the mapping $x \to xg$ satisfies: $x(gh) = (xg)h$ and $x1 = x$. An action is *effective* if the only element of G which acts as the identity is the identity $1 \in G$. The following is a useful criterion for right-orderability:

Lemma 6.3. *If the group G acts effectively on $\mathbb{R}$ by order-preserving homeomorphisms, then G is right-orderable.*

By way of a proof, consider a well-ordering of the rational numbers. Define, for g and $h \in G$,

$$g < h \Leftrightarrow xg < xh \quad \text{at the first } x \in \mathbb{Q} \text{ such that } xg \neq xh.$$

It is routine to verify that this defines a right-invariant strict total ordering of G. (By the way, we could have used any ordered set in place of $\mathbb{R}$.)

The universal cover $\tilde{D}_n$ of D_n can be embedded in the hyperbolic plane $\mathbb{H}^2$ in such a way that the covering translations are isometries. This gives a hyperbolic structure on D_n. It also gives a beautiful tiling of $\mathbb{H}^2$, illustrated in Fig. 9 for the case $n = 2$.

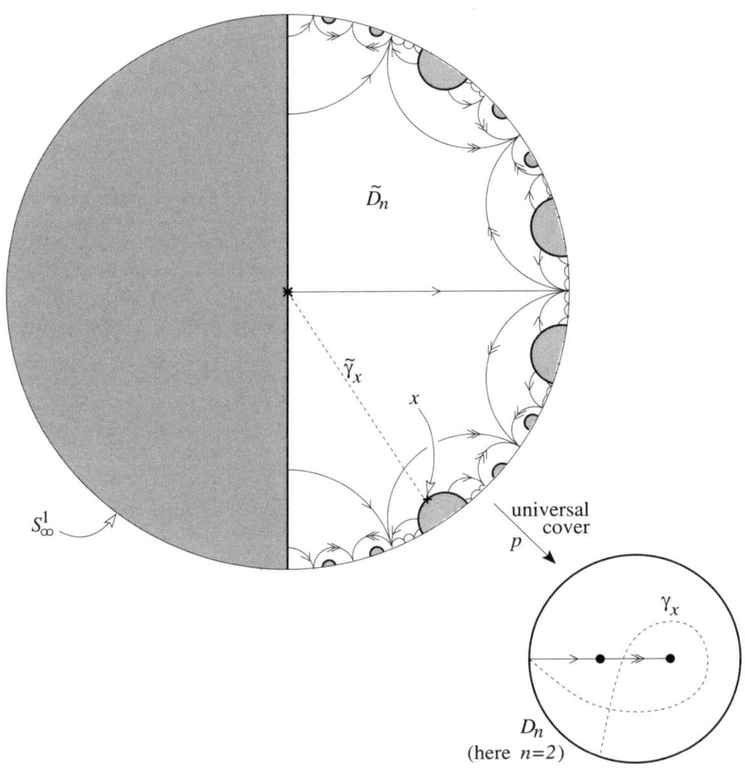

Fig. 9. The universal cover of a twice-punctured disk, with a lifted geodesic. (Courtesy of H. Short and B. Wiest [29].)

Choose a basepoint $* \in \partial D$ and a specific lift $\tilde{*} \in \mathbb{H}^2$. Now a braid is represented by a homeomorphism of D_n, which fixes $*$. This homeomorphism lifts to a homeomorphism of $\tilde{D}_n$, unique if we specify that it fixes $\tilde{*}$. In turn, this homeomorphism extends to a homeomorphism of the boundary of $\tilde{D}_n$ But in fact, this homeomorphism fixes the interval of $\partial\tilde{D}_n$ containing

$\tilde{*}$, and if we identify the complement of this interval with the real line $\mathbb{R}$, a braid defines a homeomorphism of $\mathbb{R}$. This defines an action of B_n upon $\mathbb{R}$ by order-preserving homeomorphisms, and hence a right-invariant ordering of B_n.

6.4. *Two-sided invariance*

Any right-invariant ordering of a group can be converted to a left-invariant ordering, by comparing inverses of elements, but that ordering is in general different from the given one. We will say that a group G with strict total ordering $<$ is fully-ordered, or *bi-ordered*, if

$$x < y \Rightarrow xz < yz \quad \text{and} \quad zx < zy, \quad \forall\, x, y, z \in G.$$

There are groups which are right-orderable but not bi-orderable — in fact the braid groups!

Proposition 6.4 (N. Smythe). *For $n > 2$ the braid group B_n cannot be bi-ordered.*

The reason for this is that there exists a nontrivial element which is conjugate to its inverse: take $x = \sigma_1 \sigma_2^{-1}$ and $y = \sigma_1 \sigma_2 \sigma_1$ and note that $yxy^{-1} = x^{-1}$. In a bi-ordered group, if $1 < x$ then $1 < yxy^{-1} = x^{-1}$, contradicting the other conclusion $x^{-1} < 1$. If $x < 1$ a similar contradiction arises.

Exercise 6.5. Show that in a bi-ordered group $g < h$ and $g' < h'$ imply $gg' < hh'$. Conclude that if $g^n = h^n$ for some $n \neq 0$, then $g = h$. That is, roots are unique. Use this to give an alternative proof that B_n is not bi-orderable if $n \geq 3$.

Theorem 6.6. *The pure braid groups P_n can be bi-ordered.*

This theorem was first noticed by J. Zhu, and the argument appears in [27], based on the result of Falk and Randall [14] that the pure braid groups are "residually torsion-free nilpotent". Later, in joint work with Djun Kim, we discovered a really natural, and we think beautiful, way to define a bi-invariant ordering of P_n. That ordering has the special property that, restricted to the pure braids which are also positive in the sense of Garside (expressible in the standard braid generators with no negative exponents), it is a well-ordering. We have already done half the work, by discussing Artin combing. Now we need to discuss ordering of free groups.

6.5. *Bi-ordering free groups*

Lemma 6.7. *For each $n \geq 1$, the free group F_n has a bi-invariant ordering $<$ with the further property that it is invariant under any automorphism $\phi : F_n \to F_n$ which induces the identity upon abelianization:*

$$\phi_{ab} = id : \mathbb{Z}^n \to \mathbb{Z}^n .$$

The construction depends on the Magnus expansion of free groups into rings of formal power series. Let F be a free group with free generators $x_1, \ldots, x_n$. Let $Z[[X_1, \ldots, X_n]]$ denote the ring of formal power series in the *non-commuting* variables $X_1, \ldots, X_n$.

Each term in a formal power series has a well-defined (total) degree, and we use $O(d)$ to denote terms of degree $\geq d$. The subset $\{1 + O(1)\}$ is actually a multiplicative subgroup of $Z[[X_1, \ldots, X_n]]$. Moreover, there is a multiplicative homomorphism

$$\mu : F \to Z[[X_1, \ldots, X_n]]$$

defined by

$$\mu(x_i) = 1 + X_i$$
$$\mu(x_i^{-1}) = 1 - X_i + X_i^2 - X_i^3 + \cdots . \tag{6.1}$$

There is a very nice proof that μ is injective in [21], as well as discussion of some of its properties. One such property is that commutators have zero linear terms. For example (dropping the μ)

$$[x_1, x_2] = x_1 x_2 x_1^{-1} x_2^{-1}$$
$$= (1 + X_1)(1 + X_2)(1 - X_1 + X_1^2 - \cdots)(1 - X_2 + X_2^2 - \cdots)$$
$$= 1 + X_1 X_2 - X_2 X_1 + O(3). \tag{6.2}$$

Now there is a fairly obvious ordering of $Z[[X_1, \ldots, X_n]]$. Write a power series in ascending degree, and within each degree list the monomials lexicographically according to subscripts. Given two series, order them according to the coefficient of the first term (when written in the standard form just described) at which they differ. Thus, for example, 1 and $[x_1, x_2]$ first differ at the $X_1 X_2$ term, and we see that $1 < [x_1, x_2]$. It is not difficult to verify that this ordering, restricted to the group $\{1 + O(1)\}$ is invariant under both left- and right-multiplication.

Exercise 6.8. Write x_1, $x_2 x_1 x_2^{-1}$ and $x_2^{-1} x_1 x_2$ in increasing order, according to the ordering just described.

6.6. Ordering P_n

If α and β are pure n-braids, compare their Artin coordinates (as described earlier)

$$(\alpha_1, \alpha_2, \ldots, \alpha_{n-1}) \quad \text{and} \quad (\beta_1, \beta_2, \ldots, \beta_{n-1})$$

lexicographically, using within each F_k the Magnus ordering described above. This all needs choices of conventions, for example, for generators of the free groups, described in detail in [18]. The crucial fact is that the action associated with the semidirect product, by automorphisms φ, has the property mentioned in Lemma 6.7. We recall the definition of a positive braid according to Garside: a braid is Garside-positive if it can be expressed as a word in the standard generators σ_i with only positive exponents.

Theorem 6.9 (Kim-Rolfsen). *P_n has a bi-ordering with the property that Garside-positive pure braids are greater than the identity, and the set of all Garside-positive pure braids is well-ordered by the ordering.*

6.7. Algebraic consequences

The orderability of the braid groups has implications beyond what we already knew — e.g. that they are torsion-free. In the theory of representations of a group G, it is important to understand the group algebra $\mathbb{C}G$ and the group ring $\mathbb{Z}G$. These rings also play a role in the theory of Vassiliev invariants. A basic property of a ring would be whether it has (nontrivial) zero divisors. If a group G has an element g of finite order, say $g^p = 1$, but no smaller power of g is the identity, then we can calculate in $\mathbb{Z}G$

$$(1 - g)(1 + g + g^2 + \cdots + g^{p-1}) = 1 - g^p = 0$$

and we see that both terms of the left-hand side of the equation are (nonzero) divisors of zero. A long-standing question of algebra is whether the group ring $\mathbb{Z}G$ can have zero divisors if G is torsion-free. We do know the answer for orderable groups:

Exercise 6.10. If R is a ring without zero divisors, and G is a right-orderable group, then the group ring RG has no zero divisors. Moreover, the only units of RG are the monomials rg, $g \in G$, r a unit of R.

Proposition 6.11 (Malcev, Neumann). *If G has a bi-invariant ordering, then its group ring $\mathbb{Z}G$ embeds in a division algebra, that is, an extension in which all nonzero elements have inverses.*

These results give us new information about the group rings of the braid groups.

Theorem 6.12. $\mathbb{Z}B_n$ *has no zero divisors. Moreover,* $\mathbb{Z}P_n$ *embeds in a division algebra.*

A proof of the theorem of Malcev and Neumann [22] can be found in [25].

Exercise 6.13. Which subgroups of $Aut(F_n)$ are right-orderable?

Of course, $Aut(F_n)$ itself is not right-orderable, because it has elements of finite order, e.g. permuting the generators.

6.8. *Incompatibility of the orderings*

A final note regarding orderings: As we have seen, the methods we have used for ordering B_n and P_n are quite different. One might hope there could be compatible orderings: a bi-ordering of P_n which extends to a right-invariant ordering of B_n. But, in recent work with Akbar Rhemtulla [26], we showed this is hopeless! This result was found independently by Dubrovina and Dubrovin [9].

Theorem 6.14 (Rhemtulla, Rolfsen). *For* $n \geq 5$, *there is no right-invariant ordering of* B_n, *which, upon restriction to* P_n, *is also left-invariant.*

Acknowledgments

The author wishes to thank the organizers of the summer school and conference at NUS, firstly for the honor of being invited to give the lectures upon which these notes are based, and secondly for their wonderful hospitality and efficiency which made it a memorable scientific event.

References

1. E. Artin, Theorie der Zöpfe, *Abh. Math. Sem. Hamburg. Univ.* 4 (1926), 47–72.
2. E. Artin, Theory of braids, *Annals of Math. (2)* 48 (1947), 101–126.
3. G. Baumslag, Automorphism groups of residually finite groups, *J. London Math. Soc.* 38 (1963), 117–118.
4. S. J. Bigelow, Braid groups are linear, *J. Amer. Math. Soc.* 14 (2001), no. 2, 471–486 (electronic).
5. J. Birman, *Braids, Links and Mapping Class Groups*, Annals of Math. Studies 82, Princeton Univ. Press, 1974.

6. J. Birman and H. Wenzl, Braids, link polynomials and a new algebra, *Trans. Amer. Math. Soc.* 313 (1989), 249–273.

7. G. Burde and H. Zieschang, *Knots*, de Gruyter, 1985.

8. P. Dehornoy, Braid groups and left distributive operations, *Trans Amer. Math. Soc.* 345 (1994), 115–150.

9. T. Dubrovina and N. Dubrovin, On braid groups, *Sbornik Math.* 192 (2001), 693–703.

10. D. Epstein, J. Cannon, D. Holt, S. Levy, M. Paterson and W. Thurston, *Word Processing in Groups*, Jones and Bartlett, 1992.

11. R. Fenn, An elementary introduction to the theory of braids, notes by B. Gemein, 1999, available at the author's website.

12. R. Fenn, M. Green, D. Rolfsen, C. Rourke and B. Wiest, Ordering the braid groups, *Pacific J. Math.* 191 (1999), 49–74.

13. E. Fadell and L. Neuwirth, Configuration spaces, *Math. Scand.* 10 (1962), 111–118.

14. M. Falk and R. Randell, Pure braid groups and products of free groups, *Braids (Santa Cruz 1986)*, *Contemp. Math.* 78 (1988) 217–228.

15. V. F. R. Jones, A polynomial invariant for knots via Von Neumann algebras, *Bull. Amer. Math. Soc.* 12 (1985), 103–111.

16. C. Kassell and V. Turaev, *Braids*, Springer, 2008.

17. L. Kauffman, *Knots and Physics*, World Scientific, 1993.

18. D. Kim and D. Rolfsen, An ordering for groups of pure braids and hyperplane arrangements, *Canadian J. Math.* 55 (2002), 822–838.

19. D. Krammer, The braid group B_4 is linear, *Invent. Math.* 142 (2000), no. 3, 451–486.

20. D. Long and M. Paton, The Burau representation is not faithful for $n \geq 6$, *Topology* 32 (1993), 439–447.

21. W. Magnus, A. Karrass, and D. Solitar, *Combinatorial Group Theory*, Wiley, 1966.

22. A. I. Mal'cev, On the embedding of group algebras in division algebras, *Dokl. Akad. Nauk SSSR* 60 (1948), 1944–1501.

23. J. Moody, The faithfulness question for the Burau representation, *Proc. Amer. Math. Soc.* 119 (1993), 439–447.

24. H. R. Morton, Threading knot diagrams. *Math. Proc. Cambridge Philos. Soc.* 99 (1986), no. 2, 247–260.

25. R. Mura and A. Rhemtulla, *Orderable Groups*, Lecture Notes in Pure and Applied Mathematics, vol. 27, Marcel Dekker, New York, 1977.

26. A Rhemtulla and D. Rolfsen, Local indicability in ordered groups: braids and elementary amenable groups, *Proc. Amer. Math. Soc.* 130 (2002), 2569–2577 (electronic).

27. D. Rolfsen and J. Zhu, Braids, orderings and zero divisors, *J. Knot Theory and its Ramifications* 7 (1998), 837–841.

28. D. S. Passman, *The Algebraic Structure of Group Rings*, Pure and Applied Mathematics, Wiley Interscience, 1977.

29. H. Short and B. Wiest, Orderings of mapping class groups after Thurston, *L'enseignement Mathématique* 46 (2000), 279–312.

SIMPLICIAL OBJECTS AND HOMOTOPY GROUPS

Jie Wu

Department of Mathematics
National University of Singapore
2 Science Drive 2, 117543, Singapore
E-mail: matwuj@nus.edu.sg

We discuss simplicial objects and homotopy groups. The first section covers the topics on Δ-objects and homology. The relations between Δ-sets, simplicial complexes and Δ-complexes are given. The simplicial and singular homology can be directly obtained as the derived functors of Δ-sets. The second section covers the topics on simplicial sets and homotopy. After introducing the definition of simplicial sets, the relations between simplicial sets, Δ-sets and simplicial complexes are discussed. Then, we give the connections between simplicial sets and spaces. By introducing fibrant simplicial sets, the homotopy theory on the category of simplicial sets can be set up. In particular, the homotopy groups can be combinatorially defined using simplicial sets. In the last section, we discuss simplicial group theory. The Moore chain complexes are deeply discussed. As an important example, the general homotopy groups of the 2-sphere can be described as the center of certain combinatorially given groups. We give new approaches to many classical results. Also there are many new results on these topics.

1. Δ-Objects and Homology

1.1. Δ-sets

Definition 1.1. A Δ-*set* means a sequence of sets $X = \{X_n\}_{n \geq 0}$ with *faces* $d_i \colon X_n \to X_{n-1}$, $0 \leq i \leq n$, such that

$$d_i d_j = d_j d_{i+1} \tag{1.1}$$

for $i \geq j$, which is called the Δ-*identity*.

Remark 1.2. One can use coordinate projections for catching Δ-identity:

$$d_i \colon (x_0, \ldots, x_n) \longrightarrow (x_0, \ldots, x_{i-1}, x_{i+1}, \ldots, x_n).$$

Let $\mathcal{O}^+$ be the category whose objects are finite ordered sets and whose morphisms are functions $f \colon X \to Y$ such that $f(x) < f(y)$ if $x < y$. Note that the objects in $\mathcal{O}^+$ are given by $[n] = \{0, 1, \ldots, n\}$ for $n \geq 0$ and the morphisms in $\mathcal{O}^+$ are generated by $d^i \colon [n-1] \longrightarrow [n]$ with

$$d^i(j) = \begin{cases} j & \text{if } j < i \\ j+1 & \text{if } j \geq i \end{cases}$$

for $0 \leq i \leq n$, that is d^i is the ordered embedding missing i. We may write the function d^i in matrix form:

$$d^i = \begin{pmatrix} 0 & 1 & \cdots & i-1 & i & i+1 & \cdots & n-1 \\ 0 & 1 & \cdots & i-1 & i+1 & i+2 & \cdots & n \end{pmatrix}.$$

The morphisms d^i satisfy the following identity:

$$d^j d^i = d^{i+1} d^j$$

for $i \geq j$.

Remark 1.3. For seeing that morphisms in $\mathcal{O}^+$ are generated by d^i, observe that any morphism in $\mathcal{O}^+$ means an ordered embedding, which can be written as the compositions of d^i's.

Let $\mathcal{S}$ denote the category of sets.

Proposition 1.4. *Δ-sets are one-to-one correspondent to contravariant functors from $\mathcal{O}^+$ to $\mathcal{S}$.*

Proof. Let $F \colon \mathcal{O}^+ \to \mathcal{S}$ be a contravariant functor. Define $X_n = F([n])$ and

$$d_i = F(d^i) \colon X_n = F([n]) \to X_{n-1} = F([n-1]).$$

Then X is a Δ-set.

Conversely suppose that X is a Δ-set. Define the $F \colon \mathcal{O}^+ \to \mathcal{S}$ by setting $F([n]) = X_n$ and $F(d^i) = d_i$. Then F is a contravariant functor. $\qquad\square$

A Δ-set $G = \{G_n\}_{n \geq 0}$ is called a Δ-*group* if each G_n is a group, and each face d_i is a group homomorphism. In other words, a Δ-group means a contravariant functor from $\mathcal{O}^+$ to the category of groups. More abstractly, for any category $\mathcal{C}$, a Δ-object over $\mathcal{C}$ means a contravariant functor from

$\mathcal{O}^+$ to $\mathcal{C}$. In other words, a Δ-object over $\mathcal{C}$ means a sequence of objects over $\mathcal{C}$, $X = \{X_n\}_{n \geq 0}$ with faces $d_i \colon X_n \to X_{n-1}$ as morphisms in $\mathcal{C}$.

Example 1.5 (n-simplex). The n-simplex $\Delta^+[n]$, as a Δ-set, is as follows:

$$\Delta^+[n]_k = \{(i_0, i_1, \ldots, i_k) \mid 0 \leq i_0 < i_1 < \cdots < i_k \leq n\}$$

for $k \leq n$ and $\Delta^+[n]_k = \emptyset$ for $k > n$. The face $d_j \colon \Delta^+[n]_k \to \Delta^+[n]_{k-1}$ is given by

$$d_j(i_0, i_1, \ldots, i_k) = (i_0, i_1, \ldots, \hat{i}_j, \ldots, i_k),$$

that is deleting i_j. Let $\sigma_n = (0, 1, \ldots, n)$. Then

$$(i_0, i_1, \ldots, i_k) = d_{j_1} d_{j_2} \cdots d_{j_{n-k}} \sigma_n,$$

where $j_1 < j_2 < \cdots < j_{n-k}$ with $\{j_1, \ldots, j_k\} = \{0, 1, \ldots, n\} \smallsetminus \{i_0, i_1, \ldots, i_k\}$. In other words, any elements in $\Delta^{[n]}$ can be written an iterated face of σ_n.

Definition 1.6. A Δ-*map* $f \colon X \to Y$ means a sequence of functions

$$f \colon X_n \to Y_n$$

for each $n \geq 0$ such that $f \circ d_i = d_i \circ f$, that is the diagram

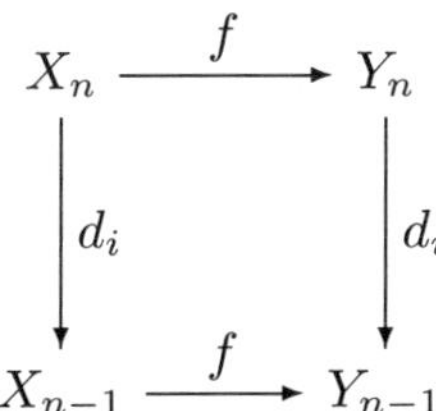

commutes. A Δ-*subset* A of a Δ-set X means a sequence of subsets $A_n \subseteq X_n$ such that

$$d_i(A_n) \subseteq A_{n-1}$$

for all $0 \leq i \leq n < \infty$. A Δ-set X is called to be *isomorphic* to a Δ-set Y, denoted by $X \cong Y$, if there is a bijective Δ-map $f \colon X \to Y$.

Let X be a Δ-set and let A be a Δ-subset. Clearly the inclusion $A \subseteq X$, that is, $A_n \subseteq X_n$ for each $n \geq 0$, is a Δ-map.

Proposition 1.7. *Let X be a Δ-set and let $x \in X_n$ be an element. Then there exists a unique Δ-map*

$$f_x \colon \Delta^+[n] \to X$$

such that $f_x(\sigma_n) = x$.

Proof. From the assumption $f_x(\sigma_n) = x$, we have

$$\begin{aligned}
f_x(i_0, i_1, \ldots, i_k) &= f_x(d_{j_1} d_{j_2} \cdots d_{j_{n-k}} \sigma_n) \\
&= d_{j_1} d_{j_2} \cdots d_{j_{n-k}} f_x(\sigma_n) \\
&= d_{j_1} d_{j_2} \cdots d_{j_{n-k}} x.
\end{aligned}$$

This defines a Δ-map f_x such that $f_x(\sigma_n) = x$. $\qquad\square$

The simplicial map $f_x \colon \Delta^+[n] \to X$ is called *representing map* of x.

Definition 1.8. Let X be a Δ-set and let $S \subseteq \bigcup_{n=0}^{\infty} X_n$. The Δ-*subset* generated by S is defined by

$$\langle S \rangle^{\Delta} = \bigcap \left\{ A \subseteq X \mid S \subseteq \bigcup_{n=0}^{\infty} A_n \ A = \{A_n\} \text{ is a } \Delta\text{-subset of } X \right\}.$$

For $x \in X_n$, $\langle \{x\} \rangle^{\Delta}$ is simply denoted by $\langle x \rangle^{\Delta}$. A Δ-set X is called *monogenic* if it is generated by a single element.

Proposition 1.9. *Let X be a Δ-set and let $S \subseteq \bigcup_{n=0}^{\infty} X_n$. Then*

$$\langle S \rangle_n^{\Delta} = (S \cap X_n) \cup \bigcup_{\substack{0 \leq j_1 < j_2 < \cdots < j_k \leq n+k \\ 1 \leq k < \infty}} d_{j_1} d_{j_2} \cdots d_{j_k}(S \cap X_{n+k})$$

for each $n \geq 0$.

Proof. Let

$$K_n = (S \cap X_n) \cup \bigcup_{\substack{0 \leq j_1 < j_2 < \cdots < j_k \leq n+k \\ 1 \leq k < \infty}} d_{j_1} d_{j_2} \cdots d_{j_k}(S \cap X_{n+k})$$

Then $d_i(K_n) \subseteq K_{n-1}$ by the Δ-identity (1.1) for $0 \leq i \leq n$. Thus $K = \{K_n\}_{n \geq 0}$ is a Δ-subset of X. If A is a Δ-subset of X with $S \subseteq \bigcup_{n=0}^{\infty} A_n$, then clearly $K_n \subseteq A_n$ for each $n \geq 0$. Thus $K = \langle S \rangle^{\Delta}$ and hence the assertion. $\qquad\square$

Problem 1.10. *Classify all monogenic Δ-sets.*

1.2. *Geometric simplicial complexes*

In this section, we give a review on (geometric) simplicial complex.

1.2.1. *Simplices*

Definition 1.11. A set $\{a^0, \ldots, a^n\}$ of $(n+1)$ points in $\mathbb{R}^m$ is called *geometrically independent* if the vectors $a^1 - a^0$, $a^2 - a^0$, $\ldots, a^n - a^0$ are linearly independent. A *geometric n-simplex*

$$\sigma^n = \left\{ x = \sum_{i=0}^{n} t_i a^i \ \Big|\ t_i \geq 0 \text{ and } \sum_{i=0}^{n} t_i = 1 \right\} \subseteq \mathbb{R}^m$$

with subspace topology, where $\{a^0, \ldots, a^n\}$ linearly independent. Sometimes we write $\sigma = a^0 a^1 \cdots a^n$ for meaning that σ is spanned by the vertices $a^0, a^1, \ldots, a^n$. The numbers t_i are uniquely determined by the point x, which are called *barycentric coordinates* of x of σ with respect to $a^0, \ldots, a^n$. The points a^i are called the *vertices* of σ^n. If $\{a^{i_0}, a^{i_1}, \ldots, a^{i_r}\}$ is a subset of $\{a^0, \ldots, a^n\}$ with $0 \leq i_0 < \cdots < i_r \leq n$, then the subspace

$$t^r = \left\{ \sum_{j=0}^{r} t_j a^{i_j} \ \Big|\ \lambda_j \geq 0 \text{ and } \sum_{i=0}^{n} t_j = 1 \right\}$$

is called a *face* of σ^n. The number n is called the *dimension* of σ. A face of σ different from σ itself is called a *proper face*. The union of all proper faces of σ is called the *boundary* of σ, denoted by $\partial \sigma$. The *interior* $\mathrm{Int}(\sigma)$ of σ is defined by $\mathrm{Int}\,\sigma = \sigma \smallsetminus \partial \sigma$, sometimes called *open simplex*.

Exercise 1.1.

(1). The barycentric coordinates $t_i(x)$ of x with respect to $a^0, \ldots, a^n$ are continuous on x.

(2). $x \in \partial \sigma \Longleftrightarrow t_i(x) = 0$ for some $0 \leq i \leq n$. Thus $x \in \mathrm{Int}\,\sigma \Longleftrightarrow t_i(x) > 0$ for all $0 \leq i \leq n$.

(3). σ is compact and convex in $\mathbb{R}^m$, where a subset A of $\mathbb{R}^m$ is called *convex* if for each pairs $x, y \in A$ the line segment joining x and y lies in A.

(4). Given a simplex, there is one and only one geometrically independent set of points spanning σ.

(5). σ equals to the union of all line segments joining a_0 to points of the simplex τ spanned by $a_1, \ldots, a_n$. Int σ is convex and is open in plane spanned by the points $a^0, a^1, \ldots, a^n$; its closure is σ. Furthermore Int σ is the union of all open line segments joining a^0 to the points of Int τ.

(6). Recall that in $\mathbb{R}^m$, the *norm* of a point $x = (x_1, \ldots, x_m)$ is defined to be

$$\|x\| = \sqrt{\sum_{i=1}^{m} x_i^2}.$$

The m-dimensional *unit ball* D^m is defined by $D^n = \{x \in \mathbb{R}^m \mid \|x\| \leq 1\}$ and the $(m-1)$-dimensional unit sphere S^{m-1} is defined to be

$$S^{m-1} = \{x \in \mathbb{R}^m \mid \|x\| = 1\}.$$

There is a homeomorphism of σ with the unit ball D^n that carries $\partial\sigma$ onto the unit sphere S^{n-1}.

1.2.2. *Geometric simplicial complex*

For having (geometric) simplicial complexes with arbitrary many simplices and arbitrary dimension of its simplices, we need to extend our m-dimensional Euclidean space to an infinite dimensional vector space as follows:

Let J be an arbitrary index set, and let $\mathbb{R}^J$ be the J-fold product of $\mathbb{R}$ with itself. An element in $\mathbb{R}^J$ is a function from J to $\mathbb{R}$ denoted by $x = (x_\alpha)_{\alpha \in J}$. The product $\mathbb{R}^J$ is vector space with addition given by the usual component-wise addition and multiplication by scalars, that is, $(x + y)_\alpha = x_\alpha + y_\alpha$ and $(cx)_\alpha = cx_\alpha$ for $\alpha \in J$.

Let $\mathbb{E}^J$ be the subset of $\mathbb{R}^J$ consisting of all points $(x_\alpha)_{\alpha \in J}$ such that $x_\alpha = 0$ for all but finitely many values of α. Then $\mathbb{E}^J$ is a vector subspace of $\mathbb{R}^J$ with a basis given by e^α for $\alpha \in J$, where e^α is the map from J to $\mathbb{R}$ with $e^\alpha(\alpha) = 1$ and $e^\alpha(\beta) = 0$ for $\beta \neq \alpha$. (**Note.** $\{e^\alpha\}_{\alpha \in J}$ does not form a basis for $\mathbb{R}^J$ if J is an infinite set. Since $\{e^\alpha_{\alpha \in J}\}$ is a basis, we can also consider that $\mathbb{E}^J$ is the vector subspace of $\mathbb{R}^J$ generated by e^α for $\alpha \in J$.) The space $\mathbb{E}^J$ has a *metric* defined by setting

$$|x - y| = \max\{|x_\alpha - y_\alpha|_{\alpha \in J}\}.$$

Definition 1.12. A *geometric simplicial complex* K is a collection of simplices, all contained in some Euclidean space $\mathbb{E}^J$ for some index set J such that

(1). if σ^n is a simplex in K and τ^p is a face of σ^n, then τ^p is in K; and

(2). if σ^n and τ^p are simplices of K, then $\sigma^n \cap \tau^p$ is either empty, or a common face of σ^n and τ^p.

The *dimension* of K is defined to be

$$\dim K = \sup\{\dim \sigma \mid \sigma \text{ a simplex of } K\}.$$

So an n-dimensional simplicial complex means a simplicial complex without simplices of dimension higher than n. If L is a sub-collection of K that contains all faces of its elements, then L is a simplicial complex in its own right, called a *subcomplex* of K. One special subcomplex of K is the collection of all simplices of K dimension at most n, called the *n-skeleton* of K denoted by $\operatorname{sk}_n K$. The points of the collection $\operatorname{sk}_0 K$ are called *vertices* of K.

Exercise 1.2. A collection K of simplices is a simplicial complex if and only if the following hold:

(1). Every face of a simplex of K is in K;

(2). Every pair of distinct simplices of K have disjoint interior.

Definition 1.13. Let

$$|K| = \bigcup_{\sigma^n \in K} \sigma^n \subseteq \mathbb{R}^J$$

the union of the simplices of K. Giving each simplex its natural topology as subspace of $\mathbb{R}^N$ for some large N, we then define the topology on $|K|$ by requiring that: a subset A of $|K|$ is closed in $|K|$ if and only if $A \cap \sigma$ is closed in σ for each simplex $\sigma \in K$. It is easy to see that this defines a topology on $|K|$. The space $|K|$ is called *polyhedron* of K.

Note. In some references, a polyhedron means $|K|$ for a finite simplicial simplex K, that is, K only has finitely many simplices. In such a case, since we can choose the index set J to be a finite set, $|K| \subseteq \mathbb{R}^m$ for some large m and our *re-defined topology* on $|K|$ coincides with the subspace topology of $|K|$ in $\mathbb{R}^m$. If K has infinitely many simplices, then our redefined topology could be different from the subspace topology. For instance, let K be the collection of all line segments from the origin to the points in the unit circle S^1. The underline set $|K|$ is the same as the unit disk D^2, but one can check that the topology on $|K|$ is different from D^2. In fact, one can show that the redefined topology on $|K|$ has more open sets (and so more closed sets) than the subspace topology, that is, the topology of $|K|$ is a finer (larger) than the topology $|K|$ inherits as a subspace of $\mathbb{E}^J$, because if A is closed in $|K|$ in subspace topology then $A \cap \sigma$ is closed in σ for every $\sigma \in K$ and so A is closed in $|K|$. The redefined topology on $|K|$ is important for making simplicial maps, which will be defined later, to be continuous.

Exercise 1.3. Prove the following statements:

(1). If L is a subcomplex of K, then $|L|$ is a closed subspace of $|K|$. In particular, if $\sigma \in K$, then σ is a closed subspace of $|K|$.

(2). A map $f \colon |K| \to X$ is continuous if and only if $f|_\sigma$ is continuous for every $\sigma \in K$.

(3). $|K|$ is Hausdorff.

(4). If K is finite, then $|K|$ is compact. Conversely if a subset A of $|K|$ is compact, then $A \subseteq |L|$ for some finite subcomplex L of K.

Exercise 1.4. Prove the following statements:

(1). If K is a simplicial complex, then the intersection of any collection of subcomplex of K is a subcomplex of K.

(2). If $\{K_\alpha\}$ is a collection of simplicial complexes in $\mathbb{E}^J$, and if the intersection of every pair $|K_\alpha| \cap |K_\beta|$ is the polyhedron of a simplicial complex which is a subcomplex of both K_α and K_β, then the union $\bigcup_\alpha K_\alpha$ is a simplicial complex.

Exercise 1.5. Let K be a simplicial complex. Show that for any point $x \in |K|$ there is a unique simplex σ of K such that $x \in \mathrm{Int}\, \sigma$.

Definition 1.14. Given simplicial complexes K and L, a function

$$f \colon |K| \to |L|$$

is called a *simplicial map* if it satisfies the following conditions:

(1). If a is a vertex of K, then $f(a)$ is a vertex of L.

(2). If $a^0 a^1 \cdots a^n$ is a simplex of K, then $f(a^0), f(a^1), \ldots, f(a^n)$ span a simplex of L (possibly with repeats).

(3). If $x = \sum_{i=0}^{n} t_i a^i$ is a point in a simplex $a^0 a^1 \cdots a^n$ of K, then

$$f(x) = \sum_{i=0}^{n} t_i f(a^i).$$

That is f is *linear* on each simplex.

From the definition, for having a simplicial map, one needs:

(1). A function f which sends the vertices of K to vertices of L such that
(2). Whenever $a^0, a^1, \ldots, a^n$ span a simplex of K, $f(a^0), f(a^1), \ldots, f(a^n)$ spans a simplex of L.

Proposition 1.15. *A simplicial map $f\colon |K| \to |L|$ is continuous.*

Proof. The assertion follows from that f restricted to each simplex is continuous. $\qquad\square$

Exercise 1.6. Suppose that $f\colon \mathrm{sk}_0\, K \to \mathrm{sk}_0\, L$ is a bijective correspondence such that the vertices $a^0, a^1, \ldots, a^n$ spanned a simplex in K if and only

$$f(a^0), f(a^1), \ldots, f(a^n)$$

spanned a simplex in L. Then the induced simplicial map $f\colon |K| \to |L|$ is a homeomorphism, called *linear isomorphism* or *simplicial homeomorphism* of K with L.

An important concept is the star of a vertex in a simplicial complex.

Definition 1.16. Let K be a simplicial complex and let v be a vertex of K. The *star* of v in K, denoted by $\mathrm{St}\, v$ or $\mathrm{St}(v, K)$, is the union of the interior of those simplices of K that have v as vertex. Its closure, denoted by $\bar{\mathrm{St}}v$, is called the *closed star* of v in K. Note that $\bar{\mathrm{St}}\, v$ is the union of all simplices of K having v as a vertex. The set $\bar{\mathrm{St}}\, v \smallsetminus \mathrm{St}\, v$ is called the *link* of v in K, denoted by $\mathrm{Lk}\, v$.

A picture for the link of v is as follows:

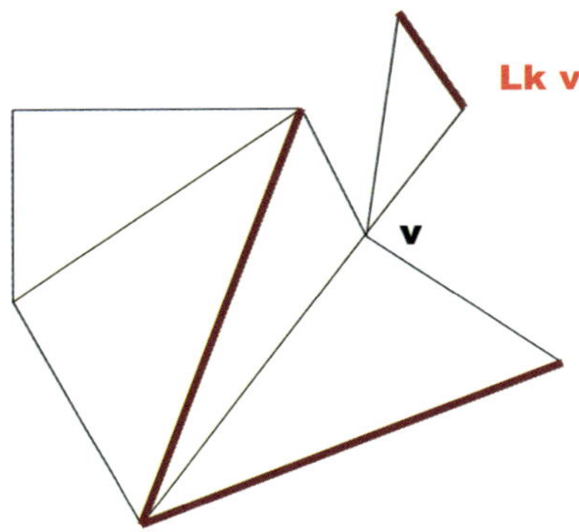

The star can be measured by continuous functions defined as follows: Let v be a vertex in K and let x be a point in $|K|$. Then x is interior to precisely one simplex of K, whose vertices are (say) $a^0, a^1, \ldots, a^n$. Then

$$x = \sum_{i=0}^{n} t_i a^i$$

with $t_i > 0$ and $\sum_{i=0}^{n} t_i = 1$. We define the *barycentric coordinate* $t_v(x)$ of x with respect to v by setting

$$t_v(x) = \begin{cases} t_i & \text{if } v = a^i \text{ for some } 0 \leq i \leq n \\ 0 & \text{otherwise.} \end{cases}$$

Proposition 1.17. *Let K be a simplicial complex and let v be a vertex. Then $t_v \colon |K| \to \mathbb{R}$ is continuous.*

Proof. Given any simplex σ of K, $t_v(x)|_\sigma$ is either identically 0 or the barycentric coordinate of x with respect to the vertex v of σ. Thus $t_v(x)|_\sigma$ is continuous. By the definition of the topology on $|K|$, $t_v(x)$ is continuous on $|K|$. $\qquad\square$

Proposition 1.18. *Let K be a simplicial complex and let v be a vertex. Then*

$$\operatorname{St} v = \{x \in |K| \mid t_v(x) > 0\}.$$

Thus $\operatorname{St} v$ is an open neighborhood of v in $|K|$.

Proof. Exercise. $\qquad\square$

Proposition 1.19 (Star covering). *Let K be a simplicial complex. Then*

$$|K| = \bigcup_{v \in \mathrm{sk}_0 K} \operatorname{St} v.$$

Proof. For any $x \in |K|$, there exists a unique simplicial σ such that $x \in \operatorname{Int} \sigma$. Let v be a vertex of σ. Then $x \in \operatorname{St} v$ and hence the result. $\qquad\square$

Proposition 1.20. *Let $x \in \bar{\operatorname{St}} v$. Then the line segment xv lies in $\bar{\operatorname{St}} v$. Moreover the line segment starting from v meets $\operatorname{Lk} v$ exactly one point.*

Proof. Exercise. $\qquad\square$

1.2.3. *Subdivision*

Definition 1.21. Let K be a geometric simplicial complex in $\mathbb{E}^J$. A complex K' is called to be a *subdivision* of K if

(1). Each simplex of K' is contained in a simplex of K.

(2). Each simplex of K equals to the union of **finitely many** simplices of K'.

These conditions imply that the union of the simplices of K' is equal to the union of the simplices of K, that is, $|K| = |K'|$ as sets. The finiteness part of condition (2) guarantee that $|K'|$ and $|K|$ are equal as topological spaces.

Definition 1.22. Suppose that K is a simplicial complex in $\mathbb{E}^J$, and w is a point in $\mathbb{E}^J$ such that each ray emanating from w intersects $|K|$ in at most one point. We define the *cone on K with vertex w* to be the collection of all simplices of the form $a^0 a^1 \cdots a^p w$, where $a^0 a^1 \cdots a^p$ is a simplex in K, along all faces of such simplices. Denote the cone by $K * w$.

The cone $K * w$ is pictured as follows:

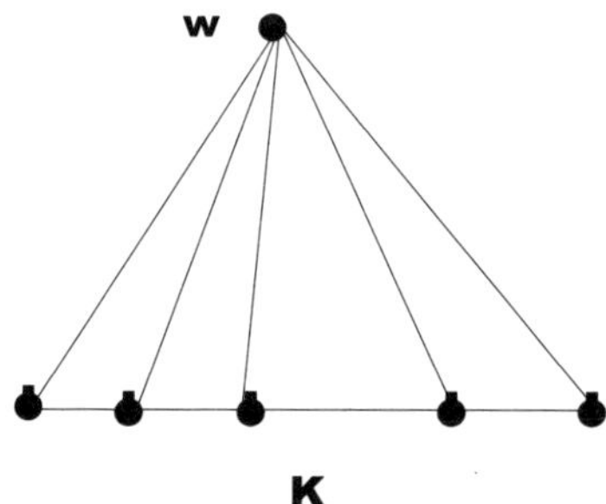

Definition 1.23. Let K be a simplicial complex. Suppose that L_p is a subdivision of $\mathrm{sk}_p K$. Let σ be a $(p+1)$-simplex of K. Note that $|\partial\sigma|$ is a polyhedron of a subcomplex of $\mathrm{sk}_p K$ and so it is a polyhedron of a subcomplex, denoted by L_σ, of L_p. If w_σ is an interior point of σ, then the cone $L_\sigma * w_\sigma$ is a simplicial complex whose underlying space is σ. We define L_{p+1} to be the union of L_p and the simplicial complexes $L_\sigma * w_\sigma$ as σ runs over all $(p+1)$-simplices of K. Then L_{p+1} is a simplicial complex, (check this!), called *subdivision of* $\mathrm{sk}_{p+1} K$ *obtained by starring L_p from the points w_σ.*

Definition 1.24. Let $\sigma = v^0 v^1 \cdots v^n$ be an n-simplex. The *barycenter* of σ is the point

$$\hat{\sigma} = \sum_{i=0}^{n} \frac{1}{n+1} v^i,$$

that is $\hat{\sigma}$ is the point of $\operatorname{Int} \sigma$ all of those barycentric coordinates with respect to the vertices are equal.

If σ is 1-simplex, then $\hat{\sigma}$ is the midpoint. If σ is a 0-simplex, then $\hat{\sigma} = \sigma$. In general, $\hat{\sigma}$ is the centroid of σ.

Definition 1.25. Let K be a simplicial complex. We define a sequence of subdivisions of the skeletons of K as follows: Let $L_0 = \operatorname{sk}_0 K$. Assume that L_p is defined as a subdivision of $\operatorname{sk}_p K$. Let L_{p+1} be the subdivision of $\operatorname{sk}_{p+1} K$ obtained by starring L_p from the barycenter of the $(p+1)$-simplices of K. The union $\bigcup_{p=0}^{\infty} L_p$ is a subdivision of K, called *barycentric subdivision* of K, denoted by $\operatorname{sd} K$. Define *the iterated barycentric subdivision* recursively by $\operatorname{sd}^n K = \operatorname{sd}^{n-1}(\operatorname{sd} K)$ for $n > 1$.

Let σ be a 2-simplex. Then $\operatorname{sd} \sigma$ is shown in the picture below:

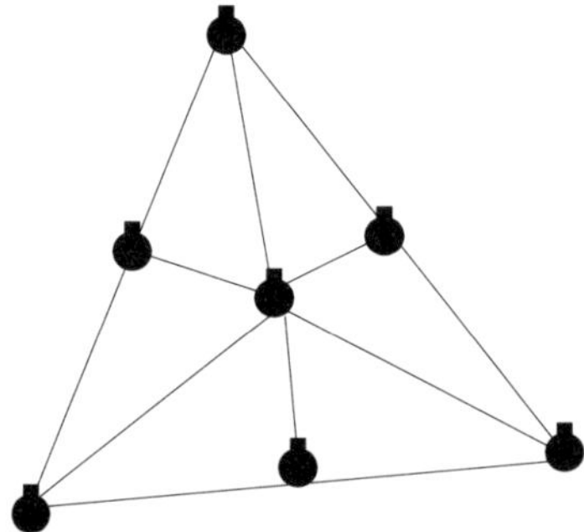

The simplices in $\operatorname{sd} K$ can be described as follows. Define a partial order on the simplices of K by setting $\sigma_1 < \sigma_2$ if σ_1 is a proper face of σ_2.

Proposition 1.26. *The simplicial complex* $\operatorname{sd} K$ *equals to the collection of all simplices of the form*

$$\hat{\sigma}_1 \hat{\sigma}_2 \cdots \hat{\sigma}_n,$$

where $\sigma_1 < \sigma_2 < \cdots < \sigma_n$ *in* K.

Proof. The proof is given by induction on p that the assertion holds for $\mathrm{sk}_p\, K$ for each $p \geq 0$. The assertion holds for $\mathrm{sk}_0\, K$ as $\mathrm{sd}\,\mathrm{sk}_0\, K = \mathrm{sk}_0\, K$. Suppose that the assertion holds for $\mathrm{sk}_p\, K$. By definition of barycentric subdivision, the assertion holds for $\mathrm{sk}_{p+1}\, K$. Since $\mathrm{sd}\, K = \bigcup_p \mathrm{sd}\,\mathrm{sk}_p\, K$, the assertion follows. $\qquad\square$

An important property of barycentric subdivision is as follows, which plays a key role for proving the simplicial approximation theorem.

Theorem 1.27. *Given a finite simplicial complex K, and given any $\epsilon > 0$, there is a positive integer N such that each simplex of $\mathrm{sd}^N K$ has diameter less than ϵ.*

Proof. 1. *If $\sigma = a^0 a^1 \cdots a^n$ is a simplex, then the diameter $\mathrm{diam}\,\sigma = \max\{|a^i - a^j| \mid 0 \leq i \leq j \leq n\}$, the maximum distance between vertices.*

Let $l = \max\{|a^i - a^j| \mid 0 \leq i < j \leq n\}$. For each vertex a^i, let $D(a^i, l) = \{x \mid |x - a^i| \leq l\}$ be the ball of radius l centered at a^i. Then $D(a^i, l)$ is convex. Since $E(a^i, l)$ contains all vertices of σ, $\sigma \subseteq D(a^i, l)$. Thus

$$|x - a^i| \leq l$$

for any $x \in \sigma$ and each $0 \leq i \leq n$.

Now given any $x \in \sigma$, consider the ball $D(x, l) = \{y \mid |y - x| \leq l\}$. Then $\sigma \subseteq D(x, l)$ because each $a^i \in D(x, l)$ as $|a^i - x| \leq l$. Thus, for any $z \in \sigma$, $|x - z| \leq l$. Hence $\mathrm{diam}\,\sigma = l$.

2. *If $\sigma = a^0 a^1 \cdots a^n$ is a simplex, then for any $x \in \sigma$*

$$|\hat{\sigma} - x| \leq \frac{n}{n+1}\mathrm{diam}\,\sigma.$$

Note that for each a^t

$$|a^t - \hat{\sigma}| = \left| a^t - \sum_{i=0}^{n} \frac{1}{n+1} a^i \right|$$

$$= \left| \sum_{\substack{0 \leq i \leq n \\ i \neq t}} \frac{1}{n+1} (a^t - a^i) \right|$$

$$\leq \sum_{\substack{0 \leq i \leq n \\ i \neq t}} \frac{1}{n+1} |a^t - a^i|$$

$$\leq \frac{n}{n+1}\mathrm{diam}\,\sigma.$$

Let $l' = \max\{|a^t - \hat{\sigma}| \mid 0 \leq t \leq n\}$. Then $l' \leq \frac{n}{n+1}\operatorname{diam}\sigma$. The ball

$$D(\hat{\sigma}, l') = \{x \mid |x - \hat{\sigma}| \leq l'\}$$

contains all vertices of σ and so $\sigma \subseteq D(\hat{\sigma}, l')$. It follows that

$$|x - \hat{\sigma}| \leq l' \leq \frac{n}{n+1}\operatorname{diam}\sigma$$

for any $x \in \sigma$.

3. For any simplicial complex L, let

$$\operatorname{mesh} L = \sup\{\operatorname{diam}\sigma \mid \sigma \text{ is a simplex of } L\}.$$

Let K be an n-dimensional finite simplicial complex. Then

$$\operatorname{mesh}\operatorname{sd} K \leq \frac{n}{n+1}\operatorname{mesh} K.$$

The proof is given by induction on the skeleton $\operatorname{sk}_p K$. For $\operatorname{sk}_0 K$, we have $\operatorname{mesh}\operatorname{sd}\operatorname{sk}_0 K = \operatorname{mesh}\operatorname{sk}_0 K = 0$. Suppose that

$$\operatorname{mesh}\operatorname{sd}\operatorname{sk}_p K \leq \frac{p}{p+1}\operatorname{mesh}\operatorname{sk}_p K.$$

Consider $\operatorname{sk}_{p+1} K$. By the definition, $\operatorname{sd}\operatorname{sk}_{p+1} K$ is the union of $\operatorname{sd}\operatorname{sk}_p K$ and the simplices of the form $\tau * \hat{\sigma}$ for $(p+1)$-simplices σ of K, where τ is a simplex of $\operatorname{sd}\partial\sigma$. If σ' is a simplex of $\operatorname{sd}\operatorname{sk}_p K$, then

$$\operatorname{diam}\sigma' \leq \frac{p}{p+1}\operatorname{mesh}\operatorname{sk}_p K \leq \frac{p}{p+1}\operatorname{mesh}\operatorname{sk}_{p+1} K \leq \frac{p+1}{p+2}\operatorname{mesh}\operatorname{sk}_{p+1} K.$$

If $\sigma' = \tau * \hat{\sigma}$ with τ a simplex of $\operatorname{sd}\partial\sigma$, then, by Step 2,

$$\operatorname{diam}\sigma' \leq \frac{p+1}{p+2}\operatorname{diam}\sigma \leq \frac{p+1}{p+2}\operatorname{mesh}\operatorname{sk}_{p+1} K.$$

The induction is finished and hence the statement.

4. Since K is a finite simplicial complex, let $\dim K = n$ and let $d = \operatorname{mesh} K$. Then

$$\operatorname{mesh}\operatorname{sd}^N K \leq \left(\frac{n}{n+1}\right)^N d.$$

Thus $\operatorname{mesh}\operatorname{sd}^N K \to 0$ as $N \to \infty$ and hence the result. $\square$

Remark 1.28. By inspecting the proof, the assertion also holds if K is a finite dimensional simplicial complex such that $\operatorname{mesh} K < +\infty$.

In practice it may be necessary to subdivide only part of a simplicial complex K, so as to leave alone a given subcomplex A. For doing subdivision in this case, we have the relative skeleton filtration defined as follows: Let $\mathrm{sk}^A_{-1} K = A$ and let $\mathrm{sk}^A_n K$ be the union of A and the simplices σ of K with $\dim \sigma \leq n$.

Definition 1.29. Let K be a geometric simplicial complex in $\mathbb{E}^J$ and let A be a subcomplex of K. A complex K' is called to be a *subdivision of K relative A* if

(1). Each simplex of K' is contained in a simplex of K.
(2). Each simplex of K equals to the union of **finitely many** simplices of K'.
(3). A is a subcomplex of K'.

The last condition requires that the simplices of the subcomplex A do not get subdivision.

Definition 1.30. Let K be a simplicial complex and let A be a subcomplex of K. We define a sequence of subdivisions of the relative skeletons $\mathrm{sk}^A_n K$ of K as follows: Let $L_{-1} = A$ and let $L_0 = \mathrm{sk}^A_0 K$. Assume that L_p is defined as a subdivision of $\mathrm{sk}^A_p K$. Let L_{p+1} be the subdivision of $\mathrm{sk}_{p+1} K$ obtained by starring L_p from the barycenter of the $(p+1)$-simplices of K NOT in A. The union $\bigcup_{p=0}^{\infty} L_p$ is a subdivision of K, called *barycentric subdivision of K relative to A*, denoted by $\mathrm{sd}(K, A)$. Define the *iterated barycentric subdivision* recursively by $\mathrm{sd}^n(K, A) = \mathrm{sd}^{n-1}(\mathrm{sd}(K, A))$ for $n > 1$.

The barycentric subdivision relative to a subcomplex is shown by the picture:

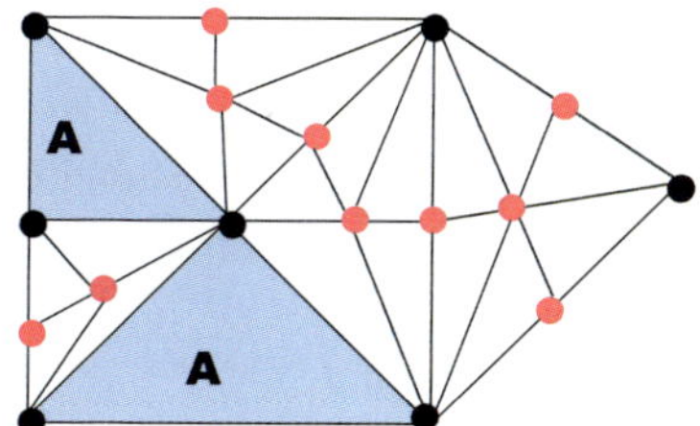

Recall that $\tau < \sigma$ if and only if τ is a proper face of σ.

Proposition 1.31. *Let K be a simplicial complex and let A be a subcomplex. Then the vertices of $\mathrm{sd}(K, A)$ are the barycenters of the simplices of*

$K \smallsetminus A$, *together with the vertices of A. Distinct points*

$$a^1, a^2, \ldots, a^q, \hat{\sigma}_1, \hat{\sigma}_2, \ldots, \hat{\sigma}_m$$

(with $\dim \sigma_i \leq \dim \sigma_{i+1}$ *for each i) span a simplex of* $\mathrm{sd}(K, A)$ *if and only if* $a^1, a^2, \ldots, a^q$ *span a simplex σ of A, σ_j is a simplex of $K \smallsetminus A$ for* $1 \leq j \leq m$ *with* $\sigma < \sigma_1 < \sigma_2 < \cdots < \sigma_m$.

Note. If $m = 0$, then it just requires $a^1, \ldots, a^q$ that span a simplex of A. If $q = 0$, it just requires $\sigma_m > \sigma_{m-1} > \cdots > \sigma_1$ and in this case $\hat{\sigma}_1, \hat{\sigma}_2, \ldots, \hat{\sigma}_m$ forms a simplex disjoint from A.

Proof. The proof follows by induction on the relative skeleton $\mathrm{sk}_n^A K$. $\quad\square$

1.3. *Abstract simplicial complexes and Δ-sets*

1.3.1. Definition and geometric realizations

Definition 1.32. An *abstract simplicial complex* $\mathcal{K}$ is a collection of finite nonempty sets, such that if A is an element in $\mathcal{K}$, so is every nonempty subset of A.

The element A of $\mathcal{K}$ is called a *simplex* of $\mathcal{K}$; its *dimension* is one less than the number of its elements. Each nonempty subset of A is called a *face* of A. The *dimension* of $\mathcal{K}$ is the supremum of the dimensions of its simplices. The vertex set $V(\mathcal{K})$ is the union of the one-point elements of $\mathcal{K}$; we shall make no distinction between the vertex v and the 0-simplex $\{v\}$. A sub collection of $\mathcal{K}$ that is itself a complex is called a *subcomplex* of $\mathcal{K}$. Two abstract simplicial complexes $\mathcal{K}$ and $\mathcal{K}'$ are called to be *isomorphic* if there exists a bijective correspondence f mapping the vertex set of $\mathcal{K}$ to the vertex set of $\mathcal{K}'$ such that $\{a_0, a_1, \ldots, a_n\} \in \mathcal{K}$ if and only if $\{f(a_0), f(a_1), \ldots, f(a_n)\} \in \mathcal{K}'$.

Definition 1.33. Let K be a geometric simplicial complex. Let V be the vertex set of K. Let $\mathcal{K}$ be the collection of all subsets $\{a_0, a_1, \ldots, a_n\}$ of V such that $a_0, a_1, \ldots, a_n$ span a simplex of K. The collection $\mathcal{K}$ is called *vertex scheme* of K, or *abstraction* of K.

Theorem 1.34. *A relation between abstract simplicial complexes and geometric simplicial complexes is as follows:*

(1). *Every abstract simplicial complex $\mathcal{K}$ is isomorphic to the vertex scheme of some geometric simplicial complex.*

(2). *Two geometric simplicial complexes are linearly isomorphic if and only if their vertex schemes are isomorphic as abstract simplicial complexes.*

Proof. We leave (2) as an exercise. To prove (1), we proceed as follows: Given an index set J, let Δ^J be the collection of all simplices in $\mathbb{E}^J$ spanned by finite subsets of the standard basis $\{e_\alpha\}_{\alpha \in J}$ for $\mathbb{E}^J$. It is easy to see that Δ^J is a simplicial complex. Moreover if σ and τ are two simplices of Δ^J, then their combined vertex set is geometrically independent and spans a simplex of Δ^J.

Now let $\mathcal{K}$ be an abstract simplicial complex with vertex set V. Choose the index set $J = V$. We specify a subcomplex K of Δ^J by the condition that for each abstract simplex $\{a_0, \ldots, a_n\} \in \mathcal{K}$, the geometric simplex spanned by $e_{a_0}, e_{a_1}, \ldots, e_{a_n}$ is to be in K. It is immediate that K is a geometric simplicial complex and $\mathcal{K}$ is isomorphic to the vertex scheme of K. $\qquad\qquad\square$

1.3.2. *Subdivision of abstract simplicial complexes*

Let V be a set with a partial order $<$. Let

$$\mathcal{S}(V) = \{\{a_0, a_1, \ldots, a_n\} \mid a_i \in V \; a_0 < a_1 < \cdots < a_n\}.$$

Then clearly $\mathcal{S}(V)$ is an abstract simplicial complex with V as its vertex set because if $\{a_0, a_1, \ldots, a_n\}$ is well-ordered, then any nonempty subset of $\{a_0, a_1, \ldots, a_n\}$ is well-ordered.

The subdivision of abstract simplicial complexes follows the ideas of barycentric subdivision of geometric simplicial complexes. Let $\mathcal{K}$ be an abstract simplicial complex. Let $A, B \in \mathcal{K}$. Let $\mathcal{K}$ be partially ordered by

$$A < B$$

if A is a proper face of B. Define

$$\mathrm{sd}(\mathcal{K}) = \mathcal{S}(\mathcal{K})$$

called *subdivision* of $\mathcal{K}$. If K is the geometric realization of $\mathcal{K}$, then, by definition, the geometric realization of $\mathrm{sd}(\mathcal{K})$ is $\mathrm{sd}(K)$.

1.3.3. *Abstract simplicial complexes and Δ-sets*

Let $\mathcal{K}$ be an abstract simplicial complex. Let $\mathcal{K}_n$ be the set of n-simplices of $\mathcal{K}$. Define the faces

$$d_i \colon \mathcal{K}_n \to \mathcal{K}_{n-1}, \quad 0 \le i \le n,$$

as follows:

> Let the vertices of K be ordered. If $\{a^0, a^1, \ldots, a^n\}$ is an n-simplex of K with $a^0 < a^1 < \cdots < a^n$, then define
>
> $$d_i\{a^0, a^1, \ldots, a^n\} = \{a^0, a^1, \ldots, a^{i-1}, a^{i+1}, \ldots, a^n\}.$$

Proposition 1.35. *Let $\mathcal{K}^\Delta = \{\mathcal{K}_n\}_{n \geq 0}$ with faces defined as above. Then $\mathcal{K}^\Delta$ is a Δ-set.*

Proof. Exercise. $\qquad\qquad\qquad\qquad\qquad\qquad\qquad\qquad\qquad\qquad\qquad$ $\square$

Definition 1.36. A Δ-set X is called *polyhedral* if there exists an abstract simplicial complex K such that $X \cong \mathcal{K}^\Delta$.

In general, a Δ-set may not be polyhedral.

Exercise 1.7. Let $X = \Delta^+[1] \cup_{\Delta^+[1]_0} \Delta^+[1]$ be the union of two copies of $\Delta^+[1]$ by identifying the vertices. Show that X is not polyhedral.

Now let X be a Δ-set and let 2^{X_0} be the set of all subsets of X_0. Define

$$\phi \colon \coprod_{n \geq 0} X_n \longrightarrow 2^{X_0}$$

by setting $\phi(x) = \{f_x(0), f_x(1), \ldots, f_x(n)\}$ for $x \in X_n$.

Theorem 1.37. *Let X be a Δ-set. Then X is polyhedral if and only if the following holds:*

(1). There exists an order of X_0 such that, for each $x \in X_n$,

$$f_x(0) \leq f_x(1) \leq \cdots \leq f_x(n).$$

(2). The function $\phi \colon \bigcup_{n \geq 0} X_n \longrightarrow 2^{X_0}$ is one-to-one.

Proof. If $X \cong \mathcal{K}^\Delta$ for some abstract simplicial complex $\mathcal{K}$, then each n-simplex is uniquely determined by its vertices. Thus ϕ is one-to-one. From the construction, we have $f_x(0) < f_x(1) < \cdots < f_x(n)$.

Conversely suppose that X is a Δ set satisfies the two conditions in the statement. First we show that, for $x \in X_n$, the cardinal of $\{f_x(0), f_x(1), \ldots, f_x(n)\}$ is $n + 1$. Otherwise there exists $0 \leq i < j \leq n$ such that $f_x(i) = f_x(j)$. Then

$$d_i(x) = f_x d_i(0, 1, \ldots, n) = f_x(0, 1, \ldots, i-1, i+1, \ldots, n)$$

and so

$$\phi(d_i(x)) = \{f_x(0), f_x(1), \ldots, f_x(i-1), f_x(i+1), \ldots, f_x(n)\}$$
$$= \{f_x(0), f_x(1), \ldots, f_x(i-1), f_x(i), f_x(i+1), \ldots, f_x(n)\}$$
$$= \phi(x).$$

Thus $d_i(x) = x$, which contradicts to that $X_n \cap X_{n-1} = \emptyset$.

Let the vertices of $\mathcal{K}$ be the elements of X_0. Define a subset

$$\{a^0, a^1, \ldots, a^n\} \subseteq X_0$$

to be an n-simplex of $\mathcal{K}$ if and only if $\{a^0, a^1, \ldots, a^n\} \in \mathrm{Im}(\phi)$, where the elements are given in the order that $a^0 < a^1 < \cdots < a^n$.

For checking that $\mathcal{K}$ is an abstract simplicial complex, let $\{a^0, a^1, \ldots, a^n\}$ be an n-simplex of $\mathcal{K}$ such that $\{a^0, a^1, \ldots, a^n\} = \phi(x)$. From the above arguments, $x \in X_n$ and so we may assume that $a^i = f_x(i)$ for $0 \leq i \leq n$. Let $\{a^{i_0}, a^{i_1}, \ldots, a^{i_p}\}$ be a subset of $\{a^0, a^1, \ldots, a^n\}$ with $0 \leq i_0 < i_1 < \cdots < i_p \leq n$. Let

$$\{j_0, j_1, \ldots, j_{n-p-1}\} = \{0, 1, \ldots, n\} \smallsetminus \{i_0, i_1, \ldots, i_p\}$$

with $0 \leq j_0 < j_1 < \cdots < j_{n-p-1}$. Then

$$d_{j_0} d_{j_1} \cdots d_{j_{n-p-1}} x = f_x d_{j_0} d_{j_1} \cdots d_{j_{n-p-1}} (0, 1, \ldots, n)$$
$$= f_x(i_0, i_1, \ldots, i_p).$$

Thus

$$\phi(d_{j_0} d_{j_1} \cdots d_{j_{n-p-1}} x) = \{a^{i_0}, a^{i_1}, \ldots, a^{i_p}\}$$

and so $\{a^{i_0}, a^{i_1}, \ldots, a^{i_p}\}$ is a p-simplex of $\mathcal{K}$.

Finally the function

$$g \colon X \longrightarrow \mathcal{K}^{\Delta}$$

with $g(x) = \{f_x(0), \ldots, f_x(n)\}$ for $x \in X_n$ is bijective Δ-map. This proves that $X \cong \mathcal{K}^{\Delta}$. $\qquad\square$

1.4. Δ-complexes and the geometric realization of Δ-sets

The standard *geometric n-simplex* Δ^n is defined by

$$\Delta^n = \left\{ (t_0, t_1, \ldots, t_n) \mid t_i \geq 0 \text{ and } \sum_{i=0}^{n} t_i = 1 \right\}.$$

Define $d^i \colon \Delta^{n-1} \to \Delta^n$ by setting

$$d^i(t_0, t_1, \ldots, t_{n-1}) = (t_0, \ldots, t_{i-1}, 0, t_i, \ldots, t_{n-1}).$$

Exercise 1.8. Show that the maps d^i satisfy the following identity:

$$d^j d^i = d^{i+1} d^j$$

for $i \geq j$.

The *boundary*

$$\partial \Delta^n = \bigcup_{i=0}^{n} d^i (\Delta[n-1])$$

is the union of all faces of Δ^n. Let $\mathrm{Int}(\Delta^n) = \Delta^n \smallsetminus \partial \Delta^n$ be the interior of Δ^n, called *open simplex*.

Definition 1.38. A Δ-*complex structure* on a space X is a collection of maps

$$\mathcal{C}(X) = \{\sigma_\alpha \colon \Delta^n \to X \mid \alpha \in J_n \ n \geq 0\}$$

such that

(1). $\sigma_\alpha|_{\mathrm{Int}(\Delta^n)} \colon \mathrm{Int}(\Delta^n) \to X$ is injective, and each point of X is in the image of exactly one such restriction $\sigma_\alpha|_{\mathrm{Int}(\Delta^n)}$.

(2). For each $\sigma_\alpha \in \mathcal{C}(X)$, each face

$$\sigma_\alpha \circ d^i \in \mathcal{C}(X).$$

(3). A set $A \subseteq X$ is open if and only if $\sigma_\alpha^{-1}(A)$ is open in Δ^n for each $\sigma_\alpha \in \mathcal{C}(X)$.

Define

$$C_n^\Delta(X) = \{\sigma_\alpha \colon \Delta^n \to X \mid \alpha \in J_n\} \subseteq \mathcal{C}(X)$$

with $d_i \colon C_n^\Delta(X) \to C_{n-1}^\Delta(X)$ given by

$$d_i(\sigma_\alpha) = \sigma_\alpha \circ d^i$$

for $0 \leq i \leq n$.

Exercise 1.9. $C^\Delta(X) = \{C_n^\Delta(X)\}_{n \geq 0}$ is a Δ-set.

Let K be a Δ-set. The *geometric realization* $|K|$ of K is defined to be

$$|K| = \coprod_{\substack{x \in K_n \\ n \geq 0}} (\Delta^n, x) / \sim = \coprod_{n=0}^{\infty} \Delta^n \times K_n / \sim,$$

where (Δ^n, x) is Δ^n labeled by $x \in K_n$ and $\sim$ is generated by

$$(z, d_i x) \sim (d^i z, x)$$

for any $x \in K_n$ and $z \in \Delta^{n-1}$ labeled by $d_i x$. For any $x \in K_n$, let $\sigma_x \colon \Delta^n = (\Delta^n, x) \to |K|$ be the canonical characteristic map. The topology on K is defined by $A \subseteq |K|$ is open if and only if the pre-image $\sigma_x^{-1}(A)$ is open in Δ^n for any $x \in K_n$ and $n \geq 0$.

Proposition 1.39. *Let K be a Δ-set. Then $|K|$ is Δ-complex.*

Proof. Let $\mathrm{sk}_n K = \{K_j\}_{0 \leq j \leq n}$ be the n-skeleton of K. Then $\mathrm{sk}_n K$ is a Δ-subset of K. Let $\partial\Delta^+[n] = \mathrm{sk}_{n-1}\Delta^+[n]$ be the boundary of $\Delta^+[n]$. Observe that

$$|\Delta^+[n]| \cong \Delta^n \quad |\partial\Delta^+[n]| \cong \partial\Delta^n.$$

From the push-out diagram

$$
\begin{array}{ccc}
\displaystyle\bigsqcup_{x \in K_n} \partial\Delta^+[n] & \longrightarrow & \mathrm{sk}_{n-1} K \\
\Big\downarrow & \text{push} & \Big\uparrow \\
\displaystyle\bigsqcup_{x \in K_n} \Delta^+[n] & \longrightarrow & \mathrm{sk}_n K,
\end{array}
$$

there is a push-out diagram

$$
\begin{array}{ccc}
\displaystyle\bigsqcup_{x \in K_n} \partial\Delta^n & \longrightarrow & |\mathrm{sk}_{n-1} K| \\
\Big\downarrow & \text{push} & \Big\uparrow \\
\displaystyle\bigsqcup_{x \in K_n} \Delta^n & \longrightarrow & |\mathrm{sk}_n K|.
\end{array}
$$

Thus $|\mathrm{sk}_n K|$ is obtained from $|\mathrm{sk}_{n-1} K|$ by attaching cells with labels in K_n. By induction, it follows that $|K|$ is a Δ-complex. $\qquad\square$

1.5. *Homology of Δ-sets*

Recall that a chain complex of groups means a sequence $C = \{C_n\}$ of groups with differential $\partial_n \colon C_n \to C_{n-1}$ such that $\partial_n \circ \partial_{n+1}$ is trivial, that is $\mathrm{Im}(\partial_{n+1}) \subseteq \mathrm{Ker}(\partial_n)$ and so the homology is defined by

$$H_n(C) = \mathrm{Ker}(\partial_n)/\mathrm{Im}(\partial_{n+1}),$$

which is a coset in general. A chain complex C is called *normal* if $\mathrm{Im}(\partial_{n+1})$ is a normal subgroup of $\mathrm{Ker}(\partial_n)$ for each n. In this case $H_n(C)$ is a group for each n.

Proposition 1.40. *Let G be a Δ-abelian group. Define*

$$\partial_n = \sum_{i=0}^{n} (-1)^i d_i \colon G_n \to G_{n-1}.$$

Then $\partial_{n-1} \circ \partial_n = 0$, that is, G is a chain complex under ∂_*.

Proof.

$$\partial_{n-1} \circ \partial_n = \sum_{i=0}^{n-1} (-1)^i d_i \sum_{j=0}^{n} (-1)^j d_j$$

$$= \sum_{0 \le i < j \le n} (-1)^{i+j} d_i d_j + \sum_{0 \le j \le i \le n-1} (-1)^{i+j} d_i d_j$$

$$= \sum_{0 \le i < j \le n} (-1)^{i+j} d_i d_j + \sum_{0 \le j < i+1 \le n} (-1)^{i+j} d_j d_{i+1}$$

$$= \sum_{0 \le i < j \le n} (-1)^{i+j} d_i d_j + \sum_{0 \le j < i \le n} (-1)^{i+j-1} d_j d_i$$

$$= 0. \qquad\qquad \square$$

Let X be a Δ-set. The homology $H_*(X; G)$ of X with coefficients in an abelian group G is defined by

$$H_*(X; G) = H_*(\mathbb{Z}(X) \otimes G, \partial_*),$$

where $\mathbb{Z}(X) = \{\mathbb{Z}(X_n)\}_{n \ge 0}$ and $\mathbb{Z}(X_n)$ is the free abelian group generated by X_n.

Proposition 1.41. *Let*

$$1 \longrightarrow C' \overset{i}{\longrightarrow} C \overset{p}{\longrightarrow} C'' \longrightarrow 1$$

be any short exact sequence of chain complexes of (possibly non-commutative) groups. Then there is a long exact sequence

$$\cdots \longrightarrow H_{k+1}(C'') \overset{\partial_{k+1}}{\longrightarrow} H_k(C') \overset{i_*}{\longrightarrow} H_k(C) \overset{p_*}{\longrightarrow} H_k(C'') \longrightarrow \cdots.$$

Moreover if C' and C'' are normal chain complexes, then ∂_{k+1} is a group homomorphism for each k.

Proof. Consider the commutative diagram

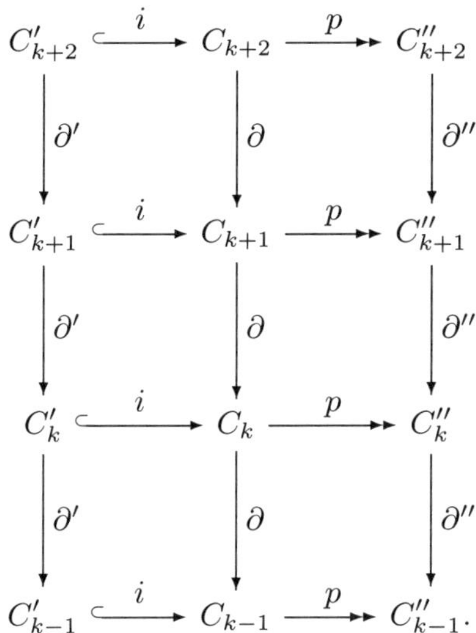

Let $x \in C''_{k+1}$ with $\partial''(x) = 1$. There exists $\tilde{x} \in C_{k+1}$ such that $p(\tilde{x}) = x$. Since

$$p(\partial(\tilde{x})) = \partial''(p(\tilde{x})) = \partial''(x) = 1,$$

there exists $\bar{x} \in C'_k$ such that $i(\bar{x}) = \partial(\tilde{x})$. Now

$$i(\partial'(\bar{x})) = \partial(i(\bar{x})) = \partial \circ \partial(\tilde{x}) = 1.$$

Thus $\bar{x}$ is a circle in C' and so $\{\bar{x}\}$ defines an element in $H_k(C')$.

Let $\hat{x}$ be another element in C_{k+1} such that $p(\hat{x}) = x$. Then

$$p(\tilde{x}\hat{x}^{-1}) = 1$$

and so there exists an element $z \in C'_{k+1}$ such that $i(z) = \tilde{x}^{-1}\hat{x}$. Now

$$i(\bar{x}\partial'(z)) = \partial(\tilde{x})(\partial(\tilde{x}))^{-1}\partial(\hat{x}) = \partial(\hat{x}).$$

Thus $\{\bar{x}\} \in H_k(C')$ is independent on the choice of the pre-image of x in C_{k+1}.

Suppose that $x' = x\partial''(y)$ with $\partial''(x) = 1$ for some $y \in C''_{k+2}$. There exists $\tilde{y} \in C_{k+2}$ such that $p(\tilde{y}) = y$. Then

$$x' = p(\tilde{x}\partial(\tilde{y}))$$

with

$$\bar{x}' = \partial(\tilde{x}\partial(\tilde{y})) = \partial(\tilde{x}) = \bar{x}.$$

This shows that

$$\partial_{k+1} \colon H_{k+1}(C'') \to H_k(C') \quad \{x\} \mapsto \{\bar{x}\}$$

is well-defined. Assume that C' and C'' are normal chain complexes. For $x, x' \in C''_{k+1}$ with $\partial''(x) = \partial''(x') = 1$. Then $p(\tilde{x}\tilde{x}') = xx'$ and so

$$\partial_{k+1}(\{x\}\{x'\}) = \partial_{k+1}(\{x\})\partial_{k+1}(\{x'\})$$

provided that C' and C'' are normal.

The composite $i_* \circ \partial_{k+1}$ is trivial because $i(\bar{x}) = \partial(\tilde{x})$. Let $y \in C'_k$ with $\partial'(y) = 1$ and $i_*(y)$ is trivial in $H_k(C)$. Then there exists $\tilde{y} \in C_{k+1}$ such that

$$i(y) = \partial(\tilde{y}).$$

By the construction of ∂_{k+1}, $\partial_{k+1}(p(\tilde{y})) = y$. This shows that

$$H_{k+1}(C'') \xrightarrow{\partial_{k+1}} H_k(C') \xrightarrow{i_*} H_k(C)$$

is exact.

Now we show that

$$H_{k+1}(C) \xrightarrow{p_*} H_{k+1}(C'') \xrightarrow{\partial_{k+1}} H_k(C')$$

is exact. Let $y \in C_{k+1}$ such that $\partial(y) = 1$. Then by the construction of ∂_{k+1}, $\partial_{k+1}(p(y)) = 1$. Thus the composite $\partial_{k+1} \circ p_*$ is trivial. Suppose that $x \in C''_{k+1}$ with $\partial''(x) = 1$ and $\bar{x} = \partial_{k+1}(x)$ is trivial in $H_k(C')$. There exists an element $z \in C'_{k+1}$ such that

$$\partial'(z) = \bar{x}.$$

Let $\hat{x} = i(z)^{-1}\tilde{x}$. Then

$$p(\hat{x}) = p(i(z)^{-1}\tilde{x}) = p(\tilde{x}) = x$$

with

$$\partial(\hat{x}) = \partial(i(z)^{-1}\tilde{x}) = i(\partial'(z)^{-1}\bar{x}) = 1.$$

Thus $\hat{x}$ defines an elements in $H_{k+1}(C)$ with $p_*(\{\hat{x}\}) = \{x\}$.

Finally we show that

$$H_k(C') \xrightarrow{i_*} H_k(C) \xrightarrow{p_*} H_k(C'')$$

is exact. Since $p \circ i$ is trivial, so is $p_* \circ i_*$. Let $y \in C_k$ with $\partial(y) = 1$ and $p_*(y)$ is trivial in $H_k(C'')$. There exists an element $z \in C''_{k+1}$ such that

$$p(y) = \partial''(z).$$

Let $\tilde{z} \in C_{k+1}$ such that $p(\tilde{z}) = z$. Then

$$\begin{aligned}
p(y\partial(\tilde{z}^{-1})) &= \partial''(z)p(\partial(\tilde{z}^{-1})) \\
&= \partial''(z)\partial''(p(\tilde{z})^{-1}) \\
&= \partial''(z)\partial''(z^{-1}) \\
&= 1.
\end{aligned}$$

Thus there exists $w \in C_k'$ such that $i(w) = y\partial(\tilde{z}^{-1})$ with

$$i(\partial'(w)) = \partial(i(w)) = \partial(y\partial(\tilde{z}^{-1})) = 1$$

and so $\partial'(w) = 1$. Hence $i_*(\{w\}) = \{y\}$. The proof is finished now. $\qquad\square$

Let X' be a Δ-subset of X. The relative homology $H_*(X, X'; G)$ is defined by

$$H_*(X, X'; G) = H_*((\mathbb{Z}(X)/\mathbb{Z}(X')) \otimes G, \partial_*).$$

Corollary 1.42. *Let X' be a Δ-subset of X. Then there is a long exact sequence*

$$\cdots \longrightarrow H_{k+1}(X, X'; G) \xrightarrow{\partial_{k+1}} H_k(X'; G) \xrightarrow{i_*} H_k(X; G) \xrightarrow{p_*} H_k(X, X'; G) \longrightarrow \cdots$$

for abelian group G. $\qquad\square$

1.6. *Simplicial homology and singular homology*

Definition 1.43. Let X be a Δ-complex. Then *simplicial homology* of X with coefficients in an abelian group G is defined by

$$H_*^\Delta(X; G) = H_*(C_*^\Delta(X); G).$$

For any space X, define

$$S_n(X) = \mathrm{Map}(\Delta^n, X)$$

be the set of all continuous maps from Δ^n to X with

$$d_i = d^{i*} \colon S_n(X) = \mathrm{Map}(\Delta^n, X) \longrightarrow S_{n-1}(X) = \mathrm{Map}(\Delta^{n-1}, X).$$

for $0 \le i \le n$. Then $S_*(X) = \{S_n(X)\}_{n \ge 0}$ is a Δ-set. This allows us to define singular homology:

Definition 1.44. For a pair of spaces (X, A), the *singular homology* $H_*(X, A; G)$ with coefficients in an abelian group G is defined by

$$H_*(X, A; G) = H_*(S_*(X), S_*(A); G).$$

See Hatcher's book [17] for details.

2. Simplicial Sets and Homotopy

The notion of simplicial sets was arisen from establishing combinatorial models for spaces. Recall that, given any topological space X, there is a singular simplicial set $S_*(X)$, where $S_n(X)$ is the set of all continuous maps from the n-simplex to X. Let $C_n(X)$ be the free abelian group generated by $S_n(X)$. Then one gets the chain complex $C_*(X)$ with its homology the singular homology $H_*(X; \mathbb{Z})$. In other words, the singular homology of X is obtained from the free abelian groups generated by $S_*(X)$. One may ask whether the homotopy groups $\pi_*(X)$ can be obtained in a similar way. The answer is actually *yes*. In fact $\pi_*(X)$ can be obtained directly from $S_*(X)$.

2.1. *Simplicial sets*

2.1.1. *Definitions*

Definition 2.1. A *simplicial set* means a Δ-set X together with a collection of *degeneracies* $s_i\colon X_n \to X_{n+1}$, $0 \le i \le n$, such that

$$d_j d_i = d_{i-1} d_j \tag{2.1}$$

for $j < i$,

$$s_j s_i = s_{i+1} s_j \tag{2.2}$$

for $j \le i$ and

$$d_j s_i = \begin{cases} s_{i-1} d_j & j < i \\ \mathrm{id} & j = i, i+1 \\ s_i d_{j-1} & j > i+1. \end{cases} \tag{2.3}$$

The three identities for $d_i d_j$, $s_j s_i$ and $d_i s_j$ are called the *simplicial identities*.

Remark 2.2. One can use *deleting-doubling* for catching simplicial identities:

$$d_i\colon (x_0, \ldots, x_n) \longrightarrow (x_0, \ldots, x_{i-1}, x_{i+1}, \ldots, x_n).$$

$$s_i\colon (x_0, \ldots, x_n) \longrightarrow (x_0, \ldots, x_{i-1}, x_i, x_i, x_{i+1}, \ldots, x_n).$$

Let $\mathcal{O}$ be the category whose objects are finite ordered sets and whose morphisms are functions $f\colon X \to Y$ such that $f(x) \le f(y)$ if $x < y$. The objects in $\mathcal{O}$ are given by $[n] = \{0, \ldots, n\}$ for $n \ge 0$, which are the same as the objects in $\mathcal{O}^+$. The morphisms in $\mathcal{O}$ are generated by d^i, which is defined in $\mathcal{O}^+$, and the following morphism

$$s^i\colon [n+1] \to [n] \quad s^i = \begin{pmatrix} 0 & 1 & \cdots & i-1 & i & i+1 & i+2 & \cdots & n+1 \\ 0 & 1 & \cdots & i-1 & i & i & i+1 & \cdots & n \end{pmatrix}$$

for $0 \leq i \leq n$, that is, s^i hits i twice.

Exercise 2.1. Show that simplicial sets are one-to-one correspondent to contravariant functors from $\mathcal{O}$ to $\mathcal{S}$.

More abstractly we have the definition of simplicial objects over any category.

Definition 2.3. Let $\mathcal{C}$ be a category. A *simplicial object* over $\mathcal{C}$ means a contravariant functor from $\mathcal{O}$ to $\mathcal{C}$. Equivalently a simplicial object $\mathcal{C}$ means a sequence of objects $X = \{X_n\}_{n\geq 0}$ with face morphisms $d_i\colon X_n \to X_{n-1}$ and degeneracy morphisms $s_i\colon X_n \to X_{n+1}$, $0 \leq i \leq n$, such that the three simplicial identities (2.1), (2.2) and (2.3) hold. A *simplicial group* means a simplicial object over the category of groups. In other words, a simplicial set X is a simplicial group if and only if (1) each X_n is a group and (2) all faces and degeneracies are group homomorphisms.

Definition 2.4. A *simplicial map* (*simplicial morphism*) $f\colon X \to Y$ means a sequence of functions (morphisms) $f\colon X_n \to Y_n$ for each $n \geq 0$ such that $f \circ d_i = d_i \circ f$ and $f \circ s_i = s_i \circ f$, that is the diagram

$$
\begin{array}{ccccc}
X_{n+1} & \xleftarrow{\;s_i\;} & X_n & \xrightarrow{\;d_i\;} & X_{n-1} \\
\big\downarrow{f} & & \big\downarrow{f} & & \big\downarrow{f} \\
Y_{n+1} & \xleftarrow{\;s_i\;} & Y_n & \xrightarrow{\;d_i\;} & Y_{n-1}
\end{array}
$$

commutes. Moreover if each X_n is a subset of Y_n such that the inclusions $X_n \subseteq Y_n$ form a simplicial map, then X is called a *simplicial subset* of Y. A simplicial set X is called to be *isomorphic* to a simplicial set Y, denoted by $X \cong Y$, if there is a bijective simplicial map $f\colon X \to Y$.

2.1.2. *Basic constructions*

Example 2.5 (*n*-simplex). The n-simplex $\Delta[n]$, as a simplicial set, is as follows:

$$\Delta[n]_k = \{(i_0, i_1, \ldots, i_k) \mid 0 \leq i_0 \leq i_1 \leq \cdots \leq i_k \leq n\}$$

for $k \leq n$. The face $d_j \colon \Delta[n]_k \to \Delta[n]_{k-1}$ is given by

$$d_j(i_0, i_1, \ldots, d_k) = (i_0, i_1, \ldots, \hat{i}_j, \ldots, i_k),$$

that is deleting i_j. The degeneracy $s_j \colon \Delta[n]_k \to \Delta[n]_{k+1}$ is defined by

$$s_j(i_0, i_1, \ldots, d_k) = (i_0, i_1, \ldots, i_j, i_j, \ldots, i_k),$$

that is doubling i_j. Let $\sigma_n = (0, 1, \ldots, n) \in \Delta[n]_n$. Then any elements in $\Delta[n]$ can be written as iterated compositions of faces and degeneracies of σ_n.

Proposition 2.6. *Let X be a simplicial set and let $x \in X_n$ be an element. Then there exists a unique simplicial map*

$$f_x \colon \Delta[n] \to X$$

such that $f_x(\sigma_n) = x$.

Proof. We leave the proof as an exercise to the reader. $\square$

Definition 2.7. Let X be a simplicial set and $A = \{A_n\}_{n \geq 0}$ with $A_n \subseteq X_n$. The *simplicial subset of X generated by A* is defined by

$$\langle A \rangle = \bigcap \{A \subseteq Y \subseteq X \mid Y \text{ is a simplicial subset of } X\},$$

namely $\langle A \rangle$ consists of elements in X that can be written as iterated compositions of faces and degeneracies of the elements in A.

Example 2.8 (*n*-sphere). The simplicial n-sphere S^n is defined by

$$S^n = \Delta[n]/\partial(\Delta[n]),$$

where $\partial(\Delta[n])$ is the simplicial subset of $\Delta[n]$ generated by $\Delta[n]_k$ for $k < n$. Let's write explicitly for the elements in the simplicial circle S^1.

Now that

$$\Delta[1]_k = \{(i_0, \ldots, i_k) \mid 0 \leq i_0 \leq \cdots \leq i_k \leq 1\}$$

$$= \{(\overbrace{0, \ldots, 0}^{i}, 1, \ldots, 1 \mid 0 \leq i \leq k+1\}$$

has $k + 2$ elements. Now

$$\partial(\Delta[1])_k = \{(0,\ldots,0),(1,\ldots,1)\}.$$

By definition, $S^1 = \Delta[1]/\partial(\Delta[1])$. Thus

$$S^1_k = \{*,(i_0,\ldots,i_k) \mid 0 \le i_0 \le \cdots \le i_k \le 1\}$$

$$= \{(\overbrace{0,\ldots,0}^{i},1,\ldots,1 \mid 1 \le i \le k\}$$

has $k + 1$ elements including the basepoint $* = (0,\ldots,0) \sim (1,\ldots,1)$.

For general simplicial n-sphere S^n, we have $S^n_k = \{*\}$ for $k < n$ and

$$S^n_k = \{*,(i_0,\ldots,i_k) \mid 0 \le i_0 \le \cdots \le i_k \le n \text{ with } \{i_0,\ldots,i_k\} = \{0,1,\ldots,n\}\}$$

for $k \ge n$. $\qquad\square$

Example 2.9 (Cartesian product). Let X and Y be simplicial sets. Define $X \times Y$ by setting

$$(X \times Y)_n = X_n \times Y_n$$

with $d_i^{X \times Y} = (d_i^X, d_i^Y)$ and $s_i^{X \times Y} = (s_i^X, s_i^Y)$. Show that $X \times Y$ is a simplicial set. $\qquad\square$

Example 2.10. Let $f\colon X \to Y$ be a simplicial map. Then

$$\mathrm{Im}(f\colon X \to Y) = \{\mathrm{Im}(f\colon X_n \to Y_n)\}_{n \ge 0}$$

is a simplicial subset of Y. $\qquad\square$

For a sequence of nonnegative integers $I = (i_1, i_2, \ldots, i_k)$ of length $k = l(I)$, denote $d_I = d_{i_1} \cdots d_{i_k}$ and $s_I = s_{i_1} \cdots s_{i_k}$.

Example 2.11 (Basepoint). Let X be a simplicial set and let $x_0 \in X_0$. Then the image of the representing map

$$f_{x_0}\colon \Delta[0] \longrightarrow X$$

is a simplicial subset of X consisting of only one element $f_{x_0}(0,\ldots,0) = s_I(x_0)$ in each dimension. Thus a basepoint $*$ of X means a sequence of elements

$$\{f_{x_0}(\overbrace{0,\ldots,0}^{n+1})\}_{n \ge 0}$$

corresponds to the elements $x_0 \in X_0$.

A *pointed simplicial set* means a simplicial set with a given basepoint. A *pointed simplicial map* means a simplicial map that preserves the basepoints.

A simplicial set is called *reduced* if X_0 has only one element. For a reduced simplicial set, there is a unique choice of basepoints. Moreover any simplicial map between reduced simplicial sets is *pointed.*   □

Example 2.12 (Push-out). Let $C \xleftarrow{\;g\;} A \xrightarrow{\;f\;} B$ be simplicial maps. Define X_n to be the push-out in the diagram

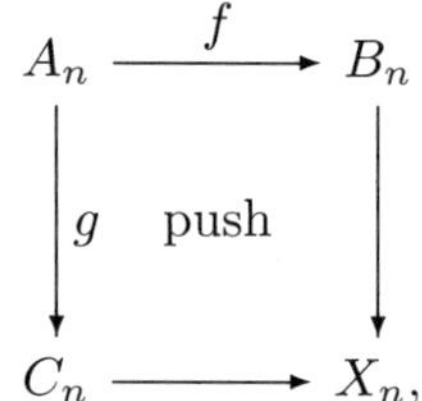

that is X_n is the quotient of the disjoint union $B_n \coprod C_n$ subject to the equivalence relation generated by $f(a) \sim g(a)$ for $a \in A_n$. Then $X = \{X_n\}_{n \geq 0}$ with faces and degeneracies induced from that in B and C forms a simplicial set with a push-out diagram of simplicial sets

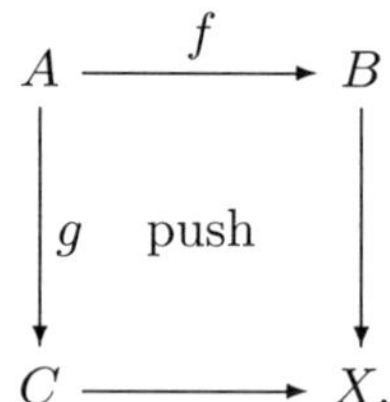

For instance, let $\partial \Delta[n] = \langle d_i(0, 1, \ldots, n) \mid 0 \leq i \leq n \rangle \subseteq \Delta[n]$ be the simplicial subset of $\Delta[n]$ generated by all of the faces of the n-simplex $\sigma_n = (0, 1, \ldots, n)$. Then

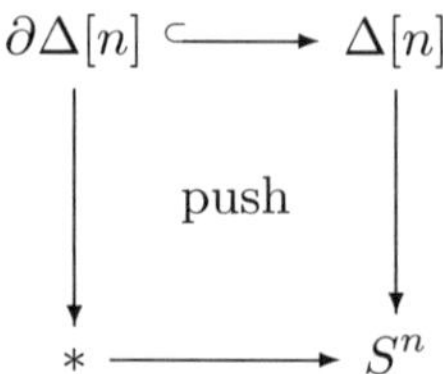

is a push-out diagram.   □

Example 2.13 (Wedge and smash). Let X and Y be pointed simplicial sets. The *wedge* $X \vee Y$ of X and Y is the simplicial set obtained by identifying the basepoint of X with the basepoint of Y. In other words, $X \vee Y$ is defined by the push-out diagram

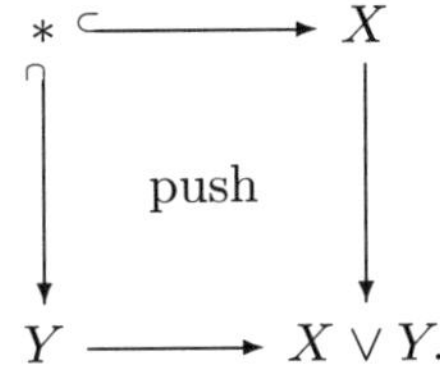

Note that $X \vee Y$ can be described as the simplicial subset of $X \times Y$ consisting of $(x, *)$ for $x \in X$ and $(*, y)$ for $y \in Y$. The *smash product* $X \wedge Y$ is defined to be the simplicial quotient $X \times Y / (X \vee Y)$.

An element $x \in X_n$ is called *nondegenerate* if $x \neq s_i(y)$ for any $y \in X_{n-1}$.

Exercise 2.2. Determine nondegenerate elements in $S^1 \wedge S^1$. In general, determine nondegenerate elements in $S^n \wedge S^m$.

Remark 2.14. As spaces, we know that $S^m \wedge S^n \cong S^{n+m}$. It should be pointed that, as simplicial sets, $S^m \wedge S^n \not\cong S^{m+n}$ although they have the same "homotopy type".

Example 2.15 (Skeleton of simplicial sets). Let X be a simplicial set. Let $\mathrm{sk}_n(X) = \langle X_j \mid j \leq n \rangle$ is the simplicial subset of X generated by X_j for $j \leq n$. Then there is a filtration

$$\mathrm{sk}_0 X \subseteq \mathrm{sk}_1 X \subseteq \cdots \mathrm{sk}_n X \subseteq \cdots$$

with $X = \bigcup_n \mathrm{sk}_n X$, called *skeleton filtration* of X. There are push-out diagrams

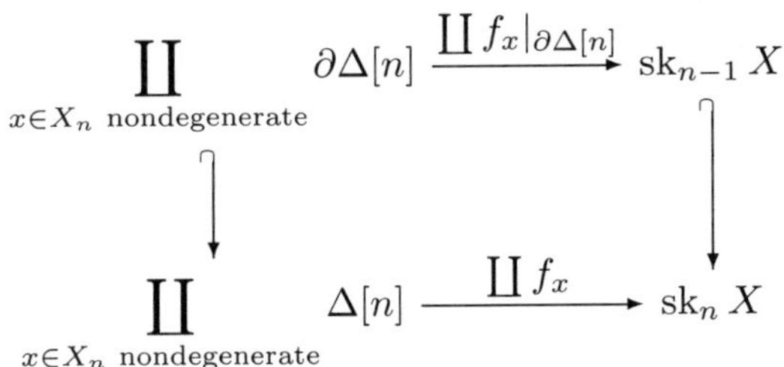

for each $n \geq 1$.

Example 2.16 (Cone and suspension). Let X be a pointed simplicial set. The *reduced cone* CX is defined by setting

$$(CX)_n = \{(x, q) \mid x \in X_{n-q}, 0 \le q \le n\} \text{ with } (*, q) \text{ all identified to } *,$$

$$
\begin{aligned}
d_i(x, q) &= \begin{cases} (x, q-1) & \text{for } 0 \le i < q, \\ (d_{i-q}x, q) & \text{for } q \le i \le n, \end{cases} \\
s_i(x, q) &= \begin{cases} (x, q+1) & \text{for } 0 \le i < q, \\ (s_{i-q}x, q) & \text{for } q \le i \le n, \end{cases}
\end{aligned}
\tag{2.4}
$$

where for $x \in X_0$, $d_1(x, 1) = *$. By identifying x with $(x, 0)$, X is a simplicial subset of CX. The *reduced suspension* ΣX is defined by $\Sigma X = CX/X$.

We give another description for the construction CX. Let X be any simplicial set without choosing basepoint. Let $\tilde{x}_0$ be a new point not in X. Let

$$
\begin{aligned}
(\tilde{C}X)_n &= X_n \bigsqcup s_{-1}(X_{n-1}) \bigsqcup \cdots \bigsqcup s_{-1}^n(X_0) \bigsqcup \{s_0^n \tilde{x}_0\} \\
&= \{s_0^n \tilde{x}_0\} \coprod_{k=0}^{n} s_{-1}^k(X_{n-k})
\end{aligned}
$$

be the disjoint union as a set, where $s_{-1}(X_j) = X_j$. Regard s_{-1} as (-1)st degeneracy operation. From the simplicial identities, we have

$$
\begin{aligned}
d_i s_{-1} &= \begin{cases} \text{id} & \text{if } i = 0 \\ s_{-1} d_{i-1} & \text{if } i \ge 1 \end{cases} \\
s_i s_{-1} &= \begin{cases} s_{-1} s_{-1} & \text{if } i = 0 \\ s_{-1} s_{i-1} & \text{if } i \ge 1, \end{cases}
\end{aligned}
\tag{2.5}
$$

which induces the operations d_i and s_i on $\tilde{C}X$ exactly given as in Formula (2.4) by identifying $s_{-1}^q x$ with (x, q), where $d_1(s_{-1}x) = \tilde{x}_0$ for $x \in X_0$. The resulting simplicial set $\tilde{C}X$ is called *unreduced cone* of X. The reduced cone CX is then the quotient given by requiring $x_0 \sim \tilde{x}_0$ and $(x_0, 1) = s_{-1}x_0 \sim (s_0 x_0, 0) = s_0 x_0 \sim s_0 \tilde{x}_0$ for the basepoint vertex $x_0 \in X_0$. $\qquad\square$

Proposition 2.17 (Contractible criterion). *Let X be a pointed simplicial set. Then the inclusion $j\colon X \hookrightarrow CX$ admits a simplicial retraction if and only if there exists a function $s_{-1}\colon X_n \to X_{n+1}$ for each $n \ge 0$ such that $s_{-1}(*) = *$ and Identities (2.5) hold.*

Proof. Suppose that there is a simplicial map $r\colon CX \to X$ such that the composite $r \circ j = \mathrm{id}_X$. Define $s_{-1}(x) = r(x, 1)$ for $x \in X_n$. Then

$$d_i s_{-1}(x) = d_i r(x, 1) = r d_i(x, 1) = r(d_i s_{-1}(x, 0))$$
$$s_i s_{-1}(x) = s_i r(x, 1) = r s_i(x, 1) = r(s_i s_{-1}(x, 0))$$

and so Identities (2.5) hold.

Conversely, if $s_{-1}\colon X_n \to X_{n+1}$ is defined with Identities (2.5), define the function $r\colon CX \to X$ by setting $r(x, 0) = x$ and $r(x, q) = s_{-1}^q x$ for $q > 0$. Then r is a simplicial map by Identities (2.5) with $r|X = \mathrm{id}_X$ and hence the result. $\square$

Theorem 2.18 (Null-homotopy criterion). *Let $f\colon X \to Y$ be a pointed simplicial map. Then the extension problem*

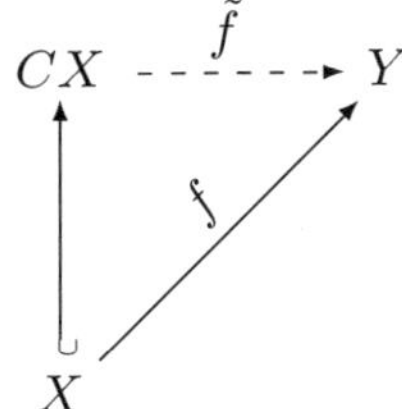

has a solution of a simplicial map $\tilde{f}$ if and only if there exists $F\colon X_n \to Y_{n+1}$ for each $n \geq 0$ such that the following conditions hold:

(1). $d_0 F = f$ and $d_i F = F d_{i-1}$ for $i > 0$;
(2). $s_i F = F s_{i-1}$ for $i > 0$.

Proof. Suppose that the solution $\tilde{f}$ exists. Define $F(x) = \tilde{f}(x, 1)$ for $x \in X_n$. Then F satisfies the conditions.

Conversely suppose that there exists $F\colon X_n \to Y_{n+1}$ satisfying the conditions. Define $\tilde{f}\colon CX \to Y$ by setting

$$\tilde{f}(x, q) = \begin{cases} f(x) & \text{if } q = 0 \\ F(x) & \text{if } q = 1 \\ s_0^{q-1} F(x) & \text{if } q \geq 2. \end{cases}$$

Note that $\tilde{f}|_X = f$. We show that $\tilde{f}$ is a simplicial map, that is, $d_j \tilde{f}(x, q) = \tilde{f} d_j(x, q)$ and $s_j \tilde{f}(x, q) = \tilde{f} s_j(x, q)$. Since $\tilde{f}|_X = f$, these identities hold for $q = 0$. Consider the case $q > 0$. Note that

$$d_j \tilde{f}(x, q) = d_j s_0^{q-1} F(x) = \begin{cases} d_0 F(x) & \text{if } j = 0,\ q = 1 \\ s_0^{q-2} F(x) & \text{if } 0 \leq j < q,\ q \geq 2 \\ s_0^{q-1} d_{j-q+1} F(x) & \text{if } j \geq q, \end{cases}$$

$$\tilde{f}(d_j(x,q)) = \begin{cases} f(x) & \text{if } j = 0,\ q = 1 \\ \tilde{f}(x, q-1) = s_0^{q-2} F(x) & \text{if } 0 \le j < q,\ q \ge 2 \\ \tilde{f}((d_{j-q}x, q)) = s_0^{q-1} F(d_{j-q}x) & \text{if } j \ge q. \end{cases}$$

Since $d_0 F(x) = f(x)$ and $d_i F(x) = F(d_{i-1}(x))$ for $i > 0$, $d_j \tilde{f}(x, q) = \tilde{f} d_j(x, q)$ for any x and q. Now

$$s_j \tilde{f}(x, q) = s_j s_0^{q-1} F(x) = \begin{cases} s_0^q F(x) & \text{if } 0 \le j < q \\ s_0^{q-1} s_{j-q+1} F(x) & \text{if } j \ge q, \end{cases}$$

$$\tilde{f} s_j(x, q) = \begin{cases} \tilde{f}(x, q+1) = s_0^q F(x) & \text{if } 0 \le j < q \\ \tilde{f}(s_{j-q}, q) = s_0^{q-1} F(s_{j-q}x) & \text{if } j \ge q. \end{cases}$$

Since $s_i F(x) = F(s_{i-1}(x))$ for $i > 0$, $s_j \tilde{f}(x, q) = \tilde{f} s_j(x, q)$ for any x and q. This finishes the proof. $\qquad\square$

Exercise 2.3. Show that $CS^0 \cong \Delta[1]$.

Proposition 2.19. $\Sigma S^n \cong S^{n+1}$ *for* $n \ge 0$.

Proof. Let σ_n be the nondegenerate element in S_n^n. Then, from the definition of CS^n, the nondegenerate element in CS^n are given by $* \in (CS^n)_0$, $(\sigma_n, 0) \in (CS^n)_n$ and $(\sigma_n, 1) \in (CS^n)_{n+1}$ with $d_0(\sigma n, 1) = \sigma_n$. Thus the nondegenerate elements in ΣS^{n+1} are given by $* \in (\Sigma S^n)_0$ and $(\sigma_n, 1) \in (\Sigma S^n)_{n+1}$. It follows that the representing map

$$f_{(\sigma_n, 1)} : \Delta[n] \longrightarrow \Sigma S^n$$

is onto and it factors through the quotient S^{n+1} because $d_i(\sigma_n, 1) = *$ in ΣS^n for all $0 \le i \le n$. The resulting simplicial map

$$\bar{f}_{(\sigma_n, 1)} : S^{n+1} \longrightarrow \Sigma S^n$$

is bijective by counting the elements in each dimension, and hence the result. $\qquad\square$

Exercise 2.4. Show that $\tilde{C}\Delta[n] \cong \Delta[n+1]$.

Let X and Y be simplicial sets. Write $\mathrm{Hom}_S(X, Y)$ for the set of simplicial maps from X to Y. The following example gives a way to construct the mapping space (as a simplicial set) from X to Y.

Example 2.20 (Mapping spaces). Let $\sigma_n = (0, 1, \ldots, n) \in \Delta[n]$ be the nondegenerate element.

$$d^i = f_{d_i \sigma_n} : \Delta[n-1] \longrightarrow \Delta[n]$$

$$s^i = f_{s_i \sigma_n} : \Delta[n+1] \longrightarrow \Delta[n]$$

be the representing maps of $d_i \sigma_n$ and $s_i \sigma_n$ for $0 \leq i \leq n$. Observe that the composites

$$\Delta[n-2] \xrightarrow{d^j} \Delta[n-1] \xrightarrow{d^i} \Delta[n],$$

$$\Delta[n+2] \xrightarrow{s^j} \Delta[n+1] \xrightarrow{d^i} \Delta[n],$$

$$\Delta[n] \xrightarrow{d^j} \Delta[n+1] \xrightarrow{s^i} \Delta[n]$$

are representation maps of the elements $d_j d_i \sigma_n$, $s_j s_i \sigma_n$ and $d_j s_i \sigma_n$, respectively. From the simplicial identities, the sequences of simplicial sets $\{\Delta[n]\}_{n \geq 0}$ with d^i and s^i is a *cosimplicial* simplicial set, namely the identities

$$
\begin{aligned}
d^j d^i &= d^{i+1} d_j \quad \text{for } i \geq j \\
s^i s^j &= s^j s^{i+1} \quad \text{for } i \geq j
\end{aligned}
$$

$$
s^i d^j = \begin{cases} d^j s^{i-1} & j < i \\ \text{id} & j = i,\ i+1 \\ d^{j-1} s^i & j > i+1. \end{cases}
\tag{2.6}
$$

Let $\mathrm{Map}(X, Y)_n = \mathrm{Hom}_S(X \times \Delta[n], Y)$ and let

$$
\begin{aligned}
d_i = (\mathrm{id}_X \times d^i)^* : \mathrm{Map}(X, Y)_n \ &= \ \mathrm{Hom}_S(X \times \Delta[n], Y) \\
&\longrightarrow \ \mathrm{Hom}_S(X \times \Delta[n-1], Y) \\
&= \ \mathrm{Map}(X, Y)_{n-1}, \\
s_i = (\mathrm{id}_X \times s^i)^* : \mathrm{Map}(X, Y)_n \ &= \ \mathrm{Hom}_S(X \times \Delta[n], Y) \\
&\longrightarrow \ \mathrm{Hom}_S(X \times \Delta[n+1], Y) \\
&= \ \mathrm{Map}(X, Y)_{n+1}
\end{aligned}
$$

for $0 \leq i \leq n$. From Identities (2.6), $\mathrm{Map}(X, Y) = \{\mathrm{Map}(X, Y)_n\}_{n \geq 0}$ with d_i and s_i is a simplicial set, which is called the *mapping space* from X to Y.

2.1.3. *Polyhedral simplicial sets*

Example 2.21 (Simplicial sets from simplicial complex). Let $\mathcal{K}$ be an abstract simplicial complex, and let the vertices of $\mathcal{K}$ be ordered. Then we obtain a simplicial set $\mathcal{K}^S = \{\mathcal{K}_n^S\}_{n \geq 0}$:

$$\mathcal{K}_n^S = \{(v_0, v_1, \ldots, v_n) \,|\, v_i \text{ are vertices of a simplex of } \mathcal{K} \text{ with } v_0 \leq v_1 \leq \cdots \leq v_n\},$$

$$
\begin{aligned}
d_i(v_0, v_1, \ldots, v_n) &= (v_0, v_1, \ldots, \hat{v}_i, \ldots, v_n) && \text{deleting } v_i, \\
s_i(v_0, v_1, \ldots, v_n) &= (v_0, v_1, \ldots, v_i, v_i, \ldots, v_n) && \text{doubling } v_i
\end{aligned}
$$

with a Δ-map $\mathcal{K}^\Delta \hookrightarrow \mathcal{K}^S$. Observe that $\mathcal{K}^S$ has the nondegenerate elements given by $\mathcal{K}^\Delta$. $\qquad\square$

Definition 2.22. A simplicial set X is called *polyhedral* if there exists an abstract simplicial complex $\mathcal{K}$ such that $X \cong \mathcal{K}^S$.

For a simplicial set X, let X_n^D denote the set of nondegenerate elements in X_n.

Lemma 2.23. *Let X and Y be simplicial sets and let $f\colon X \to Y$ be simplicial map such that*

(1). If x is a nondegenerate element of X, then $f(x)$ is a nondegenerate element of Y and

(2). The induced function $f\colon X_n^D \to Y_n^D$ is one-to-one for each $n \geq 0$.

Then f is one-to-one.

Proof. We show by induction that $f\colon X_n \to Y_n$ is one-to-one for each $n \geq 0$. When $n = 0$, we are given that $f\colon X_0 \to Y_0$ is one-to-one by the second assumption. Suppose that $f\colon X_{n-1} \to Y_{n-1}$ is one-to-one. Let $x, y \in X_n$ such that $f(x) = f(y)$. If both x and y are nondegenerate elements, then $x = y$ by the assumption. Suppose that one of x, y is a degenerate element. We may assume that $y = s_j z$ for some $z \in X_{n-1}$ and some $0 \leq j \leq n - 1$. Then $f(x) = f(y) = f(s_j z) = s_j f(z)$. By the first assumption, x must be a degenerate element. Let $x = s_i w$.

If $i = j$, then

$$f(w) = d_i s_i f(w) = d_i f(s_i w) = d_i f(x) = d_i f(y) = d_i f(s_i z) = d_i s_i f(z) = f(z).$$

By induction, $w = z$ and so $x = s_i w = s_i z = y$.

If $i \neq j$, we may assume that $i < j$. Then

$$f(w) = d_i f(s_i w) = d_i f(x) = d_i f(y) = d_i f(s_j z) = f(d_i s_j z).$$

By induction, $w = d_i s_j z$ and so

$$x = s_i w = s_i d_i s_j z = s_i s_{j-1} d_i z = s_j s_i d_i z.$$

It follows that

$$f(s_i d_i z) = d_j f(s_j s_i d_i z) = d_j f(x) = d_j f(y) = d_j f(s_j z) = d_j s_j f(z) = f(z)$$

and so $s_i d_i z = z$ by induction. Thus

$$x = s_j s_i d_i z = s_j z = y$$

and hence the result. $\qquad\qquad\square$

Now let X be a simplicial set and let 2^{X_0} be the set of all subsets of X_0. Define

$$\phi\colon \coprod_{n \geq 0} X_n^D \longrightarrow 2^{X_0}$$

by setting $\phi(x) = \{f_x(0), f_x(1), \ldots, f_x(n)\}$ for $x \in X_n$.

Theorem 2.24. *Let X be a simplicial set. Then X is polyhedral if and only if the following holds:*

(1). There exists an order of X_0 such that, for each $x \in X_n$,

$$f_x(0) \leq f_x(1) \leq \cdots \leq f_x(n).$$

(2). The function $\phi\colon \bigcup_{n \geq 0} X_n^D \longrightarrow 2^{X_0}$ is one-to-one.

Proof. $\Longrightarrow$ follows from the construction.

$\Longleftarrow$. By the proof of Theorem 1.37 of Section 1, for each $x \in X_n^D$, we have $f_x(0) < f_x(1) < \cdots < f_x(n)$ and the abstract simplicial complex $\mathcal{K}$ with vertex set X_0 is defined by requiring that: $\{a^0, \ldots, a^n\}$ is a simplex of $\mathcal{K}$ if and only if the set $\{a^0, \ldots, a^n\} \in \operatorname{Im} \phi$. Moreover by the proof of Theorem 1.37 of Section 1, the map ϕ induces a Δ-map

$$g\colon \mathcal{K}^\Delta \hookrightarrow X$$

given by

$$g(\{f_x(0), f_x(1), \ldots, f_x(n)\}) = x$$

for each $x \in X_n^D$. The Δ-map g extends to a simplicial map

$$\tilde{g}\colon \mathcal{K}^S \longrightarrow X.$$

Since g maps onto nondegenerate elements of X, the simplicial map $\tilde{g}$ is onto. Since $\tilde{g}$ sends each nondegenerate element of $\mathcal{K}^S$ (that are given by the elements in $\mathcal{K}^\Delta$) to a nondegenerate element of X and $g\colon \mathcal{K}^\Delta_n \to X^D_n$ is one-to-one, the map $\tilde{g}$ is one-to-one by the above lemma. Thus $\tilde{g}$ is a simplicial isomorphism and hence the result. $\qquad\square$

2.1.4. *Relations between Δ-sets and simplicial sets*

Let X be a simplicial set. By forgetting degeneracy, we can regard X as a Δ-set. On the other hand, given a Δ-set X, we can construct a simplicial set X^S by *"adding degeneracies"*.

Let $\sigma_n = (0, 1, \ldots, n) \in \Delta[n]$ be the nondegenerate element. Recall that we have simplicial maps

$$d^i = f_{d_i \sigma_n} \colon \Delta[n-1] \longrightarrow \Delta[n]$$

$$s^i = f_{s_i \sigma_n} \colon \Delta[n+1] \longrightarrow \Delta[n]$$

be the representing maps of $d_i \sigma_n$ and $s_i \sigma_n$ for $0 \leq i \leq n$. Given a Δ-set X, we define a sequence of sets by setting

$$X_n^S = \coprod_{k \geq 0} \Delta[k]_n \times X_k / \sim,$$

where $\sim$ is the equivalence relation generated by

$$(a, d_i x) \sim (d^i(a), x)$$

for any $a \in \Delta[k-1]_n$, $x \in X_k$, $k \geq 1$ and $0 \leq i \leq k$. Define the faces

$$d_j \colon X_n^S \to X_{n-1}^S$$

and the degeneracies

$$s_j \colon X_n^S \to X_{n+1}^S$$

by setting

$$d_j(a, x) = (d_j a, x), \quad s_j(a, x) = (s_j a, x)$$

for $0 \leq j \leq n$, that is, taking the faces and degeneracies on the first coordinate. We check that d_j and s_j preserve the equivalence relation:

$$d_j(a, d_i x) = (d_j a, d_i x)$$

$$d_j(d^i(a), x) = (d_j d^i(a), x)$$
$$= (d^i(d_j a), x) \quad \text{because } d^i \text{ is a simplicial map}$$

$$s_j(a, d_i x) = (s_j a, d_i x)$$

$$s_j(d^i(a), x) = (s_j d^i(a), x)$$
$$= (d^i(s_j a), x) \quad \text{because } d^i \text{ is a simplicial map.}$$

Thus d_j and s_j preserve the equivalence relation and so $d_j \colon X_n^S \to X_{n-1}^S$ and $s_j \colon X_n^S \to X_{n+1}^S$ are well-defined. The simplicial identities automatically hold as they are induced from the simplicial structure of $\Delta[k]$'s.

The construction looks complicated. But the output simplicial set is not so big as the equivalence relation identifies many elements. Let

$$\phi_n \colon X_n \to X_n^S$$

be the function defined by $\phi_n(x)$ to be the equivalence class $[\sigma_n, x]$ in X_n^S. Then

$$d_j(\phi_n(x)) = d_j[\sigma_n, x]$$
$$= [d_j \sigma_n, x]$$
$$= [d^j(\sigma_{n-1}, x]$$
$$= [\sigma_{n-1}, d_j x]$$
$$= \phi(d_j(x)).$$

Thus $\phi = \{\phi_n\} \colon X \to X^S$ is a Δ-map.

Lemma 2.25. *The Δ-map $\phi \colon X \to X^S$ is one-to-one.*

Proof. Observe that the equivalence relation is generated by connecting some elements in $\Delta[k]_n \times X_k$ and $\Delta[k-1]_n \times X_{k-1}$ for each k. Thus $[\sigma_n, x] \neq [\sigma_n, y]$ for $x \neq y \in X_n$, and so ϕ is one-to-one. $\qquad\square$

Theorem 2.26. *The construction $X \mapsto X^S$ is a functor from the category of Δ-sets and Δ-maps to the category of simplicial sets and simplicial maps. Moreover suppose that X is a Δ-set such that one of the faces*

$$d_i \colon X_n \to X_{n-1}$$

*is onto for each $n \geq 1$. Then the simplicial set X^S is generated by $\phi(X)$
with all nondegenerate elements given in $\phi(X)$.*

Proof. For showing the first assertion, let $f\colon X \to Y$ be a Δ-map. Then
the maps

$$\mathrm{id}_{\Delta[k]} \times f\colon \Delta[k]_n \times X_k \longrightarrow \Delta[k]_n \times Y_k$$

induces a function $f_n^S\colon X_n^S \to Y_n^S$ because from

$$\begin{aligned}
\mathrm{id}_{\Delta[k]} \times f(d^i a, x) &= (d^i a, f(x)) \\
\mathrm{id}_{\Delta[k-1]} \times f(a, d_i x) &= (a, f(d_i x)) \\
&= (a, d_i(f(x)))
\end{aligned}$$

the functions $\{\mathrm{id}_{\Delta[k]} \times f\}$ preserves the equivalence relations. Moreover
$f^S = \{f_n^S\} \to X^S \to Y^S$ is a simplicial map because the functions
$\{\mathrm{id}_{\Delta[k]} \times f\}$ commutes with faces and degeneracies. Clear $(g \circ f)^S = g^S \circ f^S$
for Δ-maps $f\colon X \to Y$ and $g\colon Y \to Z$. Thus $X \mapsto X^S$ is a functor.

For the second assertion, we first show that every equivalence class in
X_n^S can be represented by an element in $\Delta[n]_n \times X_n$. Let $(a, x) \in \Delta[k]_n \times X_k$.

If $k < n$, since there exists a face $d_i\colon X_{k+1} \to X_k$ which is onto, $x = d_i y$ for some $y \in X_{k+1}$. Thus $(a, x) = (a, d_i y) \sim (d^i a, y)$ with $(d^i, y) \in \Delta[k+1]_n \times X_{k+1}$. By iterating this procedure, $(a, x) \sim (b, z)$ for some
$(b, z) \in \Delta[n] \times X_n$.

If $k > n$, then all elements in $\Delta[k]_n$ are given by iterated faces σ_k. Thus
$a = d_{i_1} d_{i_2} \cdots d_{i_{k-n}} \sigma_k$ for some $0 \leq i_1 < i_2 < \cdots < i_{k-n} \leq k$. It follows that

$$d^{i_{k-n}} \circ \cdots \circ d^{i_2} d^{i_1}(\sigma_n) = d_{i_1} d_{i_2} \cdots d_{i_{k-n}} \sigma_k = a$$

and so

$$\begin{aligned}
(a, x) &= (d^{i_{k-n}} \circ \cdots \circ d^{i_2} d^{i_1}(\sigma_n), x) \\
&\sim (d^{i_{k-n-1}} \circ \cdots \circ d^{i_2} d^{i_1}, d_{i_{k-n}} x) \\
&\cdots \\
&\sim (\sigma_n, d_{i_1} d_{i_2} \cdots d_{i_{k-n}} x).
\end{aligned}$$

Thus every equivalence class in X_n^S can be represented by an element
in $\Delta[n]_n \times X_n$. Now we let $(a, x) \in \Delta[n]_n \times X_n$. Note that σ_n is only
nondegenerate element in

$$\Delta[n]_n = \{(i_0, i_1, \ldots, i_n) \mid 0 \leq i_0 \leq i_1 \leq \cdots \leq i_n \leq n\}.$$

If $a = \sigma_n$, then $(a, x) \in \phi(X)$. Otherwise

$$a = s_{l_1} s_{l_2} \cdots s_{l_k} d_{j_1} d_{j_2} \cdots d_{j_k} \sigma_n$$

for some $0 \leq l_k < \cdots < l_2 < l_1 \leq n - 1$ and $0 \leq j_1 < j_2 < \cdots < j_k \leq n$ with $k \geq 1$. (Recall that d_j is given by deleting the coordinates in the sequences $(i_0, \ldots, i_n)$ and s_j is given by doubling the coordinates. Since $a = (i_0, \ldots, i_n) \neq (0, 1, \ldots, n)$, we can apply d_j operations and then apply degeneracies backwards to a.) Then

$$\begin{aligned}
[a, x] &= [s_{l_1} s_{l_2} \cdots s_{l_k} d_{j_1} d_{j_2} \cdots d_{j_k} \sigma_n, x] \\
&= s_{l_1} s_{l_2} \cdots s_{l_k} [d_{j_1} d_{j_2} \cdots d_{j_k} \sigma_n, x] \\
&= s_{l_1} s_{l_2} \cdots s_{l_k} [d^{j_k} \circ \cdots \circ d^{j_1}(\sigma_{n-k}), x] \\
&= s_{l_1} s_{l_2} \cdots s_{l_k} [\sigma_{n-k}, d_{j_1} d_{j_2} \cdots d_{j_k} x]
\end{aligned}$$

with $[\sigma_{n-k}, d_{j_1} d_{j_2} \cdots d_{j_k} x] \in \phi(X)$ and hence the result. $\square$

Let $f \colon X \to Y$ be a Δ-map. Then there is a commutative diagram

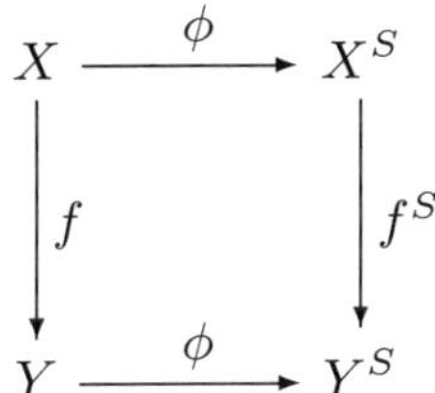

because, for any $x \in X_n$,

$$f^S \circ \phi(x) = f^S[\sigma_n, x] = [\sigma_n, f(x)] = \phi \circ f(x).$$

Thus $\phi \colon X \to X^S$ is a natural transformation. More precisely ϕ is a natural transformation from the identity functor of the category of Δ-sets to the functor $(-)^S$ composing with the forgetful functor from the category of simplicial sets to the category of Δ-sets.

Proposition 2.27. *The Δ-map $\phi \colon X \to X^S$ has the following properties:*

(1). *Let Z be any simplicial set and let $g \colon X \to Z$ be a Δ-map. Then there exists a simplicial map $\tilde{g} \colon X^S \to Z$ such that $g = \tilde{g} \circ \phi$.*

(2). *Suppose that X satisfies the property that one of the faces $d_i \colon X_n \to X_{n-1}$ is onto for each $n \geq 1$. Then $\tilde{g}$ is uniquely determined by g.*

Proof. The second assertion follows from the first assertion because X^S is generated by X in this case.

For proving assertion (1), let

$$\bar{Z}_n = \coprod_{k \geq 0} \Delta[k]_n \times Z_k / \approx$$

where $\approx$ is the equivalence relation generated by

$$(a, d_i z) \approx (d^i(a), z), \quad (b, s_i z) \approx (s^i b, z)$$

for any $a \in \Delta[k-1]_n$, $b \in \Delta[k+1]_n$, $z \in Z_k$ and $0 \leq i \leq k$. Then $\bar{Z} = \{\bar{Z}_n\}_{n \geq 0}$ is a simplicial set with the faces

$$d_j \colon \bar{Z}_n \to \bar{Z}_{n-1}$$

and the degeneracies

$$s_j \colon \bar{Z}_n \to \bar{Z}_{n+1}$$

defined by

$$d_j(a, z) = (d_j a, z), \quad s_j(a, z) = (s_j a, z)$$

for $0 \leq j \leq n$. (Similar to the construction X^S, one can check that faces and degeneracies defined above preserve the equivalence relation.)

Let $\psi \colon Z_n \to \bar{Z}_n$ be the function defined by $\psi(z) = [\sigma_n, z]$. Then the sequence of functions $\psi = \{\psi_n\} \colon Z \to \bar{Z}$ is a simplicial map.

Let $\bar{g}_n \colon X_n^S \to \bar{Z}_n$ be induced by the functions

$$\mathrm{id}_{\Delta[k]} \times g \colon \Delta[k]_n \times X_k \longrightarrow \Delta[k]_n \times Z_k.$$

From the lines in the proof of the above theorem, it follows that the function $\bar{g}_n$ is well-defined and $\bar{g} = \{\bar{g}_n\} \colon X^S \to \bar{Z}$ is a simplicial map. Now there is a commutative diagram

$$
\begin{array}{ccc}
X & \xrightarrow{\phi} & X^S \\
\downarrow{g} & & \downarrow{\bar{g}} \\
Z & \xrightarrow{\psi} & \bar{Z}
\end{array}
$$

because for any $x \in X_n$

$$\bar{g} \circ \phi(x) = \bar{g}([\sigma_n, x]) = [\sigma_n, g(x)] = \psi \circ g(x).$$

The assertion will follow by proving that $\psi \colon Z \to \bar{Z}$ is an isomorphism because if so, the composite $\tilde{g} = \psi^{-1} \circ \bar{g}$ is the desired map.

To show that $\psi \colon Z \to \bar{Z}$ is onto, from the proof of the above theorem, every equivalence class in $\bar{Z}_n$ can be represented by an element in $\Delta[n]_n \times Z_n$. Let $(a, x) \in \Delta[n]_n \times Z_n$. If $a \neq \sigma_n$, then $a = s_{i_1} s_{i_2} \cdots s_{i_k} d_{j_1} d_{j_2} \cdots d_{j_k} \sigma_n$

for some $0 \leq l_k < \cdots < l_2 < l_1 \leq n-1$ and $0 \leq j_1 < j_2 < \cdots < j_k \leq n$ with $k \geq 1$. Then

$$
\begin{aligned}
d^{j_k} \cdots d^{j_2} d^{j_1} s^{i_k} \cdots s^{i_2} s^{i_1}(\sigma_n) &= d^{j_k} \cdots d^{j_2} d^{j_1} s^{i_k} \cdots s^{i_2}\left(s_{i_1}\sigma_{n-1}\right) \\
&= d^{j_k} \cdots d^{j_2} d^{j_1} s^{i_k} \cdots s^{i_3}\left(s_{i_1} s^{i_2}(\sigma_{n-1})\right) \\
&= d^{j_k} \cdots d^{j_2} d^{j_1} s^{i_k} \cdots s^{i_3}\left(s_{i_1} s_{i_2}\sigma_{n-1}\right) \\
&\quad \cdots \\
&= d^{j_k} \cdots d^{j_2} d^{j_1}\left(s_{i_1} s_{i_2} \cdots s_{i_k}\sigma_{n-k}\right) \\
&= d^{j_k} \cdots d^{j_2}\left(s_{i_1} s_{i_2} \cdots s_{i_k} d^{i_1}(\sigma_{n-k})\right) \\
&= d^{j_k} \cdots d^{j_2}\left(s_{i_1} s_{i_2} \cdots s_{i_k} d_{i_1}\sigma_{n-k+1}\right) \\
&= d^{j_k} \cdots d^{j_3}\left(s_{i_1} s_{i_2} \cdots s_{i_k} d_{i_1} d^{j_2}(\sigma_{n-k+1})\right) \\
&= s_{i_1} s_{i_2} \cdots s_{i_k} d_{j_1} d_{j_2} \cdots d_{j_k}\sigma_n \\
&= a.
\end{aligned}
$$

Thus

$$
(a, x) = \left(d^{j_k} \cdots d^{j_2} d^{j_1} s^{i_k} \cdots s^{i_2} s^{i_1}(\sigma_n), x\right) \approx \left(\sigma_n, s_{i_1} s_{i_2} \cdots s_{i_k} d_{j_1} d_{j_2} \cdots d_{j_k} x\right)
$$

and so ψ is onto.

To see that ψ is one-to-one, similar to the above argument, every element in set $\coprod_{k=0}^{\infty} \Delta[k]_n$ is the image of the iterated composites of the functions d^i and s^j on $\sigma_n \in \Delta[n]_n$. The equivalence relation $\approx$ is to identify the elements $(a, z) \in \Delta[k]_n \times Z_k$ with an element (σ_n, z') where x' is obtained by the correspondent iterated composites of faces and degeneracies on z. Thus (σ_n, z) gives the unique representative in the equivalence class $[\sigma_n, z]$ and so ψ is one-to-one. The proof is finished now. $\qquad\square$

Note. We can also directly construct $\psi^{-1}\colon \bar{Z} \to Z$ as follows: Let $(a, z) \in \Delta[k]_n \times Z_k$. Since $a \in \Delta[k]_n$,

$$
a = s_{i_1} s_{i_2} \cdots s_{i_q} d_{j_1} d_{j_2} \cdots d_{j_t}\sigma_k,
$$

for unique sequences $0 \leq l_k < \cdots < l_2 < l_1 \leq n-1$ and $0 \leq j_1 < j_2 < \cdots < j_t \leq k$ with $k \geq 0$. Define

$$
\hat{\psi}_n(a, z) = s_{i_1} s_{i_2} \cdots s_{i_q} d_{j_1} d_{j_2} \cdots d_{j_t} z.
$$

Exercise 2.5. Let $\hat{\psi}_n\colon \Delta[k]_n \times Z_k \to Z_n$ be defined as above. Show that

(1). $\hat{\psi}_n$ induces a function $\bar{\psi}\colon \bar{Z}_n \to Z_n$.
(2). The sequence of functions $\bar{\psi} = \{\bar{\psi}_n\}\colon \bar{Z} \to Z$ is a simplicial map.
(3). $\bar{\psi}$ is inverse of $\psi\colon Z \to \bar{Z}$.

2.1.5. *More examples of simplicial sets*

Example 2.28 (Classifying spaces of categories). A category $\mathcal{C}$ is called *small* if the class of morphisms in $\mathcal{C}$ is a set. Let $\mathcal{C}$ be any small category. The simplicial set $B\mathcal{C}$ is defined by setting $B\mathcal{C}_n$ is the set of the strings

$$A_0 \xrightarrow{f_1} A_1 \xrightarrow{f_2} \cdots \xrightarrow{f_n} A_n$$

of composable arrows of length n of morphisms $f_j \in \mathrm{Mor}(\mathcal{C})$ with d_i is given by deleting A_i in the sequence and s_i is given by doubling A_i is the sequence. That is d_i on the above sequence is

$$A_0 \xrightarrow{f_1} A_1 \xrightarrow{f_2} \cdots \xrightarrow{f_{i-1}} A_{i-1} \xrightarrow{f_{i+1}\circ f_i} A_{i+1} \xrightarrow{f_{i+2}} \cdots \xrightarrow{f_n} A_n$$

and s_i on the above sequence is

$$A_0 \xrightarrow{f_1} A_1 \xrightarrow{f_2} \cdots \xrightarrow{f_i} A_i \xrightarrow{\mathrm{id}} A_i \xrightarrow{f_{i+1}} A_{i+1} \xrightarrow{f_{i+2}} \cdots \xrightarrow{f_n} A_n.$$

The simplicial set $B\mathcal{C}$ is called the *classifying space* of the category $\mathcal{C}$.

Example 2.29 (Classifying spaces of groups). Let M be a monoid. The simplicial set WM is defined by setting

$$WM_n = \{(g_0, g_1, \ldots, g_n) \mid g_i \in M\} = M^{n+1}$$

with faces and degeneracies given by

$$d_i(g_0, g_1, \ldots, g_n) = \begin{cases} (g_0, g_1, \ldots, g_{i-1}, g_i g_{i+1}, g_{i+2} \ldots, g_n) & \text{if } i < n \\ (g_0, \ldots, g_{n-1}) & \text{if } i = n \end{cases}$$

$$s_i(g_0, g_1, \ldots, g_n) = (g_0, g_1, \ldots, g_i, e, g_{i+1} \ldots, g_n).$$

For checking simplicial identities, one could describe the sequence $(g_0, g_1, \ldots, g_n)$ as a string

$$\xrightarrow{g_0} \heartsuit \xrightarrow{g_1} \heartsuit \xrightarrow{g_2} \cdots \xrightarrow{g_n} \heartsuit.$$

The face operation d_i is given by removing ith heart for $0 \leq i \leq n$. In this sense, the last face d_n deletes g_n because the heart of g_n is removed. The degeneracy operation s_i is given by doubling ith heart for $0 \leq i \leq n$.

Define the left action of M on M^{n+1} by setting

$$g \cdot (g_0, g_1, \ldots, g_n) = (g g_0, g_1, \ldots, g_n).$$

Then $d_i(g \cdot (g_0, g_1, \ldots, g_n)) = g \cdot d_i(g_0, g_1, \ldots, g_n)$ and $s_i(g \cdot (g_0, g_1, \ldots, g_n)) = g \cdot s_i(g_0, g_1, \ldots, g_n)$, that is, the action

$$M \times WM \longrightarrow WM$$

is simplicial, where M is regarded as a discrete in the sense that $M_n = M$ and $d_i = s_i = \mathrm{id}_M$. By modulo the action of M, we obtain the simplicial set $\bar{W}M = WM/M$.

Define $s_{-1} \colon WM_n \to WM_{n+1}$ by setting

$$s_{-1}(g_0, g_1, \ldots, g_n) = (e, g_0, g_1, \ldots, g_n).$$

Then Identities (2.5) hold. This means that the geometric realization $|WM|$ is contractible. If M is a group, then the geometric realization of $\bar{W}M$ is the classifying of M. $\qquad\square$

Exercise 2.6. Let M be a monoid. Using the strings

$$\diamond \xleftarrow{\; g_0 \;} \diamond \xleftarrow{\; g_1 \;} \cdots \diamond \xleftarrow{\; g_{n-1} \;} \diamond \xleftarrow{\; g_n \;}$$

show that the sequence of sets $\{M^{n+1}\}_{n \geq 0}$ with

$$d_i(g_0, g_1, \ldots, g_n) = \begin{cases} (g_1, \ldots, g_n) & \text{if } i = 0 \\ (g_0, g_1, \ldots, g_{i-2}, g_{i-1}g_i, g_{i+1} \ldots, g_n) & \text{if } i < n \end{cases}$$
$$s_i(g_0, g_1, \ldots, g_n) = (g_0, g_1, \ldots, e, g_i, g_{i+1} \ldots, g_n).$$

forms a simplicial set with a right action of M.

Example 2.30 (Another construction of classifying spaces). Let X be any set. Define EX by setting $EX_n = X^{n+1}$ with

$$d_i(x_0, x_1, \ldots, x_n) = (x_0, \ldots, x_{i-1}, x_{i+1}, \ldots, x_n)$$
$$s_i(x_0, x_1, \ldots, x_n) = (x_0, \ldots, x_{i-1}, x_i, x_i, x_{i+1}, \ldots, x_n).$$

Then EX is a simplicial set. This defines a functor from the category of sets to the category of simplicial sets. Moreover the functions

$$EX_n \times EY_n \longrightarrow E(X \times Y)_n$$
$$((x_0, \ldots, x_n)) \times ((y_0, \ldots, y_n)) \mapsto ((x_0, y_0), (x_1, y_1), \ldots, (x_n, y_n))$$

induces a simplicial isomorphism $EX \times EY \cong E(X \times Y)$. Thus, if X is a group (monoid), then EX is a simplicial group (monoid).

Let X be a pointed set with basepoint $*$. Define s_{-1} on EX by setting

$$s_{-1}(x_0, \ldots, x_n) = (*, x_0, \ldots, x_n).$$

Then Identities (2.5) hold. Thus the geometric realization $|EX|$ is contractible.

Suppose that $X = G$ is a group. Observe that $(EG)_0 = G$ and so G is a simplicial subgroup EG, where G is regarded as a discrete simplicial group. The resulting simplicial coset $BG = EG/G$ is the quotient of a contractible simplicial group EG by its simplicial subgroup G. Thus the geometric realization of BG is the classifying space of G. $\qquad\square$

Example 2.31 (Classifying spaces of simplicial monoids). Let M be a simplicial monoid. The simplicial set WM is defined by setting

$$WM_n = \{(g_n, g_{n-1}, \ldots, g_0) \mid g_i \in M_i\} = M_n \times M_{n-1} \times \cdots \times M_0$$

with faces and degeneracies given by

$$d_i(g_n, g_{n-1}, \ldots, g_0)$$
$$= \begin{cases} (d_i g_n, d_{i-1} g_{n-1}, \ldots, d_1 g_{n-i+1}, d_0 g_{n-i} g_{n-i-1}, g_{n-i-2} \ldots, g_0) & \text{if } i < n \\ (d_n g_n, d_{n-1} g_{n-1} \ldots, d_1 g_1) & \text{if } i = n \end{cases}$$

$$s_i(g_n, g_{n-1}, \ldots, g_0) = (s_i g_n, s_{i-1} g_{n-1}, \ldots, s_0 g_{n-i}, e, g_{n-i-1} \ldots, g_0).$$

The left action of M on WM given by

$$h \cdot (g_n, g_{n-1}, \ldots, g_0) = (hg_n, g_{n-1}, \ldots, g_0)$$

for $h \in M_n$ and $(g_n, g_{n-1}, \ldots, g_0) \in WM_n$ is simplicial. The classifying space $\bar{W}M = WM/M$ is the quotient of WM by the action of M. $\qquad\square$

Example 2.32 (Simplicial structure on braids). Let $X_n = B_{n+1}$ be the braid group of $(n+1)$-strands with faces and degeneracies given by:

> $d_i\beta$ is the braid obtained by removing the $(i+1)$st strand of β and $s_i\beta$ is the braid obtained by doubling the $(i+1)$st strand of β (that is the $(i+1)$st strand is replaced by two untwisted strands in its small neighborhood).

Then $X = \{X_n\}$ is a simplicial set. Since the faces and degeneracies are not group homomorphism, X is a not a simplicial group. (In fact, X is so-called *crossed simplicial group* as it satisfies $d_i(\beta\beta') = d_i\beta d_{\beta(i)}\beta'$ and $s_i(\beta\beta') = s_i\beta s_{\beta(i)}\beta'$, where $\beta(i)$ means that the ith strand goes from ith point to $\beta(i)$th point.) The sequence of pure braid groups $\{P_{n+1}\}_{n\geq 0}$ with the above faces and degeneracies is a simplicial group.

2.2. Geometric realization of simplicial sets

The standard *geometric n-simplex* Δ^n is defined by

$$\Delta^n = \left\{ (t_0, t_1, \ldots, t_n) \mid t_i \geq 0 \text{ and } \sum_{i=0}^{n} t_i = 1 \right\}.$$

Define $d^i \colon \Delta^{n-1} \to \Delta^n$ and $s^i \colon \Delta^{n+1} \to \Delta^n$ by setting

$$d^i(t_0, t_1, \ldots, t_{n-1}) = (t_0, \ldots, t_{i-1}, 0, t_i, \ldots, t_{n-1}),$$

$$s^i(t_0, t_1, \ldots, t_{n+1}) = (t_0, \ldots, t_{i-1}, t_i + t_{i+1}, \ldots, t_{n+1})$$

for $0 \leq i \leq n$.

Let X be a simplicial set. Then its *geometric realization* $|X|$ is a *CW*-complex defined by

$$|X| = \coprod_{\substack{x \in X_n \\ n \geq 0}} (\Delta^n, x)/\sim = \coprod_{n=0}^{\infty} \Delta^n \times X_n/\sim,$$

where (Δ^n, x) is Δ^n labeled by $x \in X_n$ and $\sim$ is generated by

$$(z, d_i x) \sim (d^i z, x)$$

for any $x \in X_n$ and $z \in \Delta^{n-1}$ labeled by $d_i x$, and

$$(z, s_i x) \sim (s^i z, x)$$

for any $x \in X_n$ and $z \in \Delta^{n+1}$ labeled by $s_i x$. Note that the points in $(\Delta^{n+1}, s_i x)$ and $(\Delta^{n-1}, d_i x)$ are identified with the points in (Δ^n, x).

Exercise 2.7. Prove that $|\Delta[n]| \cong \Delta^n$ and $|S^n| \cong S^n$.

Let $f \colon X \to Y$ be a simplicial map. Then its *geometric realization* $|f|$ is defined by

$$|f|(z, x) = (z, f(x))$$

for any $x \in X_n$ and $z \in \Delta^n$ labeled by x. Clearly $|f|$ is continuous.

Proposition 2.33. *Let X be a simplicial set. Then $|X|$ is a CW-complex. Thus the geometric realization gives a functor from the category of simplicial sets to the category of CW-complexes.*

Proof. From the push-out diagram

$$
\coprod_{x \in X_n \text{ nondegenerate}} \partial\Delta[n] \xrightarrow{\coprod f_x|_{\partial\Delta[n]}} \mathrm{sk}_{n-1} X
$$

$$
\coprod_{x \in X_n \text{ nondegenerate}} \Delta[n] \xrightarrow{\coprod f_x} \mathrm{sk}_n X,
$$

there is a push-out diagram

$$
\coprod_{x \in X_n \text{ nondegenerate}} |\partial\Delta[n]| = \partial|\Delta^n| \xrightarrow{\coprod |f_x|_{\partial\Delta[n]}|} |\mathrm{sk}_{n-1} X|
$$

$$
\coprod_{x \in X_n \text{ nondegenerate}} |\Delta[n]| = \Delta^n \xrightarrow{\coprod f_x} |\mathrm{sk}_n X|.
$$

Thus $|X|$ is obtained by attaching cell-by-cell and so $|X|$ is a CW-complex.
$\square$

Proposition 2.34. *Let K be an abstract simplicial complex. Then there is a homeomorphism $|K^S| \cong |K|$.*

Proof. Exercise.
$\square$

Exercise 2.8. Show that $|\Delta[m] \times \Delta[n]| \cong \Delta^m \times \Delta^n$.

Theorem 2.35. *Let X and Y be simplicial sets. Then there is a one-to-one and onto continuous map*

$$
\eta \colon |X \times Y| \to |X| \times |Y|.
$$

Proof. The coordinate projections $p_1 \colon X \times Y \to X$ and $p_2 \colon X \times Y \to Y$ induce continuous maps $|p_1| \colon |X \times Y| \to |X|$ and $|p_2| \colon |X \times Y| \to |Y|$ and so a continuous map

$$
\eta = (p_1, p_2) \colon |X \times Y| \longrightarrow |X| \times |Y|.
$$

We show that η is bijective. From the above exercise, this holds for $X = \Delta[m]$ and $Y = \Delta[n]$. By induction on the skeleton filtration of Y and using the push-out

$$
\begin{array}{ccc}
\Delta[m] \times \left(\coprod\limits_{y \in Y_n \text{ nondegenerate}} \partial\Delta[n] \right) & \longrightarrow & \Delta[m] \times \mathrm{sk}_{n-1} Y \\
\downarrow & & \uparrow \\
\Delta[m] \times \left(\coprod\limits_{y \in Y_n \text{ nondegenerate}} \Delta[n] \right) & \longrightarrow & \Delta[m] \times \mathrm{sk}_n Y,
\end{array}
$$

we have that

$$
\eta \colon |\Delta[m] \times Y| \to |\Delta[m]| \times |Y|
$$

is bijective for any simplicial set Y. The proof is then finished by induction on the skeleton filtration of X and using the push-out

$$
\begin{array}{ccc}
\left(\coprod\limits_{x \in X_m \text{ nondegenerate}} \partial\Delta[m] \right) \times Y & \longrightarrow & \mathrm{sk}_{m-1} X \times Y \\
\downarrow & & \uparrow \\
\left(\coprod\limits_{x \in X_n \text{ nondegenerate}} \Delta[m] \right) \times Y & \longrightarrow & \mathrm{sk}_m X \times Y.
\end{array}
$$
$\qquad\qquad\qquad\qquad\qquad\qquad\qquad\qquad\qquad\qquad\qquad\qquad\qquad\qquad\qquad\square$

Definition 2.36. A topological space X is called *compactly generated* if X is Hausdorff and each subset A of X with the property that $A \cap C$ is closed for any every compact subspace C of X is itself closed.

Exercise 2.9. Show that:

(1). Every locally compact Hausdorff space is compactly generated.
(2). Every Hausdorff space satisfying the first axiom of countability is compactly generated.
(3). Every metric space is compactly generated.

Given a Hausdorff space X, we can redefine a topology on X as follows: Let

$$
\mathcal{C} = \{ C \subset X \mid C \text{ compact} \}
$$

be the collection of compact spaces. Define a subset A of X to be closed if and only if $A \cap C$ is closed for every $C \in \mathcal{C}$. The new topological space is denoted by $k(X)$ which has the same underline set as X with the same compact sets. Note that the identity map $k(X) \to X$ is continuous. Namely possibly $k(X)$ has more closed sets. If X is already compactly generated, then $k(X) = X$. Let X and Y be two Hausdorff spaces. Let $X \times_{\mathrm{CG}} Y = k(X \times Y)$.

The category of compactly generated spaces means the subcategory of topological spaces, where the objects are compactly generated spaces and the morphisms are continuous maps between two compactly generated spaces. The Cartesian product in the category of compactly generated spaces is given by $X \times_{\mathrm{CG}} Y$. (**Note.** If X and Y are compactly generated, then $X \times Y$ may not be compactly generated in general and so we need to add more closed sets in $X \times Y$ to make $X \times_{\mathrm{CG}} Y$ such that the Cartesian product $X \times_{\mathrm{CG}} Y$ is still compactly generated.)

Exercise 2.10. Let X and Y be simplicial sets. Show that A is closed in $|X| \times_{\mathrm{CG}} |Y|$ if and only if

$$(f_x \times f_y)^{-1}(A)$$

is closed in $\Delta^m \times \Delta^n = |\Delta[m]| \times |\Delta[n]|$ for every $x \in X_m$ and $y \in Y_n$ with $m, n \geq 0$.

By this exercise, you can then prove the following:

Theorem 2.37. *Let X and Y be simplicial sets. Then*

$$\eta \colon |X \times Y| \longrightarrow |X| \times_{\mathrm{CG}} |Y|$$

is a homeomorphism. $\qquad\qquad\square$

Corollary 2.38. *Let G be a simplicial group. Then $|G|$ is a topological group in the category of compactly generated spaces.*

Proof. The multiplication $\mu \colon G \times G \to G$ induces a continuous multiplication

$$|G \times G| = |G| \times_{\mathrm{CG}} |G| \longrightarrow |G|$$

such that $|G|$ is a topological group. $\qquad\qquad\square$

Exercise 2.11. Prove the following statement:

(1). For any simplicial set X, $|\tilde{C}X|$ is homeomorphic to the unreduced cone $\tilde{C}|X|$ of $|X|$.

(2). If X is a pointed simplicial set, CX is homeomorphic to the reduced cone $C|X|$ of $|X|$.

(3). If X is a pointed simplicial set, $|\Sigma X| \cong \Sigma|X|$.

2.3. *Homotopy and fibrant simplicial sets*

Let $I = \Delta[1]$. As a simplicial set,

$$I_n = \{t_{\frac{i}{n+1}} = (\overbrace{0,\ldots,0}^{n+1-i}, \overbrace{1,\ldots,1}^{i}) \mid 0 \le i \le n+1\}$$

with

$$d_j t_{\frac{i}{n+1}} = \begin{cases} t_{\frac{i}{n}} & \text{if } 0 \le j < n+1-i \\ t_{\frac{i-1}{n}} & \text{if } n+1-i \le j \le n \end{cases}$$
$$= t_{\frac{i-\theta(i+j-n-1)}{n}},$$

$$s_j t_{\frac{i}{n+1}} = \begin{cases} t_{\frac{i}{n+2}} & \text{if } 0 \le j < n+1-i \\ t_{\frac{i+1}{n+2}} & \text{if } n+1-i \le j \le n \end{cases}$$
$$= t_{\frac{i+\theta(i+j-n-1)}{n+2}},$$

where

$$\theta(x) = \begin{cases} 0 & \text{if } x < 0 \\ 1 & \text{if } x \ge 0. \end{cases}$$

Given a simplicial set X, the simplicial subsets $X \times 0$ and $X \times 1$ of $X \times I$ are given by

$$(X \times 0)_n = \{(x, t_0) \mid x \in X_n\},$$

$$(X \times 1)_n = \{(x, t_1) \mid x \in X_n\}.$$

Definition 2.39. Let $f, g\colon X \to Y$ be simplicial maps. We call f *homotopic* to g if there is a simplicial map

$$F\colon X \times I \to Y$$

such that $F|_{X \times 0} = f$ and $F|_{X \times 1} = g$ in which case write $f \simeq g$. If A is a simplicial subset of X and $f, g\colon X \to Y$ are simplicial maps such that $f|_A = g|_A$, we call f *homotopic to g relative to A*, denoted by $f \simeq g$ rel A, if there is a homotopy $F\colon X \times I \to Y$ such that $F|_{X \times 0} = f$, $F|_{X \times 1} = g$ and $F|_{A \times I} = f$.

A *pointed simplicial set* means a simplicial set with a choice of a basepoint. Let X and Y be pointed simplicial sets. A pointed simplicial map $f\colon X \to Y$ means a simplicial map which sends the basepoint of X to the basepoint of Y. We usually denote by $*$ the basepoint. Let $f, g\colon X \to Y$ be pointed maps. A *pointed homotopy* means a simplicial map $F\colon X \times I \to Y$ such that $F|_{X \times 0} = f$, $F|_{X \times 1} = g$ and $F|_{* \times I} = *$ (that is $F|_{* \times I}$ is the constant map to the basepoint of Y) in which we write $f \simeq g$ rel $*$.

Exercise 2.12. Let $f, g \colon X \to Y$ be simplicial maps. Show that $f \simeq g$ if and only if there is a family of functions

$$F_{i/(n+1)} \colon X_n \to Y_n$$

for $n \geq 0$ and $0 \leq i \leq n + 1$ satisfying the following conditions:

(1). $d_j F_{i/(n+1)} = F_{(i - \theta(i+j-n-1))/n} d_j$,

(2). $s_j F_{i/(n+1)} = F_{(i + \theta(i+j-n-1))/(n+2)} s_j$,

(3). $F_0 = f$ and

(4). $F_1 = g$.

We need fibrant assumption on Y such that homotopy relation $\simeq$ is an equivalence relation on the set of simplicial maps from X to Y.

Definition 2.40. Let X be a simplicial set. The elements

$$x_0, \ldots, x_{i-1}, x_{i+1}, \ldots, x_n \in X_{n-1}$$

are called *matching faces* with respect to i if

$$d_j x_k = d_k x_{j+1}$$

for $j \geq k$ and $k, j + 1 \neq i$. A simplicial set X is called *fibrant* (or *Kan complex*) if it satisfies the following homotopy extension condition for each i:

> Let $x_0, \ldots, x_{i-1}, x_{i+1}, \ldots, x_n \in X_{n-1}$ be any elements that are matching faces with respect to i. Then there exists an element $w \in X_n$ such that $d_j w = x_j$ for $j \neq i$.

Let $\Lambda^i[n]$ be the simplicial subset of $\Delta[n]$ generated by all $d_j \sigma_n$ for $j \neq i$, where $\sigma_n = (0, 1, \ldots, n) \in \Delta[n]_n$ is the nondegenerate element.

Proposition 2.41. *Let X be a simplicial set. Then X is fibrant if and only if every simplicial map $f \colon \Lambda^i[n] \to X$ has an extension for each i.*

Proof. Suppose that X is fibrant. Let $f \colon \Lambda^i[n] \to X$. The elements

$$f(d_0 \sigma_n), f(d_1 \sigma_n), \ldots, f(d_{i-1} \sigma_n), f(d_{i+1} \sigma_n), \ldots, f(d_n \sigma_n)$$

have matching faces and so there exists an element $w \in X_n$ such that $d_j w = f(d_j \sigma_n)$ for $j \neq i$. Then the representing map $g = f_w \colon \Delta[n] \to X$ is an extension of f.

Conversely let $x_0, \ldots, x_{i-1}, x_{i+1}, \ldots, x_n \in X_{n-1}$ be any elements that are matching faces with respect to i. Then the representing maps $f_{x_j} \colon \Delta[n-1] \to X$ for $j \neq i$ defines a simplicial map $f \colon \Lambda^i[n] \to X$ such that the diagram

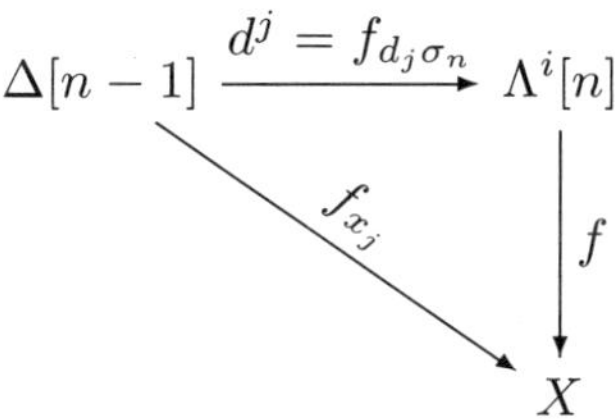

commutes for each j. (**Note.** The matching faces condition is used here for having f well-defined.) By the assumption, there exists an extension $g \colon \Delta[n] \to X$ such that $g|_{\Lambda^i[n]} = f$. Let $w = g(\sigma_n)$. Then

$$d_j w = d_j g(\sigma_n) = g(d_j \sigma_n) = f(d_j \sigma_n) = x_j$$

for $j \neq i$. Thus X is fibrant. $\qquad\square$

Lemma 2.42. *Let $A \subseteq B$ denote any of the following pairs:*

$$\Delta[n] \times \Lambda^i[m] \subseteq \Delta[n] \times \Delta[m]$$
$$(\Delta[n] \times \Lambda^i[m] \cup \partial\Delta[n] \times \Delta[m]) \subseteq \Delta[n] \times \Delta[m].$$

Let X be a fibrant simplicial set. Then any map $f \colon A \to X$ can be extended to a simplicial map $g \colon B \to X$.

Proof. For each such a pair, B can be obtained from A by successively adjoining a simplex and one of its faces, all other faces already lying in a simplicial subset. Iterated application of the extension condition gives the lemma. $\qquad\square$

Proposition 2.43. *Let Y be a fibrant simplicial set and let X be any simplicial set. Let A be a simplicial subset of X. Then any map*

$$f \colon (A \times \Delta[m]) \cup (X \times \Lambda^i[m]) \to Y$$

can be extended to a simplicial map $g \colon X \times \Delta[m] \to Y$.

Proof. Let $\tilde{\mathrm{sk}}_n(X) = A \cup \mathrm{sk}_n(X)$. First f can be extended to a simplicial map $\tilde{\mathrm{sk}}_0 X \to Y$. For $n > 0$, there is a push-out diagram

$$\coprod_{x \in \tilde{X}_n^D} \left((\partial\Delta[n] \times \Delta[m]) \cup (\Delta[n] \times \Lambda^i[m]) \right) \longrightarrow \tilde{\mathrm{sk}}_{n-1} X \times \Delta[m]$$

$$\big\downarrow \qquad\qquad\qquad \text{push} \qquad\qquad\qquad \big\downarrow$$

$$\coprod_{x \in \tilde{X}_n^D} (\Delta[n] \times \Delta[m]) \longrightarrow \tilde{\mathrm{sk}}_n X \times \Delta[m],$$

where $\tilde{X}_n^D$ is the set of nondegenerate elements in $X_n \smallsetminus A_n$. The assertion follows by induction, where the step from $\tilde{\mathrm{sk}}_{n-1} X$ to $\tilde{\mathrm{sk}}_n X$ is shown in Lemma 2.42. $\square$

Let X and Y are pointed simplicial sets. Let $\mathrm{Map}_*(X, Y)$ be the simplicial subset of $\mathrm{Map}(X, Y)$ with

$$\mathrm{Map}_*(X, Y)_n = \{ f \colon X \times \Delta[n] \to Y \mid f(s_0^n * \times \Delta[n]) = * \}$$

for each $n \geq 0$.

Proposition 2.44. *Let Y be a fibrant simplicial set (with a basepoint) and let X be any simplicial set (with a basepoint). Then the mapping space $\mathrm{Map}(X, Y)$ and $\mathrm{Map}_*(X, Y)$ are fibrant.*

Proof. Let $f_0, \ldots, f_{i-1}, f_{i+1}, \ldots, f_n \in \mathrm{Map}_*(X, Y)_{n-1}$ have matching faces, that is, each $f_j \colon X \times \Delta[n-1] \to Y$, and if we consider each such $\Delta[n-1]$ as the jth face of $\Delta[n]$, the f_j agree on $\mathrm{sk}_{n-2} \Delta[n]$ to give a simplicial map

$$f \colon (X \times \Lambda^i[n]) \cup (* \times \Delta[n]) \to Y,$$

where $f|_{* \times \Delta[n]}$ is the constant map to the basepoint of Y. By Proposition 2.43, f extends to a simplicial map $g \colon X \times \Delta[n] \to Y$, where $A = *$. Then $g \in \mathrm{Map}_*(X, Y)_n$ with $d_j g = f_j$ for $j \neq i$. Thus $\mathrm{Map}_*(X, Y)$ is fibrant. Similarly $\mathrm{Map}(X, Y)$ is fibrant. $\square$

Theorem 2.45. *Let Y be a fibrant simplicial set (with a basepoint) and let X be any simplicial set (with a basepoint). Then the homotopy relation ($\simeq$ rel $*$) is an equivalence relation. In such case, denote by $[X, Y]$ the set of the pointed homotopy classes of all pointed simplicial maps from X to Y.*

Proof. $f \simeq g$ rel $* \implies g \simeq f$ rel $*$. Let $f \simeq g$ rel $*\colon X \to Y$ under a homotopy $F\colon X \times I \to Y$ with $F|_{*\times I} = *$. Consider the mapping space $\mathrm{Map}_*(X, Y)$. Then $f, g \in \mathrm{Map}_*(X, Y)_0$ and $F \in \mathrm{Map}_*(X, Y)_1$ with $d_1 F = f$ and $d_0 F = g$. The elements $x_1 = s_0 d_1 F$ and $x_2 = F$ have matching faces with respect to 0 because

$$d_1 x_1 = d_1 s_0 d_1 F = d_1 F = d_1 x_2.$$

Since $\mathrm{Map}_*(X, Y)$ is fibrant, there exists $w \in \mathrm{Map}_*(X, Y)_2$ such that $d_1 w = s_0 d_1 F$ and $d_2 w = F$. Let $F' = d_0 w$. Then

$$d_0 F' = d_0 d_0 w = d_0 d_1 w = d_0 s_0 d_1 F = d_1 F,$$

$$d_1 F' = d_1 d_0 w = d_0 d_2 w = d_0 F$$

and so F' is a homotopy from g to f. Since $F' \in \mathrm{Map}_*(X, Y)_1$, $F'|_{*\times I}$ is the constant simplicial map. Thus F' is a pointed homotopy from g to f.

$f \overset{F}{\simeq} g$ rel $*, g \overset{G}{\simeq} h$ rel $* \implies f \simeq h$ rel $*$. We have $F, G \in \mathrm{Map}_*(X, Y)_1$ such that $d_1 F = f$, $d_0 F = d_1 G = g$ and $d_0 G = g$. The elements $x_0 = G$ and $x_2 = F$ have matching faces with respect to 1 because $d_1 x_0 = d_1 G = d_0 F = d_0 x_2$. Thus there exists $w \in \mathrm{Map}_*(X, Y)_2$ such that $d_0 w = G$ and $d_2 w = F$. Let $F' = d_1 w$. Then

$$d_1 F' = d_1 d_1 w = d_1 d_2 w = d_1 F = f,$$

$$d_0 F' = d_0 d_1 w = d_0 d_0 w = d_0 G = h$$

and so $f \simeq h$ rel $*$. $\qquad\square$

Note. The ideas in the above proof can be shown as in the picture

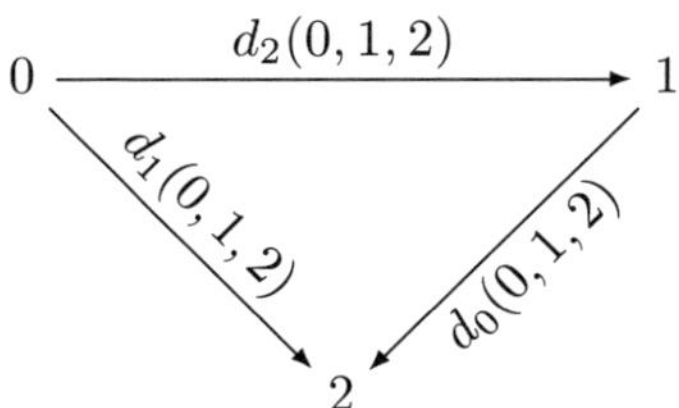

Exercise 2.13. Show that $\Delta[n], \partial\Delta[n]$ and S^n are NOT fibrant.

Exercise 2.14. Let Y be a fibrant simplicial set and let X be any simplicial set. Let A be a simplicial subset of X. Show that the homotopy relation ($\simeq$ rel A) is an equivalence relation.

Example 2.46. Let X be any topological space and let $S_n(X) = \mathrm{Map}(\Delta^n, X)$ be the set of all continuous maps from $\Delta^n \to X$. Then $S_*(X) = \{S_n(X)\}_{n \geq 0}$ is a simplicial set, called *singular simplicial set*, with faces and degeneracies

$$d_i = d^{i*} \colon \mathrm{Map}(\Delta^n, X) \to \mathrm{Map}(\Delta^{n-1}, X),$$

$$s_i = s^{i*} \colon \mathrm{Map}(\Delta^n, X) \to \mathrm{Map}(\Delta^{n+1}, X)$$

induced by the maps $d^i \colon \Delta^{n-1} \to \Delta^n$ and $s^i \colon \Delta^{n+1} \to \Delta^n$. Let $x_0, \ldots, x_{i-1}, x_{i+1}, \ldots, x_n \in S_{n-1}(X)$ have matching faces with respect to i. Namely each x_j is a continuous map

$$x_j \colon \Delta^{[n-1]} \to X.$$

Let $|\Lambda^i[n]|$ be the subcomplex of $\partial \Delta^n$ obtained by removing ith face of the n-simplex. The matching faces condition allows us to define a continuous map

$$f \colon |\Lambda^i[n]| \longrightarrow X$$

such that f restricted to jth face of the n-simplex is x_j. Observe that $|\Lambda^i[n]|$ is a retract of Δ^n. Let $r \colon \Delta^n \to |\Lambda^i[n]|$ be a retraction map. Then $g = f \circ r \colon \Delta^n \to X$ is an extension of f. Consider g as an element in $S_n(X) = \mathrm{Map}(\Delta^n, X)$. Then $d_j g = x_j$ for $j \neq i$. Thus $S_*(X)$ is fibrant. $\square$

Example 2.47. Let X be a fibrant simplicial set and let A be a simplicial retract of X. We show that A is also fibrant.

Let $r \colon X \to A$ be a simplicial retract. For any simplicial map $f \colon \Lambda^i[n] \to A$, there exists a simplicial map $g \colon \Delta[n] \to X$ such that $g|_{\Lambda^i[n]} = f$ because X is fibrant. Let $\tilde{g} = r \circ g \colon \Delta[n] \to A$. Then

$$\tilde{g}|_{\Lambda^i[n]} = r \circ g|_{\Lambda^i[n]} = r \circ f = f$$

because $f(\Lambda^i[n]) \subseteq A$. Hence A is fibrant. $\square$

An important example of fibrant simplicial sets is as follows.

Theorem 2.48 (Moore). *Any simplicial group G is a fibrant simplicial set.*

Proof. Let G be a simplicial group and let $y_0, \ldots, \hat{y}_k, \ldots, y_n \in G_{n-1}$ have matching faces. Construct the elements in G_n as follows:

$$
\begin{aligned}
w_0 &= s_0 y_0, \\
w_i &= w_{i-1}(s_i d_i w_{i-1})^{-1} s_i y_i && \text{for } 0 < i < k, \\
w_n &= w_{k-1}(s_{n-1} d_n w_{k-1})^{-1} s_{n-1} y_n, \\
w_i &= w_{i+1}(s_{i-1} d_i w_{i+1})^{-1} s_i y_i && \text{for } k < i < n.
\end{aligned}
$$

We check that $d_j w_{k+1} = y_j$ for $j \neq k$.

First we show by induction that $d_j w_i = y_j$ for $j \leq i < k$. Note that

$$
d_0 w_0 = d_0 s_0 y_0 = y_0 .
$$

Assume that $d_j w_{i-1} = y_{i-1}$ for $j \leq i - 1 < k$ with $i < k$. Then, for $j < i$

$$
\begin{aligned}
d_j w_i &= d_j \big(w_{i-1}(s_i d_i w_{i-1})^{-1} s_i y_i\big) \\
&= y_j (d_j s_i d_i w_{i-1})^{-1} d_j s_i y_i \\
&= y_j (s_{i-1} d_j d_i w_{i-1})^{-1} s_{i-1} d_j y_i \\
&= y_j (s_{i-1} d_{i-1} d_j w_{i-1})^{-1} s_{i-1} d_{i-1} y_j && \text{by matching faces condition} \\
&= y_j (s_{i-1} d_{i-1} y_j)^{-1} s_{i-1} d_{i-1} y_j && \text{by induction} \\
&= y_j
\end{aligned}
$$

and

$$
\begin{aligned}
d_i w_i &= d_i \big(w_{i-1}(s_i d_i w_{i-1})^{-1} s_i y_i\big) \\
&= d_i w_{i-1} (d_i s_i d_i w_{i-1})^{-1} d_i s_i y_i \\
&= d_i w_{i-1} (d_i w_{i-1})^{-1} y_i \\
&= y_i
\end{aligned}
$$

Hence $d_j w_i = y_j$ for $j \leq i < k$.

If $k = n$, then w_{n-1} satisfies the property that $d_j w_i = y_i$ for $i < n$. Thus we may assume that $k < n$ now. For $0 \leq t \leq n - k - 1$, we show by induction that $d_j w_{n-t} = y_j$ for $0 \leq j < k - 1$ and $n - t \leq j \leq n$. When $t = 0$, for $j < k$,

$$
\begin{aligned}
d_j w_n &= d_j \big(w_{k-1}(s_{n-1} d_n w_{k-1})^{-1} s_{n-1} y_n\big) \\
&= d_j w_{k-1} (d_j s_{n-1} d_n w_{k-1})^{-1} d_j s_{n-1} y_n \\
&= d_j w_{k-1} (s_{n-2} d_j d_n w_{k-1})^{-1} s_{n-2} d_j y_n && \text{as } j < k \leq n - 1 \\
&= d_j w_{k-1} (s_{n-2} d_{n-1} d_j w_{k-1})^{-1} s_{n-2} d_{n-1} y_j && \text{by matching faces condition} \\
&= y_j (s_{n-2} d_{n-1} y_j)^{-1} s_{n-2} d_{n-1} y_j \\
&= y_j
\end{aligned}
$$

and

$$
\begin{aligned}
d_n w_n &= d_n \big(w_{k-1}(s_{n-1} d_n w_{k-1})^{-1} s_{n-1} y_n\big) \\
&= d_n w_{k-1} (d_n s_{n-1} d_n w_{k-1})^{-1} d_n s_{n-1} y_n \\
&= d_n w_{k-1} (d_n w_{k-1})^{-1} y_n \\
&= y_n
\end{aligned}
$$

Assume that $d_j w_{n-t+1} = y_j$ for $0 \le j < k-1$ and $n-t+1 \le j \le n$ with $0 \le t \le n-k-1$. Then, for $0 \le j \le k-1$,

$$
\begin{aligned}
d_j w_{n-t} &= d_j (w_{n-t+1} (s_{n-t-1} d_{n-t} w_{n-t+1})^{-1} s_{n-t-1} y_{n-t}) \\
&= d_j w_{n-t+1} (d_j s_{n-t-1} d_{n-t} w_{n-t+1})^{-1} d_j s_{n-t-1} y_{n-t} \\
&= y_j (s_{n-t-2} d_j d_{n-t} w_{n-t+1})^{-1} s_{n-t-2} d_j y_{n-t} \\
&\qquad \text{since } j \le k-1 < k \le n-t-1 \\
&= y_j (s_{n-t-2} d_{n-t-1} d_j w_{n-t+1})^{-1} s_{n-t-2} d_{n-t-1} y_j \\
&\qquad \text{by matching faces condition} \\
&= y_j (s_{n-t-2} d_{n-t-1} y_j)^{-1} s_{n-t-2} d_{n-t-1} y_j \\
&= y_j
\end{aligned}
$$

and, for $n-t+1 \le j \le n$,

$$
\begin{aligned}
d_j w_{n-t} &= d_j (w_{n-t+1} (s_{n-t-1} d_{n-t} w_{n-t+1})^{-1} s_{n-t-1} y_{n-t}) \\
&= d_j w_{n-t+1} (d_j s_{n-t-1} d_{n-t} w_{n-t+1})^{-1} d_j s_{n-t-1} y_{n-t} \\
&= y_j (s_{n-t-1} d_{j-1} d_{n-t} w_{n-t+1})^{-1} s_{n-t-1} d_{j-1} y_{n-t} \\
&\qquad \text{since } n-t-1 < n-t \le j-1 \\
&= y_j (s_{n-t-2} d_{n-t} d_j w_{n-t+1})^{-1} s_{n-t-2} d_{n-t} y_j \\
&\qquad \text{by matching faces condition} \\
&= y_j (s_{n-t-2} d_{n-t} y_j)^{-1} s_{n-t-2} d_{n-t} y_j. \\
&= y_j
\end{aligned}
$$

Now

$$
\begin{aligned}
d_{n-t} w_{n-t} &= d_{n-t} (w_{n-t+1} (s_{n-t-1} d_{n-t} w_{n-t+1})^{-1} s_{n-t-1} y_{n-t}) \\
&= d_{n-t} w_{n-t+1} (d_{n-t} s_{n-t-1} d_{n-t} w_{n-t+1})^{-1} d_{n-t} s_{n-t-1} y_{n-t} \\
&= d_{n-t} w_{n-t+1} (d_{n-t} w_{n-t+1})^{-1} y_{n-t} \\
&= y_{n-t}.
\end{aligned}
$$

The induction is finished and so $d_j w_{k+1} = y_j$ for $j \ne k$. $\square$

2.4. *The relations between spaces and simplicial sets*

Theorem 2.49 (Simplicial extension theorem). *Let X be a simplicial set and let A be a simplicial subset of X. Let Y be a fibrant simplicial set and let $f \colon A \to Y$ be a simplicial map. Suppose that there is a continuous map $\phi \colon |X| \to |Y|$ such that $\phi|_{|A|} = |f| \colon |A| \to |Y|$. Then there exists a simplicial map $\phi' \colon X \to Y$ such that $\phi'_A = f \colon A \to Y$ and $|\phi'| \simeq \phi \colon |X| \to |Y| \operatorname{rel} |A|$.*

We state this theorem without proof. One can find a proof in [31].

Corollary 2.50. *Let X be a simplicial set with a simplicial subset A and let Y be a fibrant simplicial set. Suppose that $|A|$ is a retract of $|X|$, that is there is a continuous map $r\colon |X| \to |A|$ such that $r|_{|A|} = \mathrm{id}_{|A|}$. Then any simplicial map $f\colon A \to Y$ can be extended to a simplicial map $g\colon X \to Y$.*

Proof. Let $\phi = |f| \circ r$. Then $g = \phi'$ is an extension of f. $\qquad\square$

Let $f, g\colon X \to Y$ be pointed simplicial maps between pointed simplicial sets. Suppose that $f \simeq g$ rel $*$ under a pointed homotopy $F\colon X \times I \to Y$. By taking geometric realization, we have the commutative diagram

$$
\begin{array}{ccc}
|X \times I| \cong |X| \times I & \xrightarrow{\quad |F| \quad} & Y \\[2mm]
\big\uparrow & & \big\uparrow {\scriptstyle * \cup f \cup g} \\[2mm]
(* \times I) \cup (|X| \times 0) \cup (|X| \times 1) & \cong & |(* \times I) \cup (X \times 0) \cup (X \times 1)|
\end{array}
$$

and so $|F|$ is a pointed homotopy from $|f|$ to $|g|$, that is $|f| \simeq |g|$ rel $*$. Thus the geometric realization $f \mapsto |f|$ induces a function

$$
|\cdot|\colon [X, Y] \longrightarrow [|X|, |Y|].
$$

Theorem 2.51. *Let X be any pointed simplicial set and let Y be any pointed fibrant simplicial set. Then the geometric realization induces a one-to-one correspondence*

$$
|\cdot|\colon [X, Y] \longrightarrow [|X|, |Y|].
$$

Proof. First we show that $|\cdot|$ is onto. Let $\phi\colon |X| \to |Y|$ be any pointed map. Let $A = *$ be the simplicial subset of X generated by the basepoint. Let $f\colon A \to Y$ be the constant map to the basepoint of Y. By Theorem 2.49, there exists a simplicial map ϕ' such that $\phi'|_A = f$ and $|\phi'| \simeq \phi$ rel A. The first condition means that ϕ' is a pointed map and the second means that $|\phi'| \simeq \phi$ rel $*$. This proves that $|\cdot|$ is onto.

Now we show that $|\cdot|$ is one-to-one. Let $f, g\colon X \to Y$ be pointed maps such that $|f| \simeq |g|$ rel $*$. Let $A = (* \times I) \cup (X \times 0) \cup (X \times 1)$ and let

$$
h = * \cup f \cup g\colon A = (* \times I) \cup (X \times 0) \cup (X \times 1) \longrightarrow Y.
$$

Since $|f| \simeq |g|$ rel $*$, there exists a continuous map $\phi\colon |X| \times I = |X \times I| \to |Y|$ such that $\phi_{|A|} = |h|$. By Theorem 2.49, there exists a simplicial map

$F\colon X \times I \to Y$ such that $F|_A = h$ and so $f \simeq g$ rel $*$. The proof is finished. $\qquad\square$

Now consider the *singular simplicial functor* $S_*\colon X \mapsto S_*(X)$ for any topological space X, where $S_n(X) = \mathrm{Map}(\Delta^n, X)$ as sets with faces d_i and degeneracies s_i induced from the continuous maps

$$d^i\colon \Delta^{n-1} \to \Delta^n \quad (t_0, \ldots, t_{n-1}) \mapsto (t_0, \ldots, t_{i-1}, 0, t_i, \ldots, t_{n-1})$$

and

$$s^i\colon \Delta^{n+1} \to \Delta^n \quad (t_0, \ldots, t_{n+1}) \mapsto (t_0, \ldots, t_{i-1}, t_i + t_{i+1}, t_{i+1}, \ldots, t_{n+1}),$$

respectively.

Let X be any topological space. There is a canonical continuous map

$$e_X\colon |S_*(X)| \longrightarrow X$$

defined as follows:

Since Δ^n is compact and Hausdorff, the evaluation map

$$e_n\colon \Delta^n \times \mathrm{Map}(\Delta^n, X) \longrightarrow X$$

is continuous, where $\mathrm{Map}(\Delta^n, X)$ is given by compact-open topology. The set $\mathrm{Map}(\Delta^n, X)$ can be regarded as a topological space with discrete topology and so e_n is a continuous map by regarding $\mathrm{Map}(\Delta^n, X)$ as a set.

Lemma 2.52. *Let X be any topological space. The maps e_n induces a continuous map $e_X\colon |S_*(X)| \to X$. Moreover Φ_X is functorial with respect to X, that is, there is a commutative diagram*

$$
\begin{array}{ccc}
|S_*(X)| & \xrightarrow{\ e_X\ } & X \\[2pt]
\Big\downarrow{\scriptstyle |S_*(\phi)|} & & \Big\downarrow{\scriptstyle \phi} \\[2pt]
|S_*(Y)| & \xrightarrow{\ e_Y\ } & Y
\end{array}
$$

for any continuous map $\phi\colon X \to Y$.

Proof. Recall that

$$|S_*(X)| = \coprod_{n \geq 0} \Delta^n \times S_n(X)/\sim,$$

where $\sim$ is generated by $(t, d_i f) \sim (d^i t, f)$ for $t \in \Delta^{n-1}$ and $f \in S_n(X)$ and $(t, s_i f) \sim (s^i t, f)$ for $t \in \Delta^{n+1}$ and $f \in S_n(X)$. Note that

$$e_{n-1}(t, d_i f) = e_{n-1}(t, f \circ d^i) = f \circ d^i(t) = f(d^i t) = e_n(d^i t, f)$$

for $t \in \Delta^n$ and $f \in S_n(X)$, and

$$e_{n+1}(t, s_i f) = e_{n+1}(t, f \circ s^i) = f \circ s^i(t) = f(s^i(t)) = e_n(s^i t, f)$$

for $t \in \Delta^{n+1}$ and $f \in S_n(X)$. Thus the maps e_n induces a continuous map

$$e_X : |S_*(X)| \longrightarrow X.$$

Clearly e_X is functorial with respect to X. $\qquad\qquad\square$

Theorem 2.53. *Let X be a topological space with any choice of basepoint. The map $e_X : |S_*(X)| \to X$ induces an isomorphism on homotopy groups.*

Proof. First we show that $e_X : \pi_n(|S_*(X)|) \to \pi_n(X)$ is onto. Let $g : S^n \to X$ be a continuous map. Let $\tilde{g}$ be the composite

$$\tilde{g} : \Delta^n \xrightarrow[p]{\text{pinch}} S^n = \Delta^n / \partial D^n \xrightarrow{g} X.$$

Then $\tilde{g}$ is an element in $S_n(X) = \mathrm{Map}(\Delta^n, X)$ with $d_i \tilde{g} = \tilde{d}^i = *$ for $0 \le i \le n$. Thus the representing map

$$f_{\tilde{g}} : \Delta[n] \to S_*(X)$$

factors through the quotient simplicial set $S^n = \Delta[n]/\partial\Delta[n]$, that is there exists a simplicial map $\bar{f}_{\tilde{g}} : S^n \to S_*(X)$ such that the diagram

$$
\begin{array}{ccc}
\Delta[n] & \xrightarrow{\ f_{\tilde{g}}\ } & S_*(X) \\
\downarrow & \nearrow & \\
S^n & \bar{f}_{\tilde{g}} &
\end{array}
$$

commutes. We check the composite

$$\Delta^n = |\Delta[n]| \xrightarrow{\ |f_{\tilde{g}}|\ } |S_*(X)| \xrightarrow{\ e_X\ } X.$$

Let σ_n be the nondegenerate element in $\Delta[n]$. Note any

$$t \in \Delta^n = |\Delta[n]| = \coprod_q \Delta^q \times \Delta[n]_q / \sim$$

is represented by $(t, \sigma_n) \in \Delta^n \times \Delta[n]$. Since

$$\mathrm{id}_{\Delta^n} \times f_{\tilde{g}}(t, \sigma_n) = (t, \tilde{g}),$$

the composite

$$e_X \circ |f_{\tilde{g}}|(t) = \tilde{g}(t)$$

which gives the commutative diagram

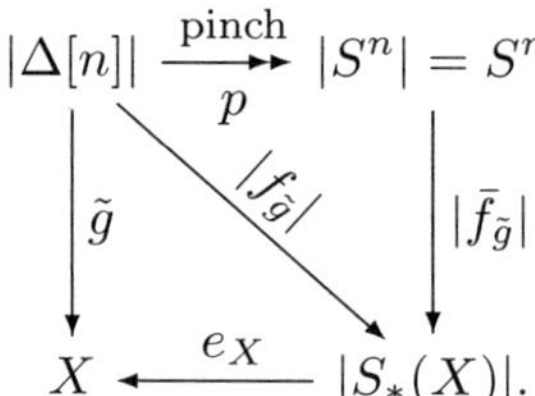

Since p is onto and

$$e_X \circ |\bar{f}_{\tilde{g}}| \circ p = \tilde{g} = g \circ p,$$

we have $e_X \circ |bar f_{\tilde{g}}| = g$ and so $\pi_n(|S_*(X)|) \to \pi_n(X)$ is onto.

Next we show that $\pi_n(|S_*(X)|) \to \pi_n(X)$ is one-to-one. By Theorem 2.51, it suffices to show that the composite

$$[S^n, S_*(X)] \xrightarrow{\;|\cdot|\;} \pi_n(|S_*(X)|) \xrightarrow{\;e_{X*}\;} \pi_n(X)$$

is one-to-one. Let $f, g \colon S^n \to S_*(X)$ be pointed maps such that

$$e_X \circ |f| \simeq e_X \circ |g| \text{ rel } *$$

under a pointed homotopy $F \colon S^n \times I \to X$. Thus $F|_{S^n \times 0} = e_x \circ |f|$, $F|_{S^n \times 1} = e_X \circ |g|$ and $F|_{* \times I} = *$. Let F' be the simplicial map given by the composite

$$F' \colon S^n \times I \xrightarrow{i_1 \times i_2} S_*(|S^n|) \times S_*(|I|) = S_*(|S^n| \times |I|) \xrightarrow{S_*(F)} S_*(X),$$

where i_1 is the representing map for the pinch map $\Delta^n \to |S^n|$ as an element in $S_n(|S^n|)$, and i_2 is the representing for the identity map of Δ^1 as an element in $S_1(|I|)$. Then $F'|_{* \times I} = *$, $F'|_{S^n \times 0}$ is the representing map for the composite

$$\Delta^n \xrightarrow{\text{pinch}} |S^n| \xrightarrow{F|_{|S^n| \times 0} = e_X \circ |f|} X, \tag{2.7}$$

and $F'|_{S^n \times 1}$ is the representing map for the composite

$$\Delta^n \xrightarrow{\text{pinch}} |S^n| \xrightarrow{F|_{|S^n| \times 1} = e_X \circ |g|} X. \tag{2.8}$$

Let $\sigma_n \in S_n^n$ be the nondegenerate element. Then $f(\sigma_n) \in S_n(X)$ is a continuous map from Δ^n to X with $f(\sigma_n) \circ d^i = *$ for $0 \le i \le n$. Consider the composite

$$\Delta^n = \coprod_q \Delta^q \times \Delta[n]_q / \sim \xrightarrow[p]{\text{pinch}} |S^n| \xrightarrow{|f|} |S_*(X)| \xrightarrow{e_X} X.$$

Let $\tilde{\sigma}_n$ be the nondegenerate element in $\Delta[n]$. For any $t \in \Delta^n$,

$$\begin{aligned}
e_X \circ |f| \circ p(t) &= e_X \circ |f| \circ p(t, \tilde{\sigma}_n) \\
&= e_X(t, f(\sigma_n)) \\
&= f(\sigma_n)(t).
\end{aligned}$$

It follows that the composite in Equation (2.7) is the representing map for the element $f(\sigma_n) \in S_n(X)$ and so $F'_{S^n \times 0} = f$. Similarly $F'_{S^n \times 1} = g$. Thus the map F' give a pointed homotopy $f \simeq g$ rel $*$. The proof is finished. $\square$

Now given a simplicial set X. There is a canonical simplicial map

$$i_X \colon X \to S_*(|X|)$$

defined as follows:

For any $x \in X_n$, let $f_x \colon \Delta[n] \to X$ be the representing map of x. Then $|f_x| \colon \Delta^n = |\Delta[n]| \to |X|$ gives an element in $S_n(|X|)$. The functions $i_X \colon X \to S_*(X)$ is defined by setting $i(x) = |f_x|$.

Lemma 2.54. *The sequence of functions $i_X \colon X \to S_*(|X|)$ is a simplicial map. Moreover i_X is functorial with respect to simplicial sets X and simplicial maps.*

Proof. The assertion follows from the following commutative diagram

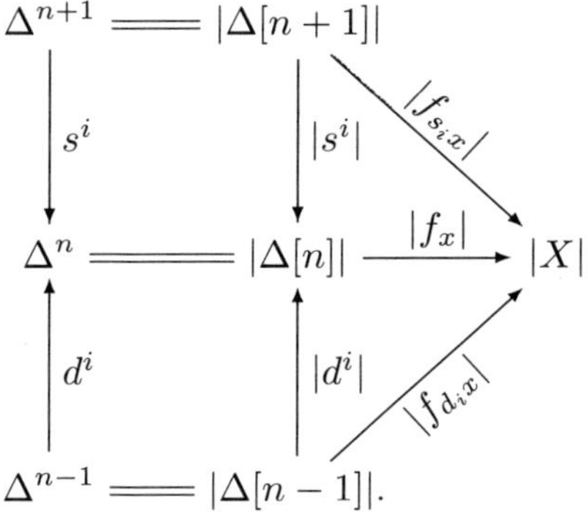

$\square$

Proposition 2.55. *The composite*

$$|X| \xrightarrow{\ |i_X|\ } |S_*(|X|)| \xrightarrow{\ e_{|X|}\ } |X|$$

is the identity map.

Proof. Recall that

$$|X| = \coprod_{n \geq 0} \Delta^n \times X_n / \sim,$$

where $(t, d_i x) \sim (d^i t, x)$ and $(t, s_i x) \sim (s^i t, x)$. Thus any element $\alpha \in |X|$ is represented by an element (t, x) with $t \in \Delta^n$ and $x \in X_n$. Let σ_n be the nondegenerate element in $\Delta[n]$. Then $f_x(\sigma_n) = x$. Thus $e_X \circ i_X(\alpha)$ is given by the point

$$e_X \circ |f_X|(t, \sigma_n) = |f_x|(t) = \{(t, x)\}$$

and hence the result. $\qquad\square$

Corollary 2.56. *Let X be a pointed simplicial set. Then the map $|i_X|$ induce an isomorphism*

$$|i_X|_* \colon \pi_n(|X|) \longrightarrow \pi_n(|S_*(|X|)|).$$

for each $n \geq 0$. $\qquad\square$

By the Whitehead Theorem which states that a pointed map between path-connected CW-complexes which induces isomorphisms on all homotopy groups is a homotopy equivalence. Thus if $\pi_0(X)$ is trivial, then $i_X \colon |X| \to |S_*(|X|)|$ is a homotopy equivalence because both $|X|$ and $|S_*(|X|)|$ are CW-complexes. The relations between simplicial sets and spaces can be given as in the picture

2.5. *Homotopy groups*

Let X be a pointed fibrant simplicial set. The homotopy group $\pi_n(X)$, as a set, is defined by

$$\pi_n(X) = [S^n, X]$$

and so $\pi_n(X) = \pi_n(|X|)$ as sets. We are going to give the combinatorial description for the group structure on $\pi_n(X)$ for $n \geq 1$ such that combinatorially defined group structure coincides with the group structure geometrically defined.

An element $x \in X_n$ is called *spherical* if $d_i x = *$ for all $0 \leq i \leq n$. Given a spherical element $x \in X_n$, then its representing map $f_x \colon \Delta[n] \to X$ factors through the simplicial quotient set $S^n = \Delta[n]/\partial\Delta[n]$. Conversely any simplicial map $f \colon S^n \to X$ gives a spherical element $f(\sigma_n) \in X_n$, where σ_n is the nondegenerate element in S^n_n. This gives a one-to-one correspondence

$$\boxed{\text{spherical elements in } X_n} \longleftrightarrow \boxed{\text{simplicial maps } S^n \to X} \ .$$

Since we have the notion of homotopy on simplicial maps, we also want to see how to give homotopy on the spherical elements in X_n.

2.5.1. *Path product and fundamental groupoids*

Let $\sigma_1 = (0,1) \in \Delta[1]_1$. A simplicial map $\lambda \colon \Delta[1] \to X$ is called a *path*. Since $\mathrm{Hom}(\Delta[1], X) \cong X_1$, the path are one-to-one correspondence to the elements in X_1 under the function given by $\lambda \mapsto x_\lambda = \lambda(\sigma_1)$. The *initial point* of λ is $\lambda(0) = \lambda(d_1\sigma_1) = d_1 x_\lambda \in X_0$, and the *ending point* of λ is $\lambda(1) = \lambda(d_0\sigma_1) = d_0 x_\lambda \in X_0$. Given two paths λ and μ such that $\lambda(1) = \mu(0)$. Then

$$d_0 x_\lambda = d_1 x_\mu$$

and so the elements $x_0 = x_\mu$ and $x_2 = x_\lambda$ have matching faces. By the fibrant assumption, there is an element $w \in X_2$ such that $d_0 w = x_\mu$ and $d_2 w = x_\lambda$, see the picture

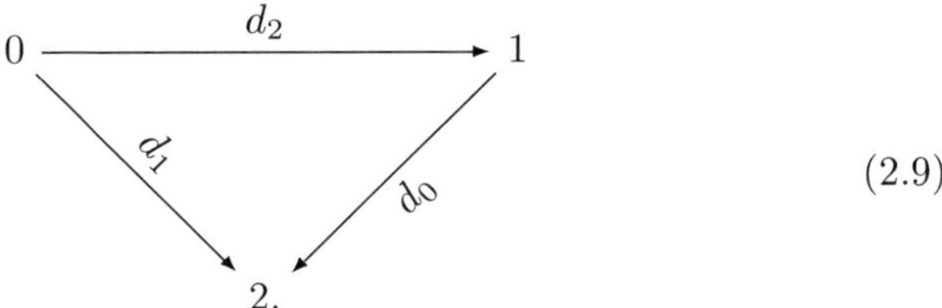

$$(2.9)$$

Define $\lambda * \mu = f_{d_1 w}$. Since w may not be unique, the product is only a *relation*. Let $\lambda \colon \Delta[1] \to X$ be a path. Let

$$[\lambda] = \{\mu \colon \Delta[1] \to X \mid \lambda \simeq \mu \text{ rel } \partial\Delta[1]\},$$

called *path homotopy class* represented by λ. We are going to show that the above product induces a well-defined product structure on path-homotopy classes.

Lemma 2.57. *Let $w, w' \in X_2$ such that $d_0 w = d_0 w'$ and $d_2 w = d_2 w'$. Then*

$$f_{d_1 w} \simeq f_{d_1 w'} \text{ rel } \partial\Delta[1].$$

Proof. Define the map

$$F \colon (\Lambda^1[2] \times I) \cup (\Delta[2] \times \partial I) \longrightarrow X$$

by setting

$$F|_{\Lambda^1[2] \times I}(x, t) = f_w|_{\Lambda^1[2]}(x) = f|_{w'}|_{\Lambda^1[2]}(x),$$
$$F|_{\Delta[2] \times 0} = f_w,$$
$$F|_{\Delta[2] \times 1} = f_{w'}.$$

Then F can be extended to a simplicial map $F' \colon \Delta[2] \times I \to X$ because the geometric realization $|(\Lambda^1[2] \times I) \cup (\Delta[2] \times \partial I)|$ is a retract of $|\Delta[2] \times I| = \Delta^2 \times I$. Now the composite

$$\Delta[1] \times I \xrightarrow{d^1 \times \mathrm{id}_I} \Delta[2] \times I \xrightarrow{F'} X$$

is a homotopy between $f_{d_1 w}$ and $f_{d_1 w'}$ relative to $\partial\Delta[1]$. $\square$

Lemma 2.58. *Let $\lambda_1, \lambda_1', \lambda_2, \lambda_2' \colon \Delta[1] \to X$ be paths such that $\lambda_i \simeq \lambda_i'$ rel $\partial\Delta[1]$ for $i = 1, 2$ and $\lambda_1 = \lambda_2$. Then $\lambda_1 * \lambda_2 \simeq \lambda_1' * \lambda_2'$ rel $\partial\Delta[1]$.*

Proof. Let $w, w' \in X_2$ such that $d_0 w = \lambda_2$, $d_2 w = \lambda_1$, $d_0 w' = \lambda_2'$ and $d_2 w' = \lambda_1'$. Let $F_i \colon \Delta[1] \times I \to X$ be a relative homotopy from λ_i to λ_i' for $i = 1, 2$. Then the map

$$F \colon (\Lambda^1[2] \times I) \cup (\Delta[2] \times \partial I) \longrightarrow X$$

by setting

$$F|_{d^0 \Delta[1] \times I} = F_2,$$
$$F|_{d^2 \Delta[1] \times I} = F_1,$$
$$F|_{\Delta[2] \times 0} = f_w,$$
$$F|_{\Delta[2] \times 1} = f_{w'},$$

is a well-defined simplicial map. So F can be extended to a simplicial map $F' \colon \Delta[2] \times I \to X$. The composite $F' \circ (d^1 \times \mathrm{id}_I)$ is a homotopy from $d_1 w$ to $d_1 w'$ relative to $\partial\Delta[1]$. $\qquad\square$

Lemma 2.59 (Associativity). *Let X be a fibrant simplicial set and let λ_1, λ_2 and λ_3 be paths in X such that $\lambda_1(1) = \lambda_2(0)$ and $\lambda_2(1) = \lambda_3(0)$. Then*

$$(\lambda_1 * \lambda_2) * \lambda_3 \simeq \lambda_1 * (\lambda_2 * \lambda_3) \ \mathrm{rel}\ \partial\Delta[1].$$

Proof. Since $d_0 x_{\lambda_2} = \lambda_2(1) = \lambda_3(0) = d_1 x_{\lambda_3}$, there exists $w_0 \in X_2$ such that $d_0 w_0 = x_{\lambda_3}$ and $d_2 w_0 = x_{\lambda_2}$. Since

$$d_2(d_1 w_0) = d_1 d_2 w = d_1 x_{\lambda_2} = \lambda_2(0) = \lambda_1(1) = d_0 x_{\lambda_1},$$

there exists $w_2 \in X_2$ such that $d_0 w_2 = d_1 w_0$ and $d_2 w = x_{\lambda_1}$. Since $d_0 x_{\lambda_1} = \lambda_1(1) = \lambda_2(0) = d_1 x_{\lambda_2}$, there exists $w_3 \in X_2$ such that $d_0 w_3 = x_{\lambda_2}$ and $d_2 w_3 = x_{\lambda_1}$. These elements can be shown in the picture

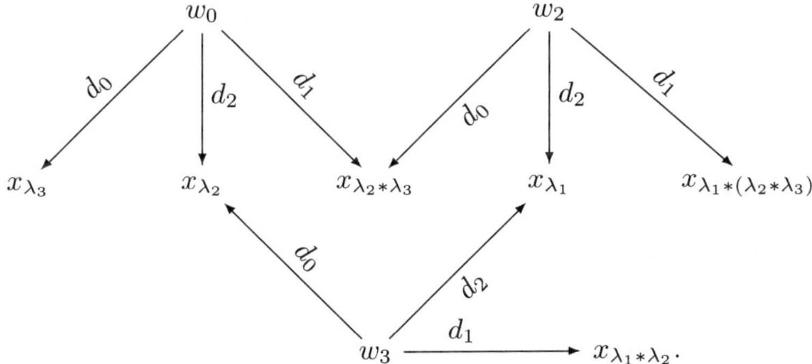

Since $d_1 w_0 = d_0 w_2$, $d_2 w_0 = d_0 w_3$ and $d_2 w_2 = d_2 w_3$, the elements w_0, w_2, w_3 have matching faces with respect to 1 and so there exists $u \in X_3$ such that $d_i u = w_i$ for $i \neq 1$. Now

$$x_{\lambda_1 * (\lambda_2 * \lambda_3)} = d_1 w_2 = d_1 d_2 u = d_1(d_1 u)$$

with

$$d_2(d_1 u) = d_1 d_3 u = d_1 w_3 = x_{\lambda_1 * \lambda_2}$$

and

$$d_0(d_1 u) = d_0 d_0 u = d_0 w_0 = x_{\lambda_3}.$$

Thus $d_1(d_1 u)$ is a representative for $(\lambda_1 * \lambda_2) * \lambda_3$. The assertion follows. $\square$

Given a point $y \in Y_0$, denote by $\epsilon_y \colon X \to Y$ for the constant simplicial map $\epsilon(x) = s_0^n(y)$ for $x \in X_n$.

Lemma 2.60 (Path inverse). *Let X be a fibrant simplicial set and let λ be a path in X. Then there exists a path λ^{-1} such that*

$$\lambda * \lambda^{-1} \simeq \epsilon_{\lambda(0)} \text{ rel } \partial\Delta[1].$$

Proof. Let $x_2 = x_\lambda$ and $x_1 = s_0\lambda(0)$. Then $d_1 x_2 = d_1 x_\lambda = \lambda(0) = d_1 x_1$. Thus x_1, x_2 have matching faces and so there exists $w \in X_2$ such that $d_2 w = x_\lambda$ and $d_1 w = s_0\lambda(0)$. Note that $f_{s_0\lambda(0)}$ is the constant path. Let $\lambda^{-1} = f_{d_0 w}$. By definition, $\lambda * \lambda^{-1} \simeq \epsilon_{\lambda(0)}$ rel $\partial\Delta[1]$. $\square$

For a fibrant simplicial set X, denote by $\mathcal{P}(X)$ the set of path homotopy classes. For a topological space X, write $\mathcal{P}(X)$ for the fundamental groupoid of X.

Theorem 2.61. *Let X be a fibrant simplicial set. Then $\mathcal{P}(X)$ is the quotient set of X_1 subject to the relation generated by $x \sim x'$ if $d_0 x = d_0 x'$, $d_1 x = d_1 x'$ and there exists $w \in X_2$ such that $d_2 w = x$, $d_1 w = x'$ and $d_0 w = \epsilon_{d_0 x}$. Moreover product structure on the pair of elements in $\mathcal{P}(X)$ which have matching faces is given by:*

> *If $x, x' \in X_1$ such that $d_0 x = d_1 x'$, then $[x] * [x'] = [d_1 w]$ for some $w \in X_2$ with the property that $d_0 w = x'$ and $d_2 w = x$.*

Moreover the map $|\cdot| \colon \mathcal{P}(X) \to \mathcal{P}(|X|)$ is an isomorphism of groupoids.

Proof. We only need to show that $f_x \simeq f_{x'}$ rel $\partial\Delta[1]$ if and only if $x \sim x'$. In geometry, one can easily see that

$$(\Delta^1 \times \Delta^1)/(1 \times \Delta^1) \cong \Delta^2$$

as in the picture:

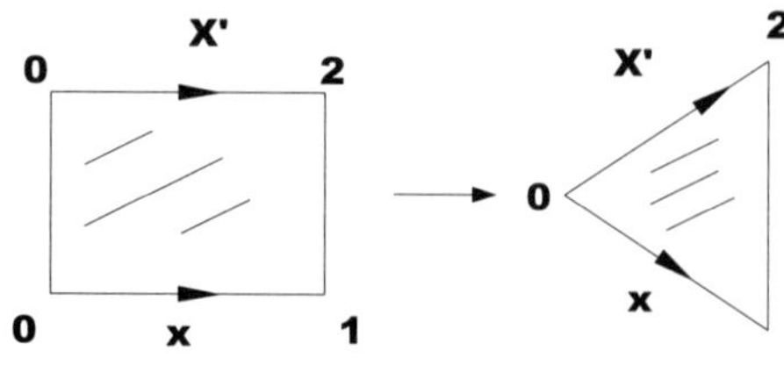

But in simplicial situation, $(\Delta[1] \times \Delta[1])/(1 \times \Delta[1]) \ncong \Delta[2]$ and so this statement is less obvious in simplicial situation. We will use simplicial extension theorem to give a proof.

Suppose that $f_x \simeq f_{x'}$ rel $\partial\Delta[1]$. Define a simplicial map

$$f \colon \partial\Delta[2] \longrightarrow X$$

by setting

$$
\begin{aligned}
f|_{d^2\Delta[1]} &= f_x, \\
f|_{d^1\Delta[1]} &= f_{x'}, \\
f|_{d^0\Delta[1]} &= \epsilon_{d_0 x}.
\end{aligned}
$$

Since $f_x \simeq f_{x'}$ rel $\partial\Delta[1]$, there is a continuous map $\phi \colon |\Delta[2]| \to |X|$ such that $\phi|_{|\partial\Delta[2]|} = |f|$ from the above picture. By simplicial extension theorem, f can be extended to a simplicial map $g \colon \Delta[2] \to X$. Let $w = g(\sigma_2)$. Then $d_2 w = x$, $d_1 w = x'$ and $d_0 w = s_0 d_0 x$ and so $x \sim x'$.

Conversely suppose that $x \sim x'$. Define a simplicial map

$$f \colon \partial(I \times I) = (\partial I \times I) \cup (I \times \partial I) \longrightarrow X$$

by setting

$$
\begin{aligned}
f|_{I \times 0} &= f_x, \\
f|_{I \times 1} &= f_{x'}, \\
f|_{0 \times I} &= \epsilon_{d_1 x}, \\
f|_{1 \times I} &= \epsilon_{d_0 x}.
\end{aligned}
$$

From the above picture, $|f| \colon |\partial(I \times I)| = \partial(|I \times I|) \to |X|$ can be extended to a continuous map from $|I \times I|$ to X. Thus, by simplicial extension theorem, f can be extended to a simplicial map $F \colon I \times I \to X$ which is a homotopy from f_x to $f_{x'}$ relative to ∂I. $\qquad\square$

Exercise 2.15. Let X be a fibrant simplicial set and let $x, x' \in X_1$. Show that $x \sim x'$ if and only if $d_0 x = d_0 x'$, $d_1 x = d_1 x'$ and there exists $w \in X_2$ such that $d_0 w = x$, $d_1 w = x'$ and $d_2 w = \epsilon_{d_1 x}$.

2.5.2. *Fundamental groups*

Whence the path product is decided, the fundamental group will follow. So we have the following:

Theorem 2.62. *Let X be a pointed fibrant simplicial set. Then the fundamental group $\pi_1(X)$ is the quotient set of the spherical elements in X_1*

subject to the relation generated by $x \sim x'$ if there exists $w \in X_2$ such that $d_2 w = x$, $d_1 w = x'$ and $d_0 w = *$. The product structure in $\pi_1(X)$ is given by:

> *$[x] \cdot [x'] = [d_1 w]$, where w is any element in X_2 with the property that $d_0 w = x'$ and $d_2 w = x$.*

Moreover the map $|\cdot|: \pi_1(X) \to \pi_1(|X|)$ preserves the product structure.

$\square$

2.5.3. *Higher homotopy groups*

The ideas of higher homotopy groups follow the same lines and so we have the following:

Theorem 2.63. *Let X be a pointed fibrant simplicial set. Then the homotopy group $\pi_n(X)$ is the quotient set of the spherical elements in X_n subject to the relation generated by $x \sim x'$ if there exists $w \in X_{n+1}$ such that $d_0 w = x$, $d_1 w = x'$ and $d_j w = *$ for $j > 1$. The product structure in $\pi_n(X)$ is given by:*

> *$[x] + [x'] = [d_1 w]$, where w is any element in X_{n+1} with the property that $d_0 w = x'$, $d_2 w = x$ and $d_j w = *$ for $j > 2$.*

Moreover the map $|\cdot|: \pi_n(X) \to \pi_n(|X|)$ preserves the product structure.

$\square$

Proposition 2.64. *Let X be a pointed fibrant simplicial set. Then $\pi_n(X)$ is commutative for $n \geq 2$.*

By using the fact that $\pi_n(|X|)$ is commutative, the proof is straightforward as $\pi_n(X) \cong \pi_n(|X|)$. A combinatorial proof is given as follows. It will take several steps: For being careful in computation, we write $[x][x']$ for the product instead of $[x] + [x']$. The definition of product is equivalent to say that:

> Let $w \in X_{n+1}$ such that all $d_i w$ are spherical elements and $d_j w = *$ for $j > 2$. Then $[d_1 w] = [d_2 w][d_0 w]$.

Recall that the simplicial identity for faces is $d_j d_i = d_i d_{j+1}$ for $j \geq i$. We can consider this relation as the symmetric matrix below the first row given by:

$$\begin{pmatrix} d_0 & d_1 & d_2 & d_3 & d_4 & \cdots \\ \Box & d_0d_1 & d_0d_2 & d_0d_3 & d_0d_4 & \cdots \\ d_0d_0 & \Box & d_1d_2 & d_1d_3 & d_1d_4 & \cdots \\ d_1d_0 & d_1d_1 & \Box & d_2d_3 & d_2d_4 & \cdots \\ d_2d_0 & d_2d_1 & d_2d_2 & \Box & d_3d_4 & \cdots \\ d_3d_0 & d_3d_1 & d_3d_2 & d_3d_3 & \Box & \cdots \\ \cdots & \cdots & \cdots & \cdots & \cdots & \cdots \end{pmatrix}.$$

For checking whether the elements $x_0, x_1, \ldots$ are matching faces, we can then look at whether the following matrix below first row is symmetric

$$\begin{pmatrix} x_0 & x_1 & x_2 & x_3 & x_4 & \cdots \\ \Box & d_0x_1 & d_0x_2 & d_0x_3 & d_0x_4 & \cdots \\ d_0x_0 & \Box & d_1x_2 & d_1x_3 & d_1x_4 & \cdots \\ d_1x_0 & d_1x_1 & \Box & d_2x_3 & d_2x_4 & \cdots \\ d_2x_0 & d_2x_1 & d_2x_2 & \Box & d_3x_4 & \cdots \\ d_3x_0 & d_3x_1 & d_3x_2 & d_3x_3 & \Box & \cdots \\ \cdots & \cdots & \cdots & \cdots & \cdots & \cdots \end{pmatrix}.$$

Proof. Step 1. *Suppose that $w \in X_{n+1}$ such that all faces d_iw are spherical elements and $d_iw = *$ for $i \neq 1, 2, 3$. Then $[d_2w] = [d_1w][d_3w]$.*

Let $x_0 = s_2d_3w$, $x_1 = w$, $x_2 = s_1d_1w$, $x_4 = s_0d_3w$ and $x_j = *$ for $j > 4$. We check that the elements x_i are matching faces with respect to 3. This is done by the matrix:

$$\begin{pmatrix} x_0 & x_1 & x_2 & \boxed{x_3} & x_4 & x_5 & \cdots \\ \Box & * & * & -- & d_3w & * & \cdots \\ * & \Box & d_1w & -- & d_3w & * & \cdots \\ * & d_1w & \Box & -- & * & * & \cdots \\ d_3w & d_2w & d_1w & \Box & * & * & \cdots \\ d_3w & d_3w & * & -- & \Box & * & \cdots \\ * & * & * & -- & * & \Box & \cdots \\ \cdots & \cdots & \cdots & -- & \cdots & \cdots \end{pmatrix}.$$

Thus there exists an element $u \in X_{n+2}$ such that $d_ju = x_j$ for $j \neq 3$. Now

$$d_0(d_3u) = d_2d_0u = d_2x_0 = d_3w$$
$$d_1(d_3u) = d_2d_1u = d_2x_1 = d_2w$$
$$d_2(d_3u) = d_2d_2u = d_2x_2 = d_1w$$

and so $[d_2w] = [d_1w][d_3w]$.

Step 2. *Suppose that $w \in X_{n+1}$ such that all faces $d_i w$ are spherical elements and $d_i w = *$ for $i \neq 0, 2, 3$. Then $[d_0 w][d_2 w] = [d_3 w]$.*

Since $a_0 = d_0 w$, $a_1 = *$, $a_3 = *$, and $a_j = *$ for $j \geq 4$ are matching faces with respect to 2, there exists $x_1 \in X_{n+1}$ such that $d_0 x_1 = d_0 w$, $d_1 x_1 = *$, and $d_j x_1 = *$ for $j \geq 3$. Let $t = d_2 x_1$.

$$1 = [d_2 x_1][d_0 w].$$

In addition to x_1, let $x_0 = s_0 d_0 w$, $x_2 = w$, $x_4 = s_2 d_3 w$ and $x_j = *$ for $j \geq 5$. From the table

$$\left(\begin{array}{cccccccc}
x_0 & x_1 & x_2 & \boxed{x_3} & x_4 & x_5 & \cdots \\
\square & d_0 w & d_0 w & -\!- & * & * & \cdots \\
d_0 w & \square & * & -\!- & * & * & \cdots \\
d_0 w & * & \square & -\!- & d_3 w & * & \cdots \\
* & d_2 x_1 & d_2 w & \square & d_3 w & * & \cdots \\
* & * & d_3 w & -\!- & \square & * & \cdots \\
* & * & * & -\!- & * & \square & \cdots \\
\cdots & \cdots & \cdots & -\!- & \cdots & \cdots &
\end{array}\right),$$

the elements x_j are matching faces and there exists $u \in X_{n+2}$ such that $d_j u = x_j$ for $j \neq 3$. From

$$\begin{aligned}
d_0(d_3 u) &= d_2 d_0 u = d_2 x_0 = * \\
d_1(d_3 u) &= d_2 d_1 u = d_2 x_1 \\
d_2(d_3 u) &= d_2 d_2 u = d_2 x_2 = d_2 w \\
d_3(d_3 u) &= d_3 d_4 u = d_3 x_4 = d_3 w,
\end{aligned}$$

we have $[d_2 w] = [d_2 x_1][d_3 w] = [d_0 w]^{-1}[d_3 w]$ and hence the statement.

Step 3. *Suppose that $w \in X_{n+1}$ such that all faces $d_i w$ are spherical elements and $d_i w = *$ for $i \neq 0, 1, 2, 3$. Then $[d_1 w][d_3 w] = [d_0 w][d_2 w]$.*

Let $x_4 = w$. Since $a_1 = *$, $a_2 = *$, $a_3 = d_0 w$ and $a_j = *$ for $j > 3$ are matching faces with respect to 0, there exists x_0 such that $d_j x_0 = a_j$ for $j > 0$. According to Step 2, $[d_0 x_0] = [d_3 x_0] = [d_0 w]$.

Since $b_0 = d_0 x_0$, $b_1 = *$, $b_3 = d_1 w$ and $b_j = *$ for $j > 3$ are matching faces with respect to 1, there exits $x_1 \in X_{n+1}$ such that $d_j x_1 = b_j$ for $j \neq 2$. From Step 2, $[d_0 x_1][d_2 x_1] = [d_3 x_1]$, that is $[d_0 x_0][d_2 x_1] = [d_1 w]$ or $[d_0 w][d_2 x_1] = [d_1 w]$.

Let $x_2 = s_2 d_2 w$ and $x_j = *$ for $j > 4$. The elements $x_0, x_1, x_2, x_4, \ldots$ are matching faces by the table

$$
\begin{pmatrix}
x_0 & x_1 & x_2 & \boxed{x_3} & x_4 & x_5 & \cdots \\
\square & d_0 x_0 & * & -- & d_0 w & * & \cdots \\
d_0 x_0 & \square & * & -- & d_1 w & * & \cdots \\
* & * & \square & -- & d_2 w & * & \cdots \\
* & d_2 x_1 & d_2 w & \square & d_3 w & * & \cdots \\
d_0 w & d_1 w & d_2 w & -- & \square & * & \cdots \\
* & * & * & -- & * & \square & \cdots \\
\cdots & \cdots & \cdots & -- & \cdots & \cdots &
\end{pmatrix}.
$$

It follows that there exists an element $u \in X_{n+2}$ such that $d_j u = x_j$ for $j \neq 3$. From

$$
\begin{aligned}
d_0(d_3 u) &= d_2 d_0 u = d_2 x_0 = * \\
d_1(d_3 u) &= d_2 d_1 u = d_2 x_1 \\
d_2(d_3 u) &= d_2 d_2 u = d_2 x_2 = d_2 w \\
d_3(d_3 u) &= d_3 d_4 u = d_3 x_4 = d_3 w,
\end{aligned}
$$

we have $[d_2 w] = [d_2 x_1][d_3 w] = [d_0 w]^{-1}[d_1 w][d_3 w]$ and hence the statement.

Final Step. Let $w \in X_{n+1}$ such that all faces $d_i w$ are spherical elements and $d_i w = *$ for $i \neq 0, 1, 2,$. From Step 3, since $[d_3 w] = 1$,

$$
[d_1 w] = [d_1 w][d_3 w] = [d_0 w][d_2 w]
$$

and by definition $[d_1 w] = [d_2 w][d_0 w]$. It follows that $[d_0 w][d_2 w] = [d_2 w][d_0 w]$. The proof is finished now. $\qquad\square$

2.5.4. *Homotopy Addition Theorem*

Lemma 2.65. *Let X be a pointed fibrant simplicial set and let x, x', x'' be spherical elements in X_n with $n \geq 2$. Suppose that there exists $y \in X_{n+1}$ such that $d_q y = x$, $d_{q+1} y = x'$, $d_{q+2} y = x''$, and $d_j y = *$ for $j \neq q$, $q + 1$, $q + 2$. Then*

$$
[x] - [x'] + [x''] = 0
$$

in $\pi_n(X)$.

Proof. The proof is given by induction on q. When $q = 0$, this is the definition of the product in $\pi_n(X)$. For $q > 0$, we use the extension condition on X to move the relation one step down.

     J. Wu

Consider the elements $x_i = *$ for $i < q - 1$, $x_{q-1} = s_{q+1}d_{q+2}y$, $x_q = y$, $x_{q+1} = s_q d_q y$, $x_{q+3} = s_{q-1}d_{q+2}y$ and $x_i = *$ for $i > q + 3$. These elements have matching faces by the table

$$
\begin{pmatrix}
\cdots & x_{q-2} & x_{q-1} & x_q & x_{q+1} & \boxed{x_{q+2}} & x_{q+3} & x_{q+4} & \cdots \\
\cdots & * & * & * & * & -- & * & * & \cdots \\
\cdots & \cdots & \cdots & \cdots & \cdots & -- & \cdots & \cdots & \cdots \\
\cdots & \square & * & * & * & -- & * & * & \cdots \\
\cdots & * & \square & d_{q-1}x_q = * & * & -- & d_{q+2}y & * & \cdots \\
\cdots & * & d_{q-1}x_{q-1} = * & \square & d_q y & -- & d_{q+2}y & * & \cdots \\
\cdots & * & * & d_q y = x & \square & -- & * & * & \cdots \\
\cdots & * & d_{q+2}y & d_{q+1}y & d_q y & \square & * & * & \cdots \\
\cdots & * & d_{q+2}y & d_{q+2}y & * & -- & \square & * & \cdots \\
\cdots & * & * & * & * & -- & * & \square & \cdots \\
\cdots & * & * & * & * & -- & * & * & \cdots \\
\cdots & \cdots & \cdots & \cdots & \cdots & -- & \cdots & \cdots & \cdots
\end{pmatrix}.
$$

So there exists $w \in X_{n+2}$ such that $d_i w = x_i$ for $i \neq q + 2$. Let $z = d_{q+2}w$. Then

$$
d_j z = d_j d_{q+2} w =
\begin{cases}
d_{q+1}d_j w = d_{q+1}x_j = * & \text{if } j < q - 1 \\
d_{q+1}d_{q-1}w = d_{q+1}x_{q-1} = d_{q+2}y = x'' & \text{if } j = q - 1 \\
d_{q+1}d_q w = d_{q+1}x_q = d_{q+1}y = x' & \text{if } j = q \\
d_{q+1}d_{q+1}w = d_{q+1}x_{q+1} = d_q y = x & \text{if } j = q + 1 \\
d_{q+2}d_{q+3}w = d_{q+2}x_{q+3} = * & \text{if } j = q + 2 \\
d_{q+2}d_{j+1}w = d_{q+2}x_{j+1} = * & \text{if } j > q + 2.
\end{cases}
$$

By induction, we have

$$[x''] - [x'] + [x] = 0$$

and hence the result. $\qquad\square$

Lemma 2.66. *Let X be a pointed fibrant simplicial set and let $x \in X_{n+1}$ with $d_i x = x_i$ for $0 \leq i \leq n+1$ such that all x_i are spherical. Let $w \in K_{n+1}$ with $d_i w = x_i$ for $q \neq i, i+1$, $d_q w = *$ and $d_{q+1}w = z$ for some $0 \leq q \leq n - 1$. Then in $\pi_n(X)$*

$$[x_q] - [x_{q+1}] + [z] = 0.$$

Proof. We check that the elements $v_i = s_{q+1}x_i$ for $i \leq q - 1$, $v_q = s_q x_q$, $v_{q+2} = x$, $v_{q+3} = w$ and $v_i = s_{q+2}x_{i-1}$ for $i \geq q + 4$ are matching faces with respect to $q+1$. Note that each x_i is spherical. We have

$$d_j v_i = \begin{cases} * & \text{if } \begin{cases} 0 \le i \le q-1 & \text{and } 0 \le j \le q \\ 0 \le i \le q-1 & \text{and } j > q+2 \\ i = q & \text{and } 0 \le j \le q-1 \\ i = q & \text{and } j > q+1 \\ i \ge q+4 & \text{and } j \le q+1 \\ i \ge q+4 & \text{and } j > q+3 \end{cases} \\ x_i & \text{if } 0 \le i \le q-1 & \text{and } j = q+1, q+2 \\ x_q & \text{if } i = q & \text{and } j = q, q+1 \\ x_j & \text{if } i = q+2 \\ d_j w & \text{if } i = q+3 \\ x_{i-1} & \text{if } i \ge q+4 & \text{and } j = q+2, q+3. \end{cases}$$

Note that $d_i w = x_i$ for $q \ne i, i+1$ and $d_q w = *$. This gives the table

$$\begin{pmatrix}
v_0 & \cdots & v_{q-1} & v_q & \boxed{v_{q+1}} & v_{q+2}=x & v_{q+3}=w & v_{q+4} & \cdots & v_{n+2} \\
* & \cdots & * & * & -\!- & x_0 & x_0 & * & \cdots & * \\
\cdots & \cdots & \cdots & \cdots & -\!- & \cdots & & & \cdots & \cdots \\
* & \cdots & \square & * & -\!- & x_{q-1} & x_{q-1} & * & \cdots & * \\
* & \cdots & * & \square & -\!- & x_q & * & * & \cdots & * \\
* & \cdots & * & x_q & \square & x_{q+1} & d_{q+1}w & * & \cdots & * \\
x_0 & \cdots & x_{q-1} & x_q & -\!- & \square & x_{q+2} & x_{q+3} & \cdots & x_{n+1} \\
x_0 & \cdots & x_{q-1} & * & -\!- & x_{q+2} & \square & x_{q+3} & \cdots & x_{n+1} \\
* & \cdots & * & * & -\!- & x_{q+3} & x_{q+3} & \square & \cdots & * \\
* & \cdots & * & * & -\!- & x_{q+4} & x_{q+4} & * & \cdots & \cdots \\
\cdots & \cdots & \cdots & \cdots & -\!- & \cdots & \cdots & \cdots & \cdots & \cdots \\
* & \cdots & * & * & -\!- & x_{n+1} & x_{n+1} & * & * & \square
\end{pmatrix}$$

Thus there exists $u \in K_{n+2}$ such that $d_i u = v_i$ for $i \ne q+1$. Now

$$d_i(d_{q+1}u) = \begin{cases} d_q d_i u = d_q v_i = * & \text{if } i < q \\ d_q d_q u = d_q v_q = x_q & \text{if } i = q \\ d_{q+1}d_{q+2}u = d_{q+1}v_{q+2} = x_{q+1} & \text{if } i = q+1 \\ d_{q+1}d_{q+3}u = d_{q+1}v_{q+3} = d_{q+1}w = z & \text{if } i = q+2 \\ d_{q+1}d_{i+1}u = d_{q+1}v_{i+1} = * & \text{if } i > q+2. \end{cases}$$

Thus $[x_q] - [x_{q+1}] + [z] = 0$ which is the assertion. $\qquad\square$

Theorem 2.67 (Homotopy Addition Theorem). *Let X be a pointed fibrant simplicial set. Let $y_i \in X_n$ be spherical elements for $0 \le i \le n+1$*

with $n \geq 2$. Then in $\pi_n(X)$ the equation

$$[y_0] - [y_1] + [y_2] - \cdots + (-1)^{n+1}[y_{n+1}] = 0$$

holds if and only if there exists $y \in K_{n+1}$ such that $d_i y = y_i$ for $0 \leq i \leq n+1$.

Note. If $n = 1$, Homotopy Addition Theorem holds by definition of the product in the fundamental group.

Proof. $\Longleftarrow$ Let $y \in X_{n+1}$ such that $d_i y = y_i$ is spherical for each $0 \leq i \leq n+1$. For each $0 \leq q \leq n-1$, since the spherical elements are always matching faces as their faces are all trivial, there exists w_q such that

$$d_i w_q = \begin{cases} * & \text{if } 0 \leq i \leq q \\ y_i & \text{if } i \geq q+2. \end{cases}$$

Let $z_q = d_{q+1} w_q$. By applying Lemma 2.66 to y and w_0 for $q = 0$, we have

$$[y_0] - [y_1] + [z_0] = 0.$$

For $q > 0$, note that $d_j w_q = d_j w_{q-1}$ for $j \neq q, q+1$, $d_q w_q = *$ and $d_{q+1} w_q = z_q$, $d_q w_{q-1} = z_{q-1}$ and $d_{q+1} w_q = y_{q+1}$, by Lemma 2.66,

$$[z_{q-1}] - [y_{q+1}] + z_q = 0$$

for $1 \leq q \leq n-1$. This gives that equations

$$[y_0] - [y_1] + [z_0] = 0$$
$$-[z_0] + [y_2] - [z_1] = 0$$
$$[z_1] - [y_3] + [z_2] = 0$$
$$\cdots$$
$$(-1)^{n-1}[z_{n-2}] + (-1)^n[y_n] + (-1)^{n+1}[z_{n-1}] = 0$$

and so

$$(-1)^{n+1}[z_{n-1}] + \sum_{i=0}^{n}(-1)^i[y_i] = 0.$$

Since

$$d_j w_{n-1} = \begin{cases} * & \text{if } j < n-1 \\ * & \text{if } j = n-1 \\ d_n w_{n-1} = z_{n-1} & \text{if } j = n \\ y_{n+1} & \text{if } j = n+1, \end{cases}$$

we have $-[z_{n-1}] + [y_{n+1}] = 0$. Thus $\sum_{i=0}^{n+1}(-1)^i[y_i] = 0$.

$\Longrightarrow$ Let $y_i \in X_n$ be spherical elements for $0 \leq i \leq n+1$ such that

$$[y_0] - [y_1] + [y_2] - \cdots + (-1)^{n+1}[y_{n+1}] = 0.$$

Since $y_1, \ldots, y_{n+1}$ are matching faces with respect to 0, there exists $w \in K_{n+1}$ such that $d_i w = y_i$ for $i \neq 0$. Note that $d_0 w$ is spherical as $d_i d_0 w = d_0 d_{i+1} w = *$. From the previous step, we have

$$\sum_{i=0}^{n+1}(-1)^i[d_i w] = 0$$

and so $[d_0 w] = [y_0]$. Let $f\colon \partial\Delta[n+1] \to X$ be the simplicial map such that $f|_{d^0\Delta[n]} = f_{d_0 w}$, $f|_{d^1\Delta[n]} = f_{y_0}$ and $f|_{d^i\Delta[n]} = \epsilon_*$ for $i > 1$. We check that the geometric realization

$$|f|\colon |\partial\Delta[n+1]| = \partial\Delta^{n+1} \longrightarrow |X|$$

can be extended to a continuous map $g\colon |\Delta[n+1]| = \Delta^{n+1} \to |X|$. Let

$$\tilde{F}\colon I \times \Delta[n] \xrightarrow{\text{pinch}} I \times S^n = I \times (\Delta[n]/\partial\Delta[n]) \xrightarrow{F} X,$$

where F is a pointed homotopy between $f_{d_0 w}$ and f_{y_0}, that is, $F|_{0 \times S^n} = f_{d_0 w}$, $F|_{1 \times S^n} = f_{y_0}$ and $F|_{I \times *} = \epsilon_*$. Consider

$$\Delta^{n+1} = \left\{ (t_0, t_1, \ldots, t_{n+2}) \in \mathbb{R}^{n+2} \mid t_i \geq 0 \ \sum_{i=0}^{n+2} t_i = 1 \right\}$$

and

$$I \times \Delta^n = \left\{ (s, t_1, t_2, \ldots, t_{n+2}) \in \mathbb{R}^{n+2} \mid t_i \geq 0 \ \sum_{i=0}^{n+1} t_i = 1 \ 0 \leq s \leq 1 \right\}.$$

Observe that the first two faces of Δ^{n+1} are the subspaces

$$d^0\Delta[n] = \{(0, t_1, t_2, \ldots, t_{n+2}) \in \mathbb{R}^{n+2}\} \cap \Delta^{n+1},$$

$$d^1\Delta[n] = \{(t_0, 0, t_2, \ldots, t_{n+2}) \in \mathbb{R}^{n+2}\} \cap \Delta^{n+1}.$$

Define the continuous map

$$\theta\colon I \times \Delta^n \longrightarrow \Delta^{n+1}$$

by setting

$$\theta(s, t_1, t_2, \ldots, t_{n+2}) = (st_1, (1-s)t_1, t_2, \ldots, t_{n+2}).$$

Then θ is onto and

$$\theta^{-1}(d^0\Delta^n) \subseteq (0 \times \Delta^n) \cup (I \times d^0\Delta^{n-1})$$

$$\theta^{-1}(d^1\Delta^n) = (1 \times \Delta^n) \cup (I \times d^0\Delta^{n-1})$$

$$\theta^{-1}(d^i\Delta^n) = I \times d^{i-1}\Delta^{n-1} \quad \text{for } i \geq 2.$$

Define a function $g\colon \Delta^{n+1} \to |X|$ by setting

$$g(x) = |\tilde{F}|(\theta^{-1}(x))$$

for $x \in \Delta^{n+1}$. We check that g is well-defined with $g|_{\partial\Delta^{n+1}} = |f|$.

Case I. $x \in \Delta^{n+1} \setminus \partial\Delta^{n+1}$. Then the pre-image of θ has exactly one point and $g(x)$ is well-defined in this case.

Case II. $x \in d^i\Delta^n$ for $i \geq 2$. Then $\theta^{-1}(x) \in I \times \partial\Delta^n$. Since $|\tilde{F}||_{I \times \partial\Delta^n} = *$,

$$g(x) = * = |f|(x).$$

Case III. $x \in d^0\Delta^n$. Consider the equation

$$\theta(s, t'_1, t'_2, \ldots, t'_{n+2}) = (st'_1, (1-s)t'_1, t'_2, \ldots, t'_{n+2}) = x = (0, t_1, t_2, \ldots, t_{n+2}).$$

If $t_i > 0$ for all $1 \leq i \leq n+2$, then $s = 0$ and $t'_i = t_i$. In this case,

$$g(x) = |\tilde{F}|(0, t_1, \ldots, t_{n+2}) = |f_{d_0w}|(x) = |f|(x).$$

If one of t_i is zero, then $\theta^{-1}(x) \in I \times \partial\Delta^n$. In this case,

$$|\tilde{F}|(\theta^{-1}(x)) = * = |f|(x).$$

Case IV. $x \in d^1\Delta^n$. Consider the equation

$$\theta(s, t'_0, t'_1, \ldots, t'_{n+1}) = (st'_0, (1-s)t'_0, t'_1, \ldots, t'_{n+1}) = x = (t_0, 0, t_2, \ldots, t_{n+2}).$$

If $t_i > 0$ for all $i \neq 1$, then $s = 1$, $t'_i = t_i$ for $0 \leq i \leq n+1$. In this case,

$$g(x) = |\tilde{F}|(1, t_0, t_2 \ldots, t_{n+1}) = |f_{y_0}|(x) = |f|(x).$$

If one of t_i is zero, then $\theta^{-1}(x) \in I \times \partial\Delta^n$. In this case,

$$|\tilde{F}|(\theta^{-1}(x)) = * = |f|(x).$$

Hence g is well-defined. To show that g is continuous, let A be a closed set of $|X|$. Then $|\tilde{F}|^{-1}(A)$ is closed in $I \times \Delta^n$. Since $I \times \Delta^n$ is Hausdorff and compact, $|\tilde{F}|^{-1}(A)$ is compact and so $g^{-1}(A) = \theta(|\tilde{F}|^{-1}(A))$ is compact. It follows that $g^{-1}(A)$ is closed and hence g is continuous.

Now by the simplicial extension theorem, the simplicial map $f\colon \partial\Delta[n+1] \to X$ can be extended to a simplicial map $f'\colon \Delta[n+1] \to X$. Let $z = f'(\sigma_{n+1})$, where σ_{n+1} is the nondegenerate element in $\Delta[n+1]$. Then $d_0 z = d_0 w$, $d_1 z = y_0$ and $d_j z = *$ for $j \geq 2$. Let $v_0 = z$, $v_1 = w$ and $v_i = s_1 d_{i-1}w = s_1 y_{i-1}$ for $i \geq 3$. These elements are matching faces by the table

$$
\begin{pmatrix}
v_0 = z & v_1 = w & \boxed{v_2} & v_3 = s_1 d_2 w & s_1 d_3 w & s_1 d_4 w & \cdots & s_1 d_{n+1} w \\
\square & d_0 w & -- & * & * & * & \cdots & * \\
d_0 w & \square & -- & y_2 & y_3 & y_4 & \cdots & y_{n+1} \\
y_0 & y_1 & \square & y_2 & y_3 & y_4 & \cdots & y_{n+1} \\
* & y_2 & -- & \square & * & * & \cdots & * \\
* & y_3 & -- & * & \square & * & \cdots & * \\
* & y_4 & -- & * & * & \square & \cdots & * \\
\cdots & \cdots & -- & \cdots & \cdots & \cdots & \cdots & \cdots \\
* & y_{n+1} & -- & * & * & * & \cdots & \square
\end{pmatrix}.
$$

Thus there exists u such that $d_i u = v_i$ for $i \neq 2$. Now

$$
d_i(d_2 u) = \begin{cases}
d_1 v_0 = y_0 & \text{if } i = 0 \\
d_1 v_1 = y_2 & \text{if } i = 1 \\
d_2 v_{i+1} = d_2 s_1 y_i = y_i & \text{if } i \geq 2.
\end{cases}
$$

The proof is finished now. $\qquad\square$

2.5.5. *A geometric proof of Homotopy Addition Theorem*

In this independent subsection, you are assumed to know some basic knowledge on algebraic topology, such as Allen Hatcher's book [17]. Observe that the boundary $\partial\Delta[n+1] = \bigcup_{i=0}^{n+1} d^i \Delta[n]$. Thus

$$
\partial\Delta[n+1]/\operatorname{sk}_{n-1}\Delta[n+1] = \bigvee_{i=0}^{n+1} (d^i\Delta[n]/d^i\partial\Delta[n]) = \bigvee_{i=0}^{n+1} S^n.
$$

Let X be a pointed simplicial set and let $y_0, y_1, \ldots, y_{n+1}$ be spherical elements in X_n. Let $f_{y_i}\colon S^n \to X$ be the representing map for y_i. Define the simplicial map g to be the composite

$$
\partial\Delta[n+1] \xrightarrow{\ q\ } \partial\Delta[n+1]/\operatorname{sk}_{n-1}\Delta[n+1] = \bigvee_{i=0}^{n+1} S^n \xrightarrow{\ \bigvee_{i=0}^{n+1} f_{y_i}\ } X.
$$

This defines a continuous map

$$
|g|\colon |\partial\Delta[n+1]| = S^n \longrightarrow |X|.
$$

We are going to show that

$$[|g|] = \sum_{i=0}^{n+1} (-1)^i [|f_{y_i}|].$$

From this, we will give a new proof for the Homotopy Addition Theorem. First we need to handle the homotopy class of the pinch map

$$|q|\colon |\partial\Delta[n+1]| = \partial\Delta^{n+1} \longrightarrow \left|\bigvee_{i=0}^{n+1} S^n\right| = \bigvee_{i=0}^{n+1} S^n.$$

Recall that

$$\Delta^n = \left\{ (t_0, t_1, \ldots, t_n) \in \mathbb{R}^{n+1} \mid t_i \geq 0 \sum_{i=0}^{n} t_i = 1 \right\}.$$

The symmetric group S_{n+1} acts on Δ^n is given by permuting coordinates. Let $\tau_{i,j}\colon \Delta^n \to \Delta^n$ be the map by switching coordinates i and j for $0 \leq i < j \leq n$, that is

$$\tau_{i,j}(t_0, t_1, \ldots, t_n) = (t_0, t_1, \ldots, t_{i-1}, t_j, t_{i+1}, \ldots, t_{j-1}, t_i, t_{j+1}, \ldots, t_n).$$

Then the boundary $\partial\Delta^n$ is invariant under the map $\tau_{i,j}$. Note that $\partial\Delta^n \cong S^{n-1}$ and so any self-map of $\partial\Delta^n$ has a degree, see Hatcher's book [17, Section 2.2].

Lemma 2.68. *The map* $\tau_{i,j}\colon \partial\Delta^n \to \partial\Delta^n$ *is of degree* -1 *for* $0 \leq i < j \leq n$.

Proof. Let

$$V_{i,j} = \{(t_0, t_1, \ldots, t_n) \in \mathbb{R}^{n+1} \mid t_i = t_j\}$$

be the n-dimensional hyperspace $\mathbb{R}^{n+1}$ with the coordinates $t_i = t_j$, consisting of the points in which the i-coordinate is the same as the jth coordinate. Then the action of $\tau_{i,j}$ is the reflection on $\partial\Delta^n$ with respect to the $(n-1)$-dimensional sphere given by $\partial\Delta^n \cap V_{i,j}$. By [17, p. 134, Section 2.2, Part (e)], the map $\tau_{i,j}$ is of degree -1. $\qquad\square$

Exercise 2.16. Let S_{n+1} act on $\partial\Delta^n$ by permuting coordinates. Show that the map $\sigma\colon \partial\Delta^n \to \partial\Delta^n$ is of degree $\mathrm{sign}(\sigma)$ for each $\sigma \in S_{n+1}$, where $\mathrm{sign}(\sigma)$ is the sign of the permutation σ, that is, $\mathrm{sign}(\sigma) = +1$ if σ is an even permutation and $\mathrm{sign}(\sigma) = -1$ if σ is an odd permutation.

For each $\sigma \in S_{n+1}$, since $\sigma(\partial \Delta^n) = \partial \Delta^n$, the action of σ on Δ^n induces an action

$$\sigma \colon \Delta^n / \partial \Delta^n \longrightarrow \Delta^n / \partial \Delta^n.$$

Note that $\Delta^n / \partial \Delta^n \cong S^n$. So any self-map of $\Delta^n / \partial \Delta^n$ has a degree.

Lemma 2.69. *For each $0 \leq i < j \leq n$, the map $\tau_{i,j} \colon \Delta^n / \partial \Delta^n \to \Delta^n / \partial \Delta^n$ is of degree -1.*

Proof. Again let

$$V_{i,j} = \{(t_0, t_1, \ldots, t_n) \in \mathbb{R}^{n+1} \mid t_i = t_j\}.$$

Then $\tau_{i,j}$ on $\Delta^n / \partial \Delta^n$ is the reflection with respect to the $(n-1)$-sphere given by $(V_{i,j} \cap \Delta^n)/(V_{i,j} \cap \partial \Delta^n)$. Thus the degree of $\tau_{i,j}$ is -1. $\quad\square$

Lemma 2.70. *Let $n \geq 2$. Then the homotopy group*

$$\pi_n \left(\bigvee_{i=1}^{q} S^n \right) = \bigoplus_{i=1}^{q} \mathbb{Z}.$$

Proof. This is a direct consequence of the Hurewicz Theorem [17, p. 366, Theorem 4.32]: Since $\pi_1 \left(\bigvee_{i=1}^{q} S^n \right) = 0$, $H_j \left(\bigvee_{i=1}^{q} S^n \right) = 0$ for $0 < j < n$,

$$\pi_n \left(\bigvee_{i=1}^{q} S^n \right) \cong H_n \left(\bigvee_{i=1}^{q} S^n \right) = \bigoplus_{i=1}^{q} \mathbb{Z}$$

by the Hurewicz Theorem. $\quad\square$

By using this fact, we can introduce the concept of *multi-degree* for the maps from S^n to $\bigvee_{i=1}^{q} S^n$. Let $f \colon S^n \to \bigvee_{i=1}^{q} S^n$ be a pointed map. Let l_j be the degree of the composite

$$S^n \xrightarrow{\ f\ } \bigvee_{i=1}^{q} S^n \xrightarrow{\ \pi_j\ } S^n,$$

where π_j is the projection to the jth copy of S^n for $1 \leq j \leq q$. Then we call that f has the multi-degree $(l_1, \ldots, l_q)$. From the above lemma, any two (pointed) maps are homotopic if they have the same multi-degree.

Lemma 2.71. *Let $n \geq 2$. Then the pinch map*

$$|q|\colon |\partial\Delta[n+1]| = \partial\Delta^{n+1} \longrightarrow \left|\bigvee_{i=0}^{n+1} S^n\right| = \bigvee_{i=0}^{n+1} S^n$$

is of multi-degree $\epsilon(1, -1, 1, -1, \ldots, (-1)^{n+1})$ for some $\epsilon = \pm 1$.

Proof. Let $|q|$ has multi-degree $(l_0, l_1, \ldots, l_{n+1})$. Observe that the composite

$$\partial\Delta^{n+1} \xrightarrow{\ q\ } \bigvee_{i=0}^{n+1} S^n \xrightarrow{\ \pi_j\ } S^n$$

is the pinch map from

$$\partial\Delta^{n+1} \twoheadrightarrow \partial\Delta^{n+1}/|\Lambda^j[n+1].$$

Since $|\Lambda^j[n+1]|$ is contractible, the composite $\pi_j \circ |q|$ is a homotopy equivalence. It follows that the degree $n_j = \pm 1$. Now we consider the symmetric group action. Observe that there is a commutative diagram

$$\begin{array}{ccc}
\partial\Delta^{n+1} \xrightarrow{\ |q|\ } \displaystyle\bigvee_{i=0}^{n+1} S^n = \bigvee_{i=0}^{n+1} d^i\Delta^n/d^i(\partial\Delta^n) \\[2mm]
\Big\downarrow{\scriptstyle \tau_{j,j+1}} \qquad\qquad\qquad\qquad \Big\downarrow{\scriptstyle \tau_{j,j+1}} \\[2mm]
\partial\Delta^{n+1} \xrightarrow{\ |q|\ } \displaystyle\bigvee_{i=0}^{n+1} S^n = \bigvee_{i=0}^{n+1} d^i\Delta^n/d^i(\partial\Delta^n)
\end{array} \qquad (2.10)$$

for $0 \leq j \leq n$. Recall that

$$d^i\Delta^n = \{(t_0, t_1, \ldots, t_{i-1}, 0, t_{i+1}, \ldots, t_{n+1}) \in \mathbb{R}^{n+2}\} \cap \Delta^{n+1}.$$

If $i \neq j, j+1$, then $\tau_{j,j+1}$ maps $d^i\Delta^n$ onto itself by switching the coordinates t_j and t_{j+1}. By Lemma 2.68, the map

$$\tau_{j,j+1}|_{d^i\Delta^n}\colon d^i\Delta^n \to d^i\Delta^n$$

is of degree -1.

If $i = j$ or $j+1$, then $\tau_{j,j+1}$ switches the faces $d^j\Delta^n$ and $d^{j+1}\Delta^n$ by keeping the same order of coordinates.

It follows that the degree

$$\deg(\tau_{j,j+1} \circ |q|) = (-l_0, -l_1, \ldots, -l_{j-1}, l_{j+1}, l_j, -l_{j+1}, -l_{j+2}, \ldots, -l_{n+1}).$$
$$(2.11)$$

On the other hand, since $\tau_{j,j+1}\colon \partial\Delta^{n+1} \to \partial\Delta^{n+1}$ is of degree -1,

$$\deg(|q| \circ \tau_{i,j}) = (-l_0, -l_1, \ldots, -l_{n+1}). \tag{2.12}$$

From Equations (2.10), (2.11) and (2.12), we obtain

$$(-l_0, -l_1, \ldots, -l_{j-1}, l_{j+1}, l_j, -l_{j+1}, -l_{j+2}, \ldots, -l_{n+1}) = (-l_0, -l_1, \ldots, -l_{n+1})$$

and so $l_{j+1} = -l_j$ for each $0 \le j \le n$. It follows that the map $|q|$ has the multi-degree $\epsilon(1, -1, \ldots, (-1)^{n+1})$ for some $\epsilon = \pm 1$. $\qquad\square$

Now we need to review how to define the product on the geometric homotopy groups $\pi_n(|X|)$. Consider S^n as the subspace of unit vectors in $\mathbb{R}^{n+1}$. Let $\mu'\colon S^n \to S^n \vee S^n$ be the map by pinching *equator* $S^{n-1} = S^n \cap \mathbb{R}^n$ to a point, where the basepoint is chosen on the equator. Let $f, g\colon S^n \to |X|$ be pointed continuous maps. Then the composite

$$S^n \xrightarrow{\ \mu'\ } S^n \vee S^n \xrightarrow{\ f \vee g\ } |X|$$

is defined as the product of f and g. The map μ' has the property that the composite

$$S^n \xrightarrow{\ \mu'\ } S^n \vee S^n \xrightarrow{\ \pi_j\ } S^n$$

is of degree 1. In other words, μ' has multi-degree $(1, 1)$. Now given pointed continuous maps $f_1, \ldots, f_q\colon S^n \to |X|$, the product $f_1 + \cdots + f_q$ is given by the composite

$$S^n \xrightarrow{\ \mu'\ } \bigvee_{i=1}^{q} S^n \xrightarrow{\ \vee_{i=1}^{q} f_i\ } |X|,$$

where μ' is of multi-degree $(1, 1, \ldots, 1)$.

Proposition 2.72. *Let $f_i\colon S^n \to |X|$ be pointed map for $0 \le i \le n+1$ with $n \ge 2$ and let g be the composite*

$$\partial\Delta^{n+1} \xrightarrow{\ |q|\ } \bigvee_{i=0}^{n+1} S^n \xrightarrow{\ \vee_{i=0}^{n+1} f_i\ } |X|.$$

Then there exists $\epsilon = \pm 1$ such that

$$\epsilon \sum_{i=0}^{n+1} (-1)^i [f_i] = [g]$$

in $\pi_n(|X|)$.

Proof. Since $|q|$ has multi-degree $\epsilon(1, -1, 1, -1, \ldots, (-1)^{n+1})$, there is a commutative diagram up to homotopy

$$
\partial\Delta^{n+1} \xrightarrow{\ |q|\ } \bigvee_{i=0}^{n+1} S^n \xrightarrow{\ \bigvee_{i=0}^{n+1} f_i\ } |X|
$$

where $\theta_i \colon S^n \to S^n$ is of degree $\epsilon(-1)^i$. The assertion follows. $\qquad\square$

Proof (Another proof of the Homotopy Addition Theorem). Let $y_0, y_1, \ldots, y_{n+1}$ be spherical elements in X_n with $n \geq 2$.

Suppose that

$$
\sum_{i=0}^{n+1} (-1)^i [y_i] = 0
$$

in $\pi_n(X)$. Then

$$
\sum_{i=0}^{n+1} (-1)^i [|f_{y_i}|] = 0
$$

in $\pi_n(|X|)$. Define the simplicial map g to be the composite

$$
\partial\Delta[n+1] \xrightarrow{\ q\ } \partial\Delta[n+1]/\operatorname{sk}_{n-1}\Delta[n+1] = \bigvee_{i=0}^{n+1} S^n \xrightarrow{\ \bigvee_{i=0}^{n+1} f_{y_i}\ } X. \quad (2.13)
$$

By Proposition 2.72, in $\pi_n(|X|)$,

$$
[|g|] = \epsilon \sum_{i=0}^{n+1} (-1)^i [|f_{y_i}|] = 0
$$

and so $|g| \colon |\partial\Delta[n+1]| \to |X|$ can be extended to a continuous map $|\Delta[n+1]| \to |X|$. By simplicial extension theorem, g can be extended to a simplicial map $g' \colon \Delta[n+1] \to X$. Let $w = g'(\sigma_{n+1})$, where σ_{n+1} is the nondegenerate element of $\Delta[n+1]$. Since

$$
g'|_{d^i\Delta[n]} = f_{y_i},
$$

we have $d_i w = d_i g'(\sigma_{n+1}) = g'(d_i \sigma_{n+1}) = f_{y_i}(d_i \sigma_{n+1}) = y_i$ for each $0 \leq i \leq n+1$.

Conversely there exists $w \in X_{n+1}$ such that $d_i w = y_i$ for each $0 \le i \le n+1$. Let g be defined as in Equation (2.13). Then

$$f_w|_{\partial\Delta[n+1]} = g$$

and so the map $g\colon \partial\Delta[n+1] \to X$ is null-homotopic, that is $[g] = 0$ in $\pi_n(X)$. By Proposition 2.72, we have

$$\sum_{i=0}^{n+1}(-1)^i[y_i] = \sum_{i=0}^{n+1}(-1)^i[|f_{y_i}|] = \epsilon[|g|] = 0$$

and hence the result. $\qquad\square$

2.5.6. *Minimal simplicial sets*

Given a space X, we can have fibrant simplicial set $S_*(X)$. But $S_*(X)$ seems too big as there are uncountable many elements in each $S_n(X)$. On the other hand, for having simplicial homotopy groups, we need fibrant assumption. This means that simplicial model for a given space cannot be too small. We hope to have a kind of *smallest* fibrant simplicial sets. This will be the concept of *minimal simplicial sets* discussed below. First we define the homotopy for the elements in X_n.

Definition 2.73. Let X be a fibrant simplicial set. Then for $x, y \in X_n$ call $x \simeq y$ if the representing maps f_x and f_y are homotopic relative to $\partial\Delta[n]$.

Proposition 2.74. *Let X be a fibrant simplicial set and let $x, y \in X_n$. Then $x \simeq y$ if and only if $d_i x = d_i y$ for all $0 \le i \le n$ and there exists $w \in X_{n+1}$ such that $d_k w = x$, $d_{k+1} w = y$, and $d_i w = d_i s_k x = d_i s_k y$ for $i \ne k, k+1$.*

Proof. Let $x, y \in X_n$ be the elements such that $d_i x = d_i y$ for all $0 \le i \le n$. Consider the simplicial map

$$\colon \Delta[n+1] \xrightarrow{\;s_k\;} \Delta[n] \xrightarrow{\;f_x\;} X.$$

Let σ_{n+1} be the nondegenerate element in $\Delta[n+1]_{n+1}$. Since

$$f_x \circ s^k(\sigma_{n+1}) = f_x(s_k\sigma_n) = s_k(f_x(\sigma_n)) = s_k x,$$

we have

$$f_x \circ s^k(d_i\sigma_{n+1}) = d_i f_x \circ s^k(\sigma_{n+1}) = d_i s_k x$$

for all $0 \le i \le n$. By restricting to $\Lambda^{k+1}[n+1]$,

$$f_x \circ s^k|_{\Lambda^{k+1}[n+1]}\colon \Lambda^{k+1}[n+1] \longrightarrow X$$

is the simplicial map such that $f_x \circ s_k|_{d^i \Delta[n]} = f_{d_i s_k x}$ for $i \neq k+1$. Since

$$\partial \Delta[n+1] = \Lambda^{k+1}[n+1] \cup d^{k+1}\Delta[n],$$

we fill in the map on the missing face $d^{k+1}\Delta[n]$ by f_y, more precisely, let

$$f = (f_x \circ s^k) \cup f_y \colon \partial\Delta[n+1] = \Lambda^{k+1}[n+1] \cup d^{k+1}\Delta[n] \longrightarrow X. \quad (2.14)$$

To see f is well-defined, note that

$$\Lambda^{k+1}[n+1] \cap d^{k+1}\Delta[n] = d^{k+1}\partial\Delta[n].$$

The map $f_x \circ s^k$ on $d^{k+1}\partial\Delta[n]$ is given by

$$f_{d_{k+1}\circ s_k x}|_{\partial\Delta[n]} = f_x|_{\partial\Delta[n]}.$$

Since $d_j x = d_j y$ for all $0 \leq j \leq n$, we have

$$f_x|_{\partial\Delta[n]} = f_y|_{\partial\Delta[n]}$$

and so the simplicial map f is well-defined. Observe that:

There exists $w \in X_{n+1}$ such that $d_k w = d_k s_k x = x$, $d_{k+1}w = y$, and $d_i w = d_i s_k x = d_i s_k y$ for $i \neq k, k+1$ if and only if f can be extended to a simplicial map $f_w \colon \Delta[n+1] \to X$, if and only if the geometric realization

$$|f| \colon |\partial\Delta[n+1]| = \partial\Delta^{n+1} \longrightarrow |X|$$

can be extended to a continuous map $\Delta^{n+1} \to X \iff [|f|] = 0$ in $\pi_n(|X|)$.

In the above statement, the first equivalence follows from the construction of the map f and the fact that simplicial maps $\Delta[n+1] \to X$ can be identified with the representing maps, and the second equivalence follows from the simplicial extension theorem.

Now construct the second simplicial map

$$f' \colon \partial(I \times \Delta[n]) \longrightarrow X \quad (2.15)$$

by requiring that $f'|_{0\times\Delta[n]} = f_x$, $f'|_{1\times\Delta[n]} = f_y$ and $f'|_{I\times\partial\Delta[n]}(s,z) = f_x(z)$. Observe that:

$f_x \simeq f_y$ rel $\partial\Delta[n]$ if and only if f' can be extended to a simplicial map $I \times \Delta[n] \to X$, if and only if the geometric realization

$$|f'| \colon |\partial(I \times \Delta[n])| = \partial(I \times \Delta^n) \longrightarrow |X|$$

can be extended to a continuous map $I \times \Delta^n \to |X| \iff [|f'|] = 0$ in $\pi_n(|X|)$.

In the above statement, the first equivalence follows from the definition of simplicial homotopy, and the second equivalence follows from simplicial extension theorem.

For having the connections between the above two statements, we need construct the third simplicial map given by defining

$$f'' = f_x \cup f_y \colon \Delta[n] \cup_{\partial \Delta[n]} \Delta[n] \colon X, \tag{2.16}$$

where $\Delta[n] \cup_{\partial \Delta[n]} \Delta[n]$ is the union of two copies of $\Delta[n]$ by identifying the elements in the boundary $\partial \Delta[n]$. Note that the geometric realization

$$|\Delta[n] \cup_{\partial \Delta[n]} \Delta[n]| = \Delta^n \cup_{\partial \Delta^n} \Delta^n \cong S^n.$$

Note that in geometry

$$\Delta^n \cup_{\partial \Delta^n} \Delta^n = \partial(I \times \Delta^n)/\sim,$$

where $\sim$ is generated by $(0, z) \sim (t, z)$ for $z \in \partial \Delta^n$ and $0 \le t \le 1$. Let

$$q_1 \colon \partial(I \times \Delta^n) \longrightarrow \Delta^n \cup_{\partial \Delta^n} \Delta^n$$

be the quotient map. Clearly q_1 is a homotopy equivalence. Since $f'(t, z) = f_x(z)$ for $(t, z) \in I \times \partial \Delta[n]$, we have $|f'| = |f''| \circ q_1$.

Let

$$q_2 = s^k \cup \mathrm{id} \colon S^n = \partial \Delta^{n+1} = |\Lambda^{k+1}[n+1]| \cup d^{k+1} \Delta^n \longrightarrow \Delta^n \cup_{\partial \Delta^n} \Delta^n = S^n.$$

From the commutative diagram

$$\partial \Delta^{n+1} = |\Lambda^{k+1}[n+1]| \cup d^{k+1} \Delta^n \xrightarrow[\simeq]{p_1} \partial \Delta^{n+1}/|\Lambda^{k+1}[n+1]| = d^{k+1}\Delta^n/d^{k+1}\partial\Delta^n$$

$$\downarrow{q_2 = s^k \cup \mathrm{id}} \qquad\qquad \|$$

$$\Delta^n \cup_{\partial \Delta^n} \Delta^n \xrightarrow[\simeq]{p_2} \Delta^n \cup_{\partial \Delta^n} \Delta^n/\Delta^n = \Delta^n/\partial\Delta^n,$$

where p_1 is given by pinching $|\Lambda^{k+1}[n+1]|$ to be the point and p_2 pinches the left Δ^n to the point. Since $|\Lambda^{k+1}[n+1]|$ and Δ^n are contractible, p_1 and p_2 are homotopy equivalences. Thus q_2 is a homotopy equivalence. Since $f = (f_x \circ s^k) \cup f_y$, we have $|f| = |f''| \circ q_2$.

Now from the commutative diagram

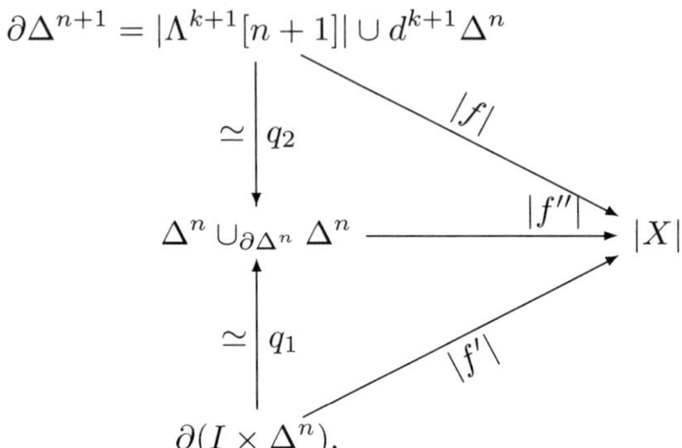

one of f, f', f'' is null homotopic implies that the other two are also null homotopic because q_1 and q_2 induce isomorphisms on $\pi_n(|X|)$. The assertion follows. $\qquad\square$

Definition 2.75. A fibrant simplicial set is called *minimal* if it has the property that $x \simeq y \Longrightarrow x = y$.

Proposition 2.76. *Let X be a fibrant simplicial set. Then X is minimal if and only if, for any $0 \leq k \leq n + 1$, whenever $v, w \in X_{n+1}$ such that $d_i v = d_i w$ for all $i \neq k$, then $d_k v = d_k w$.*

In other words, a fibrant simplicial set is minimal if and only if for any two elements with all faces but one the same, then the missed face must be the same.

Proof. $\Longrightarrow$ Let $v, w \in X_{n+1}$ such that $d_i v = d_i w$ for all $i \neq k$. Note that $\partial|\Lambda^k[n+1]| = \partial|d^k \Delta[n]|$. Since

$$|f_v|_{\Lambda^k[n+1]}| \cup |f_{d_k v}| \colon |\Lambda^k[n+1]| \cup |d^k \Delta[n]| = |\partial \Delta[n+1]| \longrightarrow |X|$$

can be extended to the continuous map $|f_v| \colon |\Delta[n+1]| \to |X|$, we have

$$|f_v|_{\Lambda^k[n+1]}| \simeq |f_{d_k v}| \text{ rel } \partial \Delta^n.$$

Similarly $|f_w|_{\Lambda^k[n+1]}| \simeq |f_{d_k w}|$ rel $\partial \Delta^n$. Since $d_i v = d_i w$ for $i \neq k$, $f_v|_{\Lambda^k[n+1]} = f_w|_{\Lambda^k[n+1]}$ and so $|f_v|_{\Lambda^k[n+1]}| = |f_w|_{\Lambda^k[n+1]}|$. It follows that

$$|f_{d_k v}| \simeq |f_{d_k w}| \text{ rel } \partial \Delta^n$$

and so $f_{d_k v} \simeq f_{d_k w}$ rel $\partial \Delta[n]$, that is $d_k v \simeq d_k w$. Hence $d_k v = d_k w$ by the minimal assumption.

$\Longleftarrow$. Let $x, y \in X_n$ such that $x \simeq y$. By Proposition 2.74, there exists $w \in X_{n+1}$ such that $d_i w = d_i s_0 x$ for $i \neq 1$ and $d_1 w = y$. Let $v = s_0 x$. Then $d_i v = d_i w$ for $i \neq 1$. By the assumption, we have $d_1 v = d_1 w$ and so $x = y$. $\qquad\square$

Lemma 2.77. *Let X be a fibrant simplicial set. Then*

(1). *Suppose that $x \in X_n$ and $v \in X_{n-1}$ such that $d_k x \simeq v$. Then there exists $z \in X_n$ such that $d_i z = d_i x$ for $i \neq k$ and $d_k z = v$.*

(2). *Let $F \colon \Delta[n] \times I \to X$ be a simplicial map and let $x = F(\sigma_n, 1)$, where σ_n is the nondegenerate element in $\Delta[n]_n$. Suppose that $x \simeq y$. Then there exists a simplicial map $G \colon \Delta[n] \times I \to X$ such that $G(\sigma_n, 1) = y$ and*

$$G|_{(\Delta[0] \times 0) \cup (\partial(\Delta[n]) \times I)} = F|_{(\Delta[0] \times 0) \cup (\partial(\Delta[n]) \times I)}.$$

Proof. (1). Since

$$|f_x|_{\Lambda^k[n]}| \simeq |f_{d_k x}| \simeq |f_v| \text{ rel } \partial\Delta^n,$$

the map

$$|f_x|_{\Lambda^k[n]}| \cup |f_v| \colon \partial\Delta^n = |\Lambda^k[n]| \cup |d^k\Delta[n]| \longrightarrow |X|$$

can be extended to a continuous map $|\Delta[n]| \to |X|$ and so

$$f_x|_{\Lambda^k[n]} \cup f_v \colon \partial\Delta[n] = \Lambda^k[n]| \cup |d^k\Delta[n] \longrightarrow X$$

can be extended to a simplicial map $g \colon \Delta[n] \to X$. Let $z = g(\sigma_n)$. Then $d_i z = d_i x$ for $i \neq k$ and $d_k z = v$.

(2). Since

$$|F|_{(\Delta[0] \times 0) \cup (\partial(\Delta[n]) \times I)}| \cup |f_x| \colon |(\Delta[0] \times 0) \cup (\partial(\Delta[n]) \times I)| \cup |\Delta[n] \times 1| \longrightarrow |X|$$

can be extended to the continuous map $|F| \colon |\Delta[n] \times I| \to |X|$,

$$|F|_{(\Delta[0] \times 0) \cup (\partial(\Delta[n]) \times I)}| \simeq |f_x| \simeq |f_y| \text{ rel } \partial\Delta^n.$$

Thus

$$|F|_{(\Delta[0] \times 0) \cup (\partial(\Delta[n]) \times I)}| \cup |f_y| \colon |(\Delta[0] \times 0) \cup (\partial(\Delta[n]) \times I)| \cup |\Delta[n] \times 1| \longrightarrow |X|$$

can be extended to a continuous map $|\Delta[n] \times I| \to |X|$ and so

$$F|_{(\Delta[0] \times 0) \cup (\partial(\Delta[n]) \times I)} \cup f_y \colon (\Delta[0] \times 0) \cup (\partial(\Delta[n]) \times I) \cup (\Delta[n] \times 1) \longrightarrow X$$

can be extended to a simplicial map $G \colon \Delta[n] \times I \to X$ which has the desired property. $\qquad\square$

Definition 2.78. Let X a simplicial set. A simplicial subset A is called a *strong deformation retract* of X if there exits a homotopy $F \colon X \times I \to X$ such that

(1). $F(a, t) = a$ for $a \in A$ and $t \in I$, that is $F|_{A \times I}$ is the composite
$$A \times I \xrightarrow{\text{proj.}} A \hookrightarrow X;$$
(2). $F|_{X \times 0} = \mathrm{id}_X$;
(3). $F(X \times 1) \subseteq A$.

A minimal simplicial set A is called a *minimal subcomplex* of X if A is a strong deformation retract of X.

Theorem 2.79 (Existence of minimal subcomplex). *Any fibrant simplicial set X contains a minimal subcomplex A.*

Proof. Let A_0 consist of a choice of one vertex in each homotopy class. Suppose that A_j is defined for all $0 \le j \le n-1$. Let A_n consist of a choice of one simplex for each homotopy class $[x]$ in X_n of simplices all of whose faces are in A_{n-1}, with requiring that if $s_i a \in [x]$ for some $a \in A_{n-1}$ with $0 \le i \le n-1$, then choose $s_i a$ in A_n.

1. *A_n is well-defined: That is $s_i a \simeq s_k a' \implies s_i a = s_k a'$ for $a, a' \in A_{n-1}$.*

If $i = k$, then $a = d_i s_i a = d_i s_i a' = a'$, because if $x \simeq y$ then $d_j x = d_j y$ for all j and so $s_i a = s_k a'$ when $i = k$.

If $i < k$, then $a = d_i s_i a = d_i s_k a' = s_{k-1} d_i a'$. Then

$$a' = d_k s_k a' = d_k s_i a = d_k s_i s_{k-1} d_i a' = s_i d_{k-1} s_{k-1} d_i a' = s_i d_i a'.$$

It follows that

$$s_i a = s_i s_{k-1} d_i a' = s_k s_i d_i a' = s_k a'$$

and so A_n is well-defined.

2. *$A = \{A_n\}_{n \ge 0}$ is a simplicial subset of X.*

By the construction, $d_j(A_n) \subseteq A_{n-1}$ and $s_j(A_{n+1})$ for $0 \le j \le n$. Thus the sequence of subsets $A = \{A_n\}_{n \ge 0}$ is a simplicial subset of X.

3. *A is a strong deformation retract of X.*

Let $\mathrm{sk}_n^A(X) = A \cup \mathrm{sk}_n(X)$. We are going to define inductively a homotopy

$$F^n \colon \mathrm{sk}_n^A(X) \times I \longrightarrow X$$

satisfying the conditions:

(1). $F^n|_{A \times I}(a, t) = a$ for $a \in A$ and $t \in I$.
(2). $F^n|_{\mathrm{sk}_n^A(X) \times 0} = \mathrm{id}_X|_{\mathrm{sk}_n^A(X)}$.
(3). $F^n|_{\mathrm{sk}_{n-1}^A(X) \times I} = F^{n-1}$.
(4). $F^n(\mathrm{sk}_n^A \times 1) \subseteq A$.

Note that every vertex $x \in X_0$ is homotopic to a vertex $w \in A_0$, that is, there is a path $\lambda_x \colon I \to X$ such that $\lambda_x(0) = x$ and $\lambda_x(1) \in A_0$. Let p be the composite $A \times I \xrightarrow{\mathrm{proj.}} A \hookrightarrow A$. Let

$$F^0 = p \cup \bigcup_{x \in X_0 \smallsetminus A_0} \lambda_x \colon \mathrm{sk}_0^A(X) \times I = (A \times I) \cup \coprod_{x \in X_0 \smallsetminus A_0} I \to X.$$

Then F^0 satisfies the required conditions. Suppose that

$$F^{n-1} \colon \mathrm{sk}_{n-1}^A(X) \times I \longrightarrow X$$

is defined satisfying the required conditions. Since $F^{n-1}|_{\mathrm{sk}_{n-1}^A(X) \times 0} = \mathrm{id}_X|_{\mathrm{sk}_{n-1}^A(X)}$,

$$\tilde{F}^{n-1} = F^{n-1} \cup (\mathrm{id}_X|_{\mathrm{sk}_n^A(X)}) \colon (\mathrm{sk}_{n-1}^A(X) \times I) \cup (\mathrm{sk}_n^A(X) \times 0) \longrightarrow X$$

is a well-defined simplicial map. Let

$$\bar{X}_n^D = \{y \in X_n \smallsetminus A_n \mid y \text{ is nondegenerate}\}.$$

Then there is a push-out diagram

$$
\begin{array}{ccc}
\coprod\limits_{y \in \bar{X}_n^D} (\partial\Delta[n] \times I) \cup (\Delta[n] \times 0) & \xrightarrow{\coprod_{y \in \bar{X}_n^D} \bar{f}_y} & (\mathrm{sk}_{n-1}^A(X) \times I) \cup (\mathrm{sk}_n^A(X) \times 0) \\
\Big\downarrow & \text{push} & \Big\downarrow{\scriptstyle j} \\
\coprod\limits_{y \in \bar{X}_n^D} \Delta[n] \times I & \xrightarrow{\coprod_{y \in \bar{X}_n^D} f_y \times \mathrm{id}_I} & \mathrm{sk}_n^A(X) \times I,
\end{array}
$$

$$(2.17)$$

where $\bar{f}_y = (f_y \times \mathrm{id}_I)|_{(\partial\Delta[n]\times I)\cup(\Delta[n]\times 0)}$. Let $y \in \bar{X}_n^D$. Since X is fibrant, there is an extension

$$
\begin{array}{ccc}
(\mathrm{sk}_{n-1}^A(X) \times I) \cup (\mathrm{sk}_n^A(X) \times 0) & \xrightarrow{\ \tilde{F}^{n-1}\ } & X \\
\uparrow{\scriptstyle \bar{f}_y} & & \uparrow{\scriptstyle F_y} \\
(\partial\Delta[n] \times I) \cup (\Delta[n] \times 0) & \hookrightarrow & \Delta[n] \times I.
\end{array}
$$

Let $x = F_y(\sigma_n, 1)$. Then all $d_j x \in A_{n-1}$ because F_y restricted to $\partial\Delta[n] \times I$ is given by the composite $\tilde{F}^{n-1} \circ \bar{f}_y$ and $\tilde{F}^{n-1}(\mathrm{sk}_{n-1}^A(X) \times 1) \subseteq A$. Thus there exists an element $z \in A_n$ such that $x \simeq z$. According to Lemma 2.77, there exists a simplicial map $G_y \colon \Delta[n] \times I \to X$ such that $G_y(\sigma_n, 1) = z \in A_n$ and

$$
G_y|_{(\partial\Delta[n]\times I)\cup(\Delta[n]\times 0)} = F_y|_{(\partial\Delta[n]\times I)\cup(\Delta[n]\times 0)}
$$

and so we have the commutative diagram

$$
\begin{array}{ccc}
(\mathrm{sk}_{n-1}^A(X) \times I) \cup (\mathrm{sk}_n^A(X) \times 0) & \xrightarrow{\ \tilde{F}^{n-1}\ } & X \\
\uparrow{\scriptstyle \bar{f}_y} & & \uparrow{\scriptstyle G_y} \\
(\partial\Delta[n] \times I) \cup (\Delta[n] \times 0) & \hookrightarrow & \Delta[n] \times I.
\end{array}
$$

By the push-out diagram in Equation (2.17), the simplicial maps $\tilde{F}^{n-1}$ and $\coprod_{y\in\bar{X}_n^D} G_y$ induces the simplicial map

$$
F^n \colon \mathrm{sk}_n^A(X) \times I \longrightarrow X
$$

such that $F^n \circ j = \tilde{F}^{n-1}$ and

$$
F^n \circ \left(\coprod_{y\in\bar{X}_n^D} f_y \times \mathrm{id}_I \right) = \coprod_{y\in\bar{X}_n^D} G_y.
$$

Thus F^n satisfies Conditions (1)–(3). For each $y \in \bar{X}_n^D$, since $G_y(\sigma_n, 1) \in A_n$, we have $G_y(\Delta[n] \times 1) \subseteq A$ and so

$$
F^n(\mathrm{sk}_n^A(X) \times 1) \subseteq A,
$$

that is F^n satisfies Condition (4). The induction is finished. Now the simplicial map

$$F = \bigcup_n F^n \colon X \times I = \bigcup_n \mathrm{sk}_n^A(X) \times I \longrightarrow X$$

makes A to be a strong deformation retract of X.

4. *A is a minimal simplicial set.*

By Step 3, A is a retract of X and so A is also a fibrant simplicial set. Suppose that $x, y \in A_n$ with $x \simeq y$. Then $x \simeq y$ in X_n with all faces in A_{n-1}. By the construction A_n, $x = y$. Thus A is a minimal simplicial set. We finish the proof now. $\qquad\square$

Let $y_1, y_2 \in X_n$ such that $d_j y_1 = d_j y_2$ for all $0 \le j \le n$. Define θ_{y_1, y_2} to be the composite

$$\theta_{y_1, y_2} \colon S^n \xrightarrow[\cong]{\theta} |\Delta[n]| \cup_{\partial|\Delta[n]|} |\Delta[n]| \xrightarrow{|f_{y_1}| \cup |f_{y_2}|} |X|,$$

where θ is given by identifying the upper hemisphere S^n_+ with the left copy $|\Delta[n]| = \Delta^n$ and the lower hemisphere S^n_- with the right copy of $|\Delta[n]|$.

Lemma 2.80. *Let X be a fibrant simplicial set and let $y_1, y_2, y_3 \in X_n$ such that $d_j y_1 = d_j y_2 = d_j y_3$ for all $0 \le j \le n$. Then*

(1). *If $y_2 \simeq y_3$, then $\theta_{y_1, y_2} \simeq \theta_{y_1, y_3}$.*
(2). *$[\theta_{y_1, y_2}] = -[\theta_{y_2, y_1}]$ in $\pi_n(|X|)$.*
(3). *$[\theta_{y_1, y_3}] = [\theta_{y_1, y_2}] + [\theta_{y_2, y_3}]$ in $\pi_n(|X|)$.*

Proof. (1). Let $F \colon \Delta[n] \times I \to X$ be a homotopy between f_{y_2} and f_{y_3} relative to $\partial\Delta[n]$. Let G be the composite

$$\Delta[n] \times I \xrightarrow{\mathrm{proj.}} \Delta[n] \xrightarrow{f_{y_1}} X.$$

Then

$$G \cup F \colon (\Delta[n] \cup_{\partial\Delta[n]} \Delta[n]) \times I \longrightarrow X$$

is a homotopy between $f_{y_1} \cup f_{y_2}$ and $f_{y_1} \cup f_{y_3}$ and so $\theta_{y_1, y_2} \simeq \theta_{y_1, y_3}$.

(2). Let τ be the self simplicial map of $\Delta[n] \cup_{\partial\Delta[n]} \Delta[n]$ by switching the left copy of $\Delta[n]$ to the right copy of $\Delta[n]$. Then there is a commutative diagram

$$
\begin{array}{ccccc}
S^n & \xrightarrow{\ \theta\ } & |\Delta[n] \cup_{\partial\Delta[n]} \Delta[n]| & \xrightarrow{\ |f_{y_1} \cup f_{y_2}|\ } & |X| \\
\Big\downarrow{\scriptstyle\tau} & & \Big\downarrow{\scriptstyle\tau} & & \Big\| \\
S^n & \xrightarrow{\ \theta\ } & |\Delta[n] \cup_{\partial\Delta[n]} \Delta[n]| & \xrightarrow{\ |f_{y_2} \cup f_{y_1}|\ } & |X|,
\end{array}
$$

where $\tau\colon S^n \to S^n$ is the reflection with respect to the equator S^{n-1} in S^n. Since $\tau\colon S^n \to S^n$ is of degree -1, we have

$$[\theta_{y_2,y_1}] = -[\theta_{y_1,y_2}].$$

(3). Let Δ_i^n be a copy of Δ^n for $i = 1, 2, 3$. Let Z be the push-out of the diagram

$$
\begin{array}{ccc}
\partial\Delta^n & \hookrightarrow & \Delta_1^n \\
\Big\downarrow & \searrow & \\
\Delta_3^n & & \Delta_2^n,
\end{array}
$$

that is Z is obtained by gluing three copies of Δ^n together by identifying their boundaries with Δ_2^n in the middle. There is a diagram

$$
\begin{array}{ccc}
\Delta_1^n \cup_{\partial\Delta^n} \Delta_3^n & \xrightarrow{\ \phi_1\ } & Z \\
\Big\| {\scriptstyle\simeq} & & \Big\uparrow{\scriptstyle\phi_2} \\
D^n \cup_{S^{n-1}} D^n & \xrightarrow{\ p\ } & (\Delta_1^n \cup_{\partial\Delta^n} \Delta_2^n) \vee (\Delta_2^n \cup_{\partial\Delta^n} \Delta_3^n),
\end{array}
$$

where p is given by pinching the common boundary S^{n-1} to the point, ϕ_1 is induced by the inclusions from Δ_1^n and Δ_2^n into Z, and ϕ_2 is the map which folds the middle two copies of Δ_2^n into one copy of Δ_2^n. Clearly this diagram is commutative up to homotopy, that is $\phi_1 \simeq \phi_2 \circ p$, see the following picture:

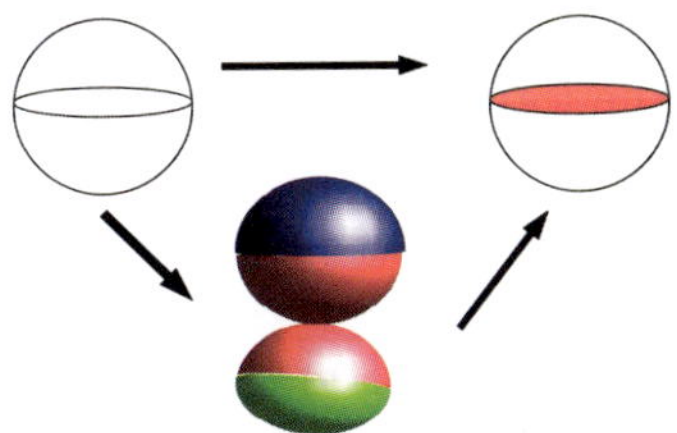

Thus

$$[\theta_{y_1,y_3}] = [|f_{y_1} \cup f_{y_2} \cup f_{y_3}| \circ \phi_1 \circ \theta]$$
$$= [|f_{y_1} \cup f_{y_2} \cup f_{y_3}| \circ \phi_2 \circ p \circ \theta]$$
$$= [(|f_{y_1} \cup f_{y_2}| \vee |f_{y_2} \cup f_{y_3}|) \circ p \circ \theta]$$
$$= [|f_{y_1} \cup f_{y_2}| \circ \theta] + [|f_{y_2} \cup f_{y_3}| \circ \theta]$$
$$= [\theta_{y_1,y_2}] + [\theta_{y_2,y_3}]$$

and hence the result. $\qquad\square$

Let X be a simplicial set. For any $x \in X_n$, let

$$V(x) = \{y \in X_n \mid d_j y = d_j x \text{ for } 0 \le j \le n\}. \qquad (2.18)$$

Let

$$\phi_x \colon V(x) \longrightarrow \pi_n(|X|)$$

be the function defined by $\phi_x(y) = [\theta_{x,y}]$ for $y \in V(x)$.

Proposition 2.81. *Let X be a fibrant simplicial set and let $y_1, y_2 \in V(x)$. Then the function ϕ_x induces a bijective function*

$$\bar{\phi}_x \colon V(x)/\simeq \; \longrightarrow \pi_n(|X|),$$

where the basepoint of $|X|$ is chosen to be a point in $|f_x|(\partial\Delta^n)$.

Proof. Let $y_1, y_2 \in V(x)$. If $y_1 \simeq y_2$, then $\phi_x(y_1) = \phi_x(y_2)$ by part (1) of Lemma 2.80. Thus $\bar{\phi}_x$ is well-defined.

Suppose that $\phi_x(y_1) = \phi_x(y_2)$. Then, by Lemma 2.80,

$$[\theta_{y_1,y_2}] = [\theta_{y_1,x}] + [\theta_{x,y_2}]$$
$$= -[\theta_{x,y_1}] + [\theta_{x,y_2}]$$
$$= -\phi_x(y_1) + \phi_x(y_2)$$
$$= 0.$$

Thus

$$|f_{y_1} \cup f_{y_2}| \colon \Delta^n \cup_{\partial\Delta^n} \Delta^n \longrightarrow |X|$$

can be extended to a continuous map $D^{n+1} \to |X|$. It follows that

$$|f_{y_1}| \simeq |f_{y_2}| \text{ rel } \partial\Delta^n$$

and so $f_{y_1} \simeq f_{y_2}$ rel $\partial\Delta[n]$ that is $y_1 \simeq y_2$. This shows that $\bar{\phi}_x$ is one-to-one.

To see that $\bar{\phi}_x$ is onto, let $g\colon S^n \to |X|$ be a pointed continuous map such that $g(ast) = |f_x|(v_0)$ for the basepoint $v_0 \in S^{n-1} = \partial\Delta^n$. Consider the lower hemisphere S^n_{-1} with $\partial S^n_{-1} \cong \partial D^n$. We may assume that $v_0 = (1, 0, \ldots, 0)$. Let $\tilde{S}^{n-1}$ be the $(n-1)$-sphere in D^n centered at $(\frac{1}{2}, 0 \ldots, 0)$ with radius $1/2$. Then $\tilde{S}^{n-1} \cap \partial D^n = \{v_0\}$. By pinching $\tilde{S}^{n-1}$ to the point v_0, we have the continuous map

$$D^n \xrightarrow{\ q\ } D^n \vee S^n.$$

Define $\phi\colon D^n \to |X|$ to be the composite

$$\Delta^n \cong D^n \xrightarrow{\ q\ } D^n \vee S^n \xrightarrow{\ |f_x| \vee g\ } |X|.$$

Since

$$\phi|_{\partial\Delta^n} = |f_x|_{\partial\Delta[n]}|,$$

there exists a simplicial map $h\colon \Delta[n] \to X$ such that

$$h|_{\partial\Delta[n]} = f_x|_{\partial\Delta[n]}$$

and

$$|h| \simeq \phi \text{ rel } \partial\Delta^n$$

by simplicial extension theorem. Let $y = h(\sigma_n) \in X_n$. Then $d_j y = d_j x$ for $0 \leq j \leq n$, that is $y \in V(x)$. Now

$$\begin{aligned}
\theta_{x,y} &= (|f_x| \cup |f_y|) \circ \theta \\
&= (|f_x| \cup |h|) \circ \theta \\
&\simeq (|f_x| \cup |\phi|) \circ \theta \quad \text{because } h \simeq \phi \text{ rel } \partial\Delta^n.
\end{aligned}$$

Thus

$$\begin{aligned}
\phi_x(y) &= [\theta_{x,y}] \\
&= [(|f_x| \cup |\phi|) \circ \theta] \\
&= [(|f_x| \cup |f_x|) \circ \theta] + [g] \\
&= [g] \qquad\qquad \text{because } [(f_x \cup f_x) \circ \theta] = 0.
\end{aligned}$$

This shows that $\bar{\phi}_x$ is onto and hence the result. $\qquad\square$

Theorem 2.82. *Let X and Y be path-connected minimal simplicial sets and let $f\colon X \to Y$ be a simplicial map such that $f_*\colon \pi_n(X) \to \pi_n(Y)$ is an isomorphism for each $n \geq 0$. Then $f\colon X \to Y$ is a simplicial isomorphism.*

Proof. Since X and Y are path-connected, the minimal assumption forces X and Y to be reduced, that is, both X_0 and Y_0 only consist of one point. Thus $f\colon X_0 \to Y_0$ is an isomorphism. Suppose that $f\colon X_j \to Y_j$ is bijective for $0 \le j < n$.

To see that $f\colon X_n \to Y_n$ is one-to-one, let $x, y \in X_n$ such that $f(x) = f(y)$. Then $f(d_j x) = d_j f(x) = d_j f(y) = f(d_j(y))$ for $0 \le j \le n$. By induction, $d_j x = d_j y$ for all $0 \le j \le n$ and so $y \in V(x)$ Now

$$|f|_*([\phi_x(y)]) = [\phi_{f(x)}(f(y))] = 0$$

in $\pi_n(|Y|)$. Thus $[\phi_x(y)] = 0$ in $\pi_n(|X|)$ and $x \simeq y$. By minimal assumption, $x = y$.

To see that $f\colon X_n \to Y_n$ is onto, let $z \in Y_n$. Let $x_j = f^{-1}(d_j z)$ for $0 \le j \le n$. Let

$$f = \bigcup_{j=0}^{n} f_{x_j} \colon \partial\Delta[n] \to X.$$

Since $|f_z|_{\partial\Delta^n}\colon \partial\Delta^n \to |Y|$ can be extended to the continuous map $|f_z|\colon \Delta^n \to |Y|$ and $f_*\colon \pi_{n-1}(|X|) \to \pi_{n-1}(|Y|)$ is an isomorphism, $|f|\colon \partial\Delta^n \to X$ can be extended to a continuous map $\Delta^n \to |X|$ and so there is a simplicial map $g\colon \Delta[n] \to X$ such that $g|_{\partial\Delta[n]} = f$. Let $x = g(\sigma_n)$. Then $d_j x = x_j$ for $0 \le j \le n$. By Proposition 2.81, there is a commutative diagram

$$
\begin{array}{ccc}
V(x)/\!\sim & \xrightarrow{\ f\ } & V(f(x))/\!\sim \\
\cong \Big\downarrow \bar{\phi}_x & & \cong \Big\downarrow \bar{\phi}_{f(x)} \\
\pi_n(|X|) & \xrightarrow[\cong]{\ f_*\ } & \pi_n(|Y|).
\end{array}
$$

Thus there exists $y \in V(x)$ such that $f(y) \simeq z$. By the minimal assumption on Y, $f(y) = z$. The proof is finished now. $\qquad\Box$

Corollary 2.83 (Uniqueness of minimal subcomplex). *Let X be a fibrant simplicial set and let A and A' be minimal subcomplex of X. Then $A \cong A'$.*

Proof. By restricting to each path-connected components, we may assume that X is path-connected. Since both of A and A' are string deformation retract of X, the composite

$$A \lhook\joinrel\longrightarrow X \xrightarrow{\ r\ } A'$$

induces an isomorphism on homotopy groups, where r is a retraction of X into A' and the base. Since A and A' are minimal, $A \cong A'$ and hence the result. $\qquad\square$

Corollary 2.84 (Whitehead Theorem). *Let X and Y be path-connected fibrant simplicial sets. Suppose that there exists a simplicial map $f\colon X \to Y$ such that $f_*\colon \pi_n(X) \to \pi_n(Y)$ is an isomorphism for each n. Then $X \simeq Y$.*

Proof. Let A and B be minimal subcomplex of X and Y, respectively. Then the composite

$$A \lhook\joinrel\longrightarrow X \xrightarrow{\ f\ } Y \xrightarrow{\ r\ } B$$

induces isomorphisms on all homotopy groups, where r is a retraction. Thus $A \cong B$. Since A is a strong deformation of X, we have $A \simeq X$. Similarly $B \simeq Y$. Thus

$$X \simeq A \cong B \simeq Y$$

and hence the result. $\qquad\square$

Read more in [40] for interesting properties of minimal simplicial sets.

2.6. *Simplicial fibration*

2.6.1. *Definition and basic properties of fibrations*

Definition 2.85. A simplicial map $p\colon E \to B$ is called a *(Kan) fibration* if for every commutative diagram of simplicial maps

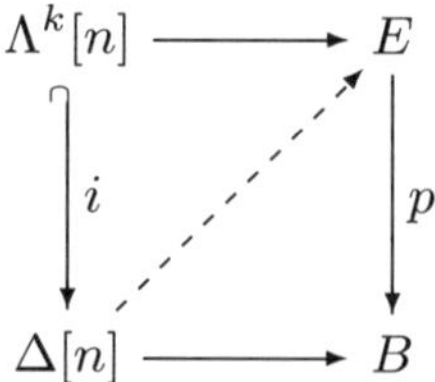

there is a simplicial map $\theta\colon \Delta[n] \to X$ (the dotted arrow) making the diagram commute, where i is the inclusion of $\Lambda^k[n]$ in $\Delta[n]$. Let $v \in B_0$ be a vertex. Then $F = p^{-1}(v)$, that is $F = \{p^{-1}(s^n v)\}_{n \geq 0}$ is called a *fiber* of p over v.

Exercise 2.17. Prove the following statements:

(1). A simplicial set X is fibrant if and only if the constant map $X \to *$ is a fibration.

(2). If $p\colon E \to B$ is a fibration, then any fiber F of p is fibrant.

(3). Let F be a fibrant simplicial set and let B be any simplicial set. Then the coordinate projection

$$p\colon F \times B \longrightarrow B \quad (x, b) \mapsto b$$

is a fibration.

(4). Let $p\colon E \to B$ be a fibration and let $f\colon A \to B$ be a simplicial map. Let $\tilde{p}\colon \tilde{E} \to A$ be given by the pull-back diagram

$$
\begin{array}{ccc}
\tilde{E} & \longrightarrow & E \\
\downarrow{\tilde{p}} \quad \text{pull} & & \downarrow{p} \\
A & \overset{f}{\longrightarrow} & B.
\end{array}
$$

Then $\tilde{p}\colon \tilde{E} \to A$ is a fibration.

From the definition, we have the following proposition.

Proposition 2.86. *Let $p\colon E \to B$ be a simplicial map. Then p is a fibration if and only if for any $x_0, x_1, \ldots, x_{k-1}, x_{k+1}, \ldots, x_n \in E_{n-1}$ and $y \in B_n$ such that*

(1). *$x_0, x_1, \ldots, x_{k-1}, x_{k+1}, \ldots, x_n$ are matching faces in X and*
(2). *$p(x_j) = d_j y$ for $j \neq k$,*

there exists $x \in E_n$ such that $p(x) = y$ and $d_j x = x_j$ for $j \neq k$. $\square$

Recall that a (left) action of a group G on a set S is a binary operation

$$G \times S \to S, \quad (g, x) \mapsto g \cdot x$$

such that $1 \cdot x = x$ and $g_1 \cdot (g_2 \cdot x) = (g_1 g_2) \cdot x$.

Definition 2.87. A *simplicial action* of a simplicial group $G = \{G_n\}_{n \geq 0}$ on a simplicial set $\{E_n\}_{n \geq 0}$ means a sequence of the actions of G_n on E_n such that the binary operations $\{G_n \times E_n \to E_n\}$ forms a simplicial map, that is,

$$d_j(g \cdot x) = d_j g \cdot d_j x \quad s_j(g \cdot x) = s_j g \cdot s_j x$$

for $g \in G_n, x \in E_n$ and $0 \le j \le n$. A simplicial action G on E is called *free* if each G_n-action on E_n is free, that is, $g \cdot x = x \Longrightarrow g = 1$.

Given a group action of G on S, let $G\backslash S$ denote the set of orbits of G on S. Given a simplicial group action of $G = \{G_n\}_{n \ge 0}$ on $E = \{E_n\}_{n \ge 0}$, we have the simplicial set $G\backslash E = \{G_n\backslash E_n\}_{n \ge 0}$ with the simplicial map $p \colon E \to G\backslash E$.

Proposition 2.88. *Let G be a simplicial group and let E be a simplicial set with a simplicial action of G. Suppose that the simplicial action of G on E is free. Then the simplicial map*

$$p \colon E \to G\backslash E$$

is a fibration.

Proof. Let $x_0, x_1, \ldots, x_{k-1}, x_{k+1}, \ldots, x_n \in E_{n-1}$ and $y \in (G\backslash E)_n = G_n\backslash E_n$ such that

(1). $x_0, x_1, \ldots, x_{k-1}, x_{k+1}, \ldots, x_n$ are matching faces in E and
(2). $p(x_j) = d_j y$ for $j \ne k$.

Let $\tilde{y} \in E_n$ such that $p(\tilde{y}) = y$. Then

$$p(d_j \tilde{y}) = d_j p(\tilde{y}) = p_j y = p(x_j)$$

for $j \ne k$. Thus there exists $z_j \in G_{n-1}$ such that

$$x_j = z_j \cdot d_j \tilde{y}$$

for each $j \ne k$. Now, for $i < j$ with $i, j \ne k$,

$$
\begin{aligned}
d_i z_j \cdot d_i d_j \tilde{y} &= d_i(z_j \cdot d_j \tilde{y}) \\
&= d_i x_j \\
&= d_{j-1} x_i \\
&\qquad \text{because } x_0, x_1, \ldots, x_{k-1}, x_{k+1}, \ldots, x_n \text{ are matching faces} \\
&= d_{j-1} z_j \cdot d_{j-1} d_i \tilde{y} \\
&= d_{j-1} z_j \cdot d_i d_j \tilde{y}.
\end{aligned}
$$

Thus $z_0, z_1, \ldots, z_{k-1}, z_{k+1}, \ldots, z_n$ are matching faces because G acts on E freely. Since G is fibrant, there exists $z \in G$ such that $d_j z = z_j$ for $j \ne k$. Now let $\bar{y} = z\tilde{y}$. Then $p(\bar{y}) = p(z\tilde{y}) = p(\tilde{y}) = y$ and

$$d_j \bar{y} = d_j z \cdot d_j \tilde{y} = x_j (d_j \tilde{y})^{-1} d_j \tilde{y} = x_j$$

for $j \ne k$. Thus $p \colon E \to G\backslash E$ is a fibration. $\qquad \square$

Problem 2.89. *Let $p\colon E \to B$ be a simplicial map. Is it true that p is a fibration if and only if the geometric realization $|p|\colon |E| \to |B|$ is a Serre fibration? (**Note.** $\Longrightarrow$ was answered to be true by Barratt (unpublished) and Quillen [29].)*

Let $\mathbf{n} = \{0, 1, \ldots, n\}$. Let $0 \le j_1 < j_2 < \cdots < j_t \le n$. For a simplicial set X, the elements $x_{j_1}, x_{j_2}, \ldots, x_{j_t}$ in X_{n-1} are called to be *matching faces* (with respect to $\mathbf{n} \setminus \{j_1, j_2, \ldots, j_t\}$) if $d_i x_j = d_{j-1} x_i$ for $0 \le i < j \le n$ with $i, j \in \{j_1, j_2, \ldots, j_t\}$.

Lemma 2.90. *Let $p\colon E \to B$ be a fibration. Assume that p is onto. Let $b \in B_n$ and let $x_{j_1}, \ldots x_{j_t} \in E_{n-1}$ with $t \le n$ such that $p(x_{j_i}) = d_{j_i}(b)$ for $1 \le i \le t$ and the elements $x_{j_1}, \ldots, x_{j_t}$ are matching faces. Then there exists $x \in E_n$ such that $d_{j_i} x = x_{j_i}$ for $1 \le i \le t$ and $p(x) = b$.*

Proof. The proof is given by induction on n. The assertion clearly holds for $n = 1$. Suppose that the assertion holds for $n - 1$. Let $b \in B_n$ and let $x_{j_1}, \ldots x_{j_t} \in E_{n-1}$ with $t \le n$ such that $p(x_{j_i}) = d_{j_i}(b)$ for $1 \le i \le t$ and the elements $x_{j_1}, \ldots, x_{j_t}$ are matching faces. We prove this by the second induction on t starting from $t = n$. If $t = n$, then the assertion follows by the definition of fibration. Assume that the assertion holds for $k < t \le n$. Let $i \in \mathbf{n} \setminus \{j_1, \ldots, j_k\}$. We are going to construct an element x_i such that $\{x_i, x_{j_1}, \ldots, x_{j_k}\}$ are matching faces and $p(x_i) = d_i b$. In other words, x_i will be a solution of the following equations

$$d_j x_i = \begin{cases} d_{i-1} x_j & \text{if } j < i \text{ and } j \in \{j_1, \ldots, j_k\}, \\ d_i x_{j+1} & \text{if } j \ge i \text{ and } j + 1 \in \{j_1, \ldots, j_k\}, \end{cases} \tag{2.19}$$
$$p(x_i) = d_i b.$$

Let s be the integer such that $j_s < i$ and $j_{s+1} > i$. Then the elements

$$z_{j_1} = d_{i-1} x_{j_1}, \ldots, z_{j_s} = d_{i-1} x_{j_s},$$
$$z_{j_{s+1}-1} = d_i x_{j_{s+1}},$$
$$z_{j_{s+2}-1} = d_i x_{j_{s+2}}, \ldots, z_{j_k-1} = d_i x_{j_k}$$

are matching faces with respect to $\mathbf{n} - \mathbf{1} \setminus \{j_1, \ldots, j_s, j_{s+1} - 1, \ldots, j_k - 1\}$ with $p(z_l) = d_l(d_i b)$ for $l \in \{j_1, \ldots, j_s, j_{s+1} - 1, \ldots, j_k - 1\}$. Since $k < n$, Equation (2.19) has a solution x_i by induction hypothesis on n. The second induction is finished and hence the result. $\qquad\square$

Lemma 2.91. *Let $p\colon E \to B$ be a fibration. Assume that p is onto. Then for any simplicial map $f\colon \Lambda^k[n] \to B$, there is simplicial map $\tilde{f}\colon \Lambda^k[n] \to E$ such that $p \circ \tilde{f} = f$.*

Proof. Let $f\colon \Lambda^k[n] \to B$ be a simplicial map and let $b_i = f(d_i\sigma_n)$ for $i \neq k$, where σ_n is the nondegenerate element in $\Delta[n]_n$. We are going to construct the elements $x_0, x_1, \ldots, x_{k-1}, x_{k+1}, \ldots, x_n$ in E_n such that $p(x_i) = b_i$ for $i \neq k$ and $d_i x_j = d_{j-1} x_i$ for $0 \le i < j \le n$ and $i, j \neq k$.

First we construct the elements $x_0, \ldots x_{k-1}$ by induction. (If $k = 0$, this is the empty case.) Since p is onto, let $x_0 \in E_n$ such that $p(x_0) = b_0$. Suppose that $x_0, \ldots, x_{j-1}$ have been constructed with the property that $d_i x_t = d_{t-1} x_i$ and $p(x_t) = b_t$ for $0 \le i < t < j$. We construct x_j to be the solution of the following equations

$$\begin{aligned} d_i x_j &= d_{j-1} x_i \quad \text{for } i < j \\ p(x_j) &= b_j. \end{aligned} \tag{2.20}$$

Since $\{d_{j-1}x_0, \ldots, d_{j-1}x_{j-1}\}$ are matching faces with respect to $\mathbf{n} \setminus \{0, 1, \ldots, j-1\}$ with $p(d_{j-1}x_i) = d_i(b_j)$. By Lemma 2.90, there exists a solution x_j to Equation (2.20). Thus we have constructed the elements $x_1, \ldots, x_{k-1}$.

Similarly we can construct inductively the elements $x_{k+1}, \ldots, x_n$ according to the required equations. (If $k = 0$, choose x_1 such that $p(x_1) = b_1$ and construct the rest elements according to the required equation.)

The constructed elements $x_0, \ldots, x_{k-1}, x_{k+1}, \ldots, x_n$ are matching faces and it induces a simplicial map $\tilde{f}\colon \Lambda^k[n] \to E$ with $p\tilde{f} = f$ because $p(x_j) = b_j$ for $j \neq k$. $\qquad\square$

Proposition 2.92. *Let $p\colon E \to B$ be a fibration. Assume that p is onto. Then B is fibrant if and only if E is fibrant.*

Proof. $\Longrightarrow$. Suppose that B is fibrant. Let $f\colon \Lambda^k[n] \to E$ be a simplicial map. Then $p \circ f\colon \Lambda^k[n] \to B$ is a simplicial map. Since B is fibrant, there exists a simplicial map $g\colon \Delta[n] \to B$ such that $g|_{\Lambda^k[n]} = p \circ p$, that is there is a commutative diagram

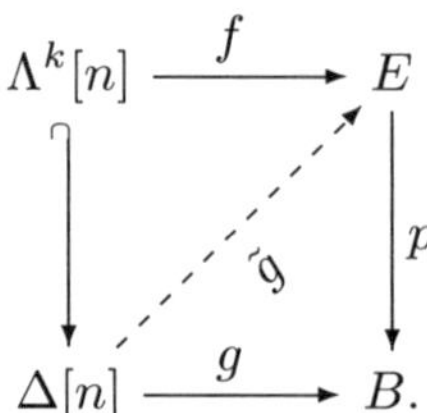

Let $\tilde{g}$ be a solution to the above commutative diagram. Then $\tilde{g}\colon \Delta[n] \to E$ is an extension of f and so E is fibrant.

$\Longleftarrow$. Suppose that E is fibrant. Let $f\colon \Lambda^k[n] \to B$ be a simplicial map. By Lemma 2.91, there exists a lifting $\tilde{f}\colon \Lambda^k[n] \to E$ such that $p \circ \tilde{f} = f$. Since E is fibrant, $\tilde{f}$ can be extended to a simplicial map $g\colon \Delta[n] \to E$. Then $p \circ g\colon \Delta[n] \to B$ is an extension of f and hence the result. $\qquad\square$

Corollary 2.93. *Let G be a simplicial group and let E be a simplicial set with a free simplicial action of G. Suppose that E is fibrant. Then $G\backslash E$ is fibrant.* $\qquad\square$

2.6.2. *Long exact sequences of homotopy groups*

An important property of fibrations is that it induces a long exact sequence on homotopy groups.

Theorem 2.94. *Let $p\colon E \to B$ be a fibration and let $F = p^*$ be the fiber. Suppose that E or B is fibrant. Then there is a (natural) boundary $\partial\colon \pi_n(B) \to \pi_{n-1}(F)$ for $n \geq 1$ which is a group homomorphism for $n > 1$ such that there is a long exact sequence*

$$\cdots \longrightarrow \pi_n(F) \xrightarrow{\ i_*\ } \pi_n(E) \xrightarrow{\ p_*\ } \pi_n(B) \xrightarrow{\ \partial\ } \pi_{n-1}(F) \longrightarrow \cdots$$

$$\cdots \xrightarrow{\ \partial\ } \pi_0(F) \xrightarrow{\ i_*\ } \pi_0(E) \xrightarrow{\ p_*\ } \pi_0(B),$$

where $i\colon F \to E$ is the inclusion. Moreover there is a (right) action of $\pi_1(B)$ on $\pi_0(F)$ such that two elements of $\pi_0(F)$ has the same image under i_* in $\pi_0(E)$ if and only if they are in the same orbit for the $\pi_1(B)$-action.

Proof. 1. *Definition of the boundary map:*

First we define the boundary $\partial\colon \pi_n(B) \to \pi_{n-1}(F)$ for each $n \geq 1$. Let b be a spherical element in B_n. Consider the commutative diagram

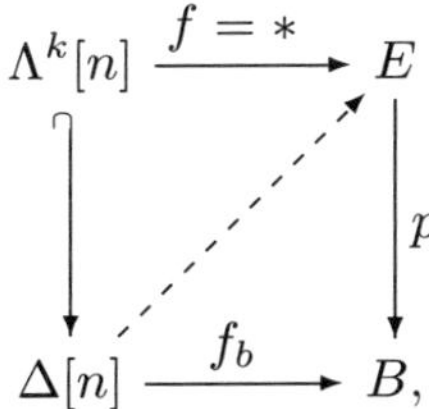

where f is the constant map to the basepoint. Then there exists an element $x \in E_n$ such that $d_j x = *$ for $j \geq 1$ and $p(x) = b$. Since

$$p(d_0 x) = d_0 p(x) = d_0 b = *,$$

$d_0 x \in F_{n-1}$ with all faces $d_j(d_0 x) = *$. Thus $d_0 x$ is a spherical element in F_{n-1}. Define

$$\partial(b) = [d_0 x] \tag{2.21}$$

in $\pi_{n-1}(F)$.

2. *The boundary map is well-defined:*

We check that $[d_0 x]$ is independent not only on the choice of x but also on the representing element b in its homotopy class. Suppose that $b \simeq b'$ and $p(x') = b'$ with $d_j x' = *$ for $j > 0$. Then there exists an element $c \in B_{n+1}$ such that $d_0 c = *$, $d_1 c = b$, $d_2 c = b'$, and $d_j c = *$ for $j > 2$. Let $y_1 = x$, $y_2 = x'$ and $y_j = *$ for $j > 2$. Then $p(y_j) = d_j(c)$ for $j \neq 0$. Moreover the elements $y_1, y_2, y_3, \ldots, y_{n+1}$ are matching faces with respect to 0 because $d_j y_0 = *$ and $d_j y_2 = *$ for $j > 0$. Thus there exists $z \in E_{n+1}$ such that $p(z) = c$ and $d_j z = y_j$ for $j \neq 0$. Now consider $d_0 z$. Since

$$p(d_0 z) = d_0 p(z) = d_0 c = *,$$

we have $d_0 z \in F_n$ with

$$d_0(d_0 z) = d_0 d_1 z = d_0 y_1 = d_0 x,$$
$$d_1(d_0 z) = d_0 d_2 z = d_0 y_2 = d_0 x',$$
$$d_j(d_0 z) = d_0 d_{j+1} z \qquad \text{for } \ j \geq 2.$$

Thus $d_0 x \simeq d_0 x'$ and so $[d_0 x] = [d_0 x']$. This proves that $[d_0 x]$ is independent on the choice of x and it induces a function

$$\partial\colon \pi_n(B) \to \pi_{n-1}(F).$$

3. *The boundary map is a homomorphism for $n \geq 2$:*

Let b, b', b'' be spherical elements in B_n such that $[b] - [b'] + [b''] = 0$ in $\pi_n(B)$, that is $[b'] = [b] + [b'']$. By Homotopy Addition Theorem, there exists $c \in B_{n+1}$ such that $d_j c = *$ for $j < n - 1$, $d_{n-1} c = b$, $d_n c = b'$ and $d_{n+1} c = b''$. Let x, x', x'' be the elements in E_n such that $d_j x = d_j x' = d_j x'' = *$ for $j > 0$. Let $y_1 = *, y_2 = *, \ldots, y_{n-1} = x, y_n = x', y_{n+1} = x''$. Then $p(y_j) = d_j c$ for $j > 0$ and the elements $y_1, \ldots, y_{n+1}$ are matching faces with respect to 0 because $d_j y_i = *$ for all $j > 0$. (**Note.** The assumption that $n \geq 2$ is used here.) Thus there exists $z \in E_{n+1}$ such that $p(z) = c$ and $d_j z = y_j$ for $j > 0$. Since $p(d_0 z) = d_0 p(z) = d_0 c = *$, $d_0 z \in F_n$. Now

$$d_j(d_0 z) = d_0 d_{j+1} z = d_0 y_{j+1} = * \qquad \text{for } j < n - 3$$
$$d_{n-2}(d_0 z) = d_0 d_{n-1} z = d_0 y_{n-1} = d_0 x,$$
$$d_{n-1}(d_0 z) = d_0 d_n z = d_0 y_n = d_0 x',$$
$$d_n(d_0 z) = d_0 d_{n+1} z = d_0 y_{n+1} = d_0 x''.$$

By the Homotopy Addition Theorem, $[d_0 x] - [d_0 x'] + [d_0 x''] = 0$, that is $[d_0 x'] = [d_0 x] + [d_0 x'']$. (If $n = 2$, then $[d_0 x'] = [d_0 x''][d_0 x]$ by the definition of the product in $\pi_1(F)$.)

4. $p_* \circ i_* = 0$ *and* $\mathrm{Ker}(p_* \colon \pi_n(E) \to \pi_n(B)) = \mathrm{Im}(i_* \colon \pi_n(F) \to \pi_n(E))$ *for* $n \geq 0$:

Since $p \circ i = *$, $p_* \circ i_* = 0$. Let $x \in E_n$ be a spherical element such that $[p(x)] = 0$. Then there exists an element $c \in B_{n+1}$ such that $d_0 c = *$, $d_1 c = p(x)$ and $d_j c = *$ for $j > 1$ by Homotopy Addition Theorem. (If $n = 0$, this is by the definition of π_0.) Let $y_1 = x$ and $y_j = *$ for $j > 1$. Then $p(y_j) = d_j c$ and the elements $y_1, \ldots, y_{n+1}$ are matching faces with respect to 0. Thus there exists $z \in E_{n+1}$ such that $p(z) = c$ and $d_j z = y_j$ for $j > 0$. Now $d_0 z \in F_n$, $d_1 z = y_1 = x$ and $d_j z = *$ for $j > 1$. Thus, by Homotopy Addition Theorem,

$$i_*([d_0 z]) - [x] = 0$$

in $\pi_n(E)$, that is, $[x] \in \mathrm{Im}(i_*)$. (If $n = 0$, then $d_0 z \in F_0$ and $d_z = x$ shows that $[x] \in \mathrm{Im}(i_*)$.)

5. $i_* \circ \partial = 0$ *and* $\mathrm{Ker}(i_* \colon \pi_n(F) \to \pi_n(E)) = \mathrm{Im}(\partial \colon \pi_{n+1}(B) \to \pi_n(F))$ *for* $n > 0$:

Let b be a spherical element in B_n. By definition, $\partial([b]) = [d_0 x]$, where $p(x) = b$ and $d_j x = *$. As an element in E_{n-1}, since $[d_0 x] = 0$ in $\pi_{n-1}(E)$, $i_*(\partial[b]) = 0$.

Now let x be a spherical element in F_n such that $i_*[x] = 0$, that is $[x] = 0$ in $\pi_n(E)$. There exists $z \in E_{n+1}$ such that $d_0 z = x$ and $d_j z = *$ for all $j > 0$. Let $b = p(z)$. By the definition of ∂, we have $\partial([b]) = [x]$.

6. $\partial \circ p_* = 0$ *and* $\mathrm{Ker}(\partial \colon \pi_n(B) \to \pi_{n-1}(F)) = \mathrm{Im}(p_* \colon \pi_n(E) \to \pi_n(B))$ *for* $n > 0$:

The composite $\partial \circ p_* = 0$ follows from the definition of ∂ as well: Let $[b] = p_*([x])$ for a spherical element $x \in E_n$. By definition, $\partial([b]) = [d_0 x] = [*] = 0$.

Now let b be a spherical element in B_n such that $\partial(b) = 0$ in $\pi_{n-1}(F)$. By definition, $\partial([b]) = [d_0 x]$, where $p(x) = b$ and $d_j x = *$ for $j > 0$. Since $[d_0 x] = 0$ in $\pi_{n-1}(F)$. There exists $x' \in F_n$ such that $d_0 x' = d_0 x$ and $d_j x' = *$ for $j > 0$. Let $y_0 = x'$, $y_1 = x$, $y_j = *$ for $2 \leq j \leq n$. Then $y_0, \ldots, y_n$ are matching faces with respect to $n + 1$. Thus there exists an element $z \in E_{n+1}$ such that $d_j z = y_j$ for $0 \leq j \leq n$. Now

$$d_j(d_{n+1} z) = d_n(d_j z) = *$$

for all $0 \leq j \leq n$. Thus $d_{n+1}z$ is a spherical element in E_n. Let $c = p(z)$. Then

$$
\begin{aligned}
d_0 c &= p(d_0 z) = p(y_0) = p(x') = * \\
d_1 c &= p(d_1 z) = p(y_1) = p(x) = b* \\
d_j c &= p(d_j z) = p(y_j) = * \qquad \text{for } 2 \leq j \leq n \\
d_{n+1} c &= p(d_{n+1} z).
\end{aligned}
$$

By the Homotopy Addition Theorem, we have

$$
[b] + (-1)^{n+1}[p(d_{n+1}z)] = [b] + (-1)^{n+1} p_*([d_{n+1}z]) = 0.
$$

(**Note.** If $n = 1$, then $[b] = p_*([d_2 z])$ by the definition of the product in the fundamental group.)

7. *The action of $\pi_1(B)$ on $\pi_0(F)$:*

Let b be a loop in B_1 and let v be any vertex in F_0. Since $p(v) = *$, there exists an element $y \in E_1$ such that $d_1 y = v$ and $p(y) = b$. Define

$$
[v] \cdot [b] = [d_0 y]. \tag{2.22}
$$

(Note that if $v = *$, then $* \cdot [b] = \partial[b]$ by definition.)

We need to check that this is well-defined. Suppose that $b \simeq b'$, $v \simeq v'$ and y, y' are given by the property $p(y) = b$, $p(y') = b'$, $d_1(y) = v$ and $d_1(y') = v'$. Then λ_y is a path with $\lambda_y(0) = d_1 y = v$, $\lambda_{y'}$ is a path with $\lambda_{y'}(0) = d_1(y') = v'$. By taking path inverse, $\lambda_{y'}^{-1}(0) = d_0(y')$ and $\lambda_{y'}^{-1}(1) = v'$ Since $v \simeq v'$, there is a path μ such that $\mu(0) = v'$ and $\mu(1) = v$. Now the path product $\lambda_{y'}^{-1} * \mu * \lambda_y$ is from $d_0(y')$ to $d_0 y$. Hence $[d_0 y] = [d_0 y']$. This proves that the binary operation

$$
\pi_0(F) \times \pi_1(B) \longrightarrow \pi_0(F)
$$

is well-defined.

If $[b] = 1$, we may assume $b = *$. Choose $y = s_0 v$. Then $p(s_0 v) = *$ and so

$$
[v] \cdot 1 = [d_0 s_0 v] = [v].
$$

Let b, b' be loops in B_1 and let $v \in F_0$. Let $y \in E_1$ such that $\lambda_y(0) = v$ and $p(y) = b$. Let $y' \in E_1$ such that $\lambda_{y'}(0) = \lambda_y(1) = d_0 y$. Let μ be the path product $\lambda_y * \lambda_{y'}$. Then $\mu(0) = \lambda_y(0) = v$ and $\mu(1) = \lambda_{y'}(1) = d_0(y')$. Now $p \colon E \to B$ induces a homomorphism of fundamental groupoids. Thus

$$
p_*([\mu]) = p_*([\lambda_y * \lambda_{y'}]) = p_*([\lambda_y]) * p_*([\lambda_{y'}]) = [b][b'].
$$

Thus shows that

$$
[v] \cdot ([b][b']) = [\mu(1)] = [d_0(y')] = [d_0 y] \cdot [b'] = ([v] \cdot [b]) \cdot [b'].
$$

8. *Two elements of $\pi_0(F)$ has the same image under i_* in $\pi_0(E)$ if and only if they are in the same orbit for the $\pi_1(B)$-action:*

Let b be a loop in B_1 and let v be any vertex in F_0. By definition, $[v] \cdot [b] = [d_0 y]$, where $d_1 y = v$ and $p(y) = b$. In E, we have $d_0 y \simeq d_1 y$ by definition of π_0. Thus $i_*([v]) = i_*([v \cdot [b])$.

Conversely let $v, v' \in F_0$ be two vertices such that $i_*([v]) = i_*([v'])$. Then there exists $y \in E_1$ such that $d_1 y = v$ and $d_0 y = v'$. Let $b = p(y)$. Then b is a loop in B_1 because $p(v) = p(v') = *$. From the definition,

$$[v] \cdot [b] = [d_0 y] = [v'].$$

The proof is finished now. $\qquad\square$

2.6.3. *Moore paths and Moore loops*

Let X be a pointed simplicial set. The *Moore Path* is defined by setting

$$(PX)_n = \{x \in X_{n+1} \mid d_1 d_2 \cdots d_{n+1} x = *\}$$

with faces

$$d_i^P = d_{i+1}^X |_{(PX)_n} : (PX)_n \to (PX)_{n-1}$$

and degeneracies

$$s_i^P = s_{i+1}^X |_{(PX)_n} : (PX)_n \to (PX)_{n+1}$$

for $0 \le i \le n$. We first need to check that d_i^P and s_i^P are well-defined. Let $x \in (PX)_n$, that is, $x \in X_{n+1}$ with $d_1 d_2 \cdots d_{n+1} x = *$. Consider $d_i x$. If $i = n + 1$, then, from $d_1 d_2 \cdots d_n (d_{n+1} x) = *$, we have $d_{n+1} \in (PX)_{n-1}$. For $1 \le i \le n$,

$$d_1 d_2 \cdots d_n(d_i x) = d_1 d_2 \cdots d_{n-1} d_i d_{n+1} x$$
$$\cdots$$
$$= d_1 d_2 \cdots d_i d_i d_{i+2} \cdots d_{n+1} x$$
$$= d_1 d_2 \cdots d_{n+1} x$$
$$= *$$

and so $d_i x \in (PX)_{n-1}$. Thus d_i^P is well-defined for $0 \le i \le n$. We can do similar examination for the degeneracies $s_i^P = s_{i+1}^X$: If $i = n + 1$, then

$$d_1 d_2 \cdots d_{n+2}(s_{n+1} x) = d_1 d_2 \cdots d_{n+1} x = *$$

and for $0 \le i \le n$

$$d_1 d_2 \cdots d_{n+2}(s_i x) = d_1 d_2 \cdots d_{n+1} s_i d_{n+1} x$$
$$= d_1 d_2 \cdots d_i d_{i+1} s_i d_{i+1} d_{i+2} \cdots d_{n+1} x$$
$$= *.$$

Thus s_i^P is well-defined for $0 \leq i \leq n$. The simplicial identities holds automatically for d_i^P and s_i^P because they are given by d_{i+1}^X and s_{i+1}^X. Thus $PX = \{(PX)_n\}_{n \geq 0}$ is a simplicial set. In addition, as we see from the above, we have one more degeneracy $s_0^X \colon (PX)_n \to (PX)_{n+1}$ for each $n \geq 0$. Define $s_{-1}^P = s_0^X|_{(PX)_n}$. By Contractible Criterion 2.17, we have the following:

Proposition 2.95. *Let X be any pointed simplicial set. Let PX be the Moore path of X. Then the geometric realization $|PX|$ is contractible.* $\square$

The first face d_0^X of X may not send $(PX)_n$ into $(PX)_{n-1}$ as we have the formula

$$d_1 d_2 \cdots d_n(d_0 x) = d_0 d_2 d_3 \cdots d_{n+1} x$$

which may not be the basepoint $*$. However,

$$d_0^X|_{(PX)_n} \colon (PX)_n \subseteq X_{n+1} \longrightarrow X_n$$

plays a role as a simplicial map. To see this, let $x \in (PX)_n$. Then

$$d_0^X(d_i^P x) = d_0^X d_{i+1}^X x$$
$$= d_i^X(d_0^X x)$$

$$d_0^X(s_i^P x) = d_0^X s_{i+1}^X x$$
$$= s_i^X(d_0^X x)$$

for $0 \leq i \leq n$. Thus we have the simplicial map

$$p_X = \{d_0^X|_{(PX)_n}\}_{n \geq 0} \colon PX \longrightarrow X.$$

Proposition 2.96. *Suppose that X is fibrant. Then*

$$p_X \colon PX \longrightarrow X$$

is a fibration.

Proof. Let

$$x_0, \ldots, x_{k-1}, x_{k+1}, \ldots, x_n \in (PX)_{n-1} \subseteq X_n$$

be matching faces with respect to k and $b \in X_n$ such that

$$d_j b = p_X(x_j) = d_0 x_j$$

for $j \neq k$. Then as elements in X_n,

$$b, x_0, x_1, \ldots, x_{k-1}, x_{k+1}, \ldots, x_n$$

are matching faces with respect to k. Thus there exists $y \in X_{n+1}$ such that $d_0 y = b$ and $d_{j+1} y = x_j$ for $j \neq k$. To see $y \in (PX)_n$, if $k \neq n$, then

$$d_1 d_2 \cdots d_n d_{n+1} y = d_1 d_2 \cdots d_n x_n = *$$

and, for $k = n$, note that $n \geq 1$ and so

$$\begin{aligned}
d_1 d_2 \cdots d_n d_{n+1} y &= d_1 d_2 \cdots d_n d_n y \\
&= d_1 d_2 \cdots d_n (d_n y) \\
&= d_1 d_2 \cdots d_n (x_{n-1}) \\
&= *.
\end{aligned}$$
$\square$

Definition 2.97. Let X be a pointed fibrant simplicial set. The *Moore loop* ΩX is defined to be the fiber of $p_X \colon PX \to X$. More precisely,

$$(\Omega X)_n = \{x \in X_{n+1} \mid d_1 d_2 \cdots d_{n+1} x = * \text{ and } d_0 x = *\}$$

with faces $d_i^\Omega = d_{i+1}^X$ and degeneracies $s_i^\Omega = s_{i+1}^X$ for $0 \leq i \leq n$.

The Moore path and Moore loop give the musical construction of path spaces and loop spaces in simplicial setting. There is a dual construction of the Moore path given as follows: Let

$$(\tilde{P}X)_n = \{x \in X_{n+1} \mid d_0^n x = d_0 d_0 \cdots d_0 x = *\}.$$

Then, similar to the case of PX, one can easily show that $\tilde{P}X = \{(\tilde{P}X)_n\}_{n \geq 0}$ is a simplicial set with faces $d_i^P = d_i^X$ and $s_i^P = s_i^X$ for $0 \leq i \leq n$. Again the geometric realization $|\tilde{P}X|$ is contractible. Moreover, the last face d_{n+1}^X induces a simplicial map

$$\tilde{p}_X = \{d_{n+1}^X\}_{n \geq 0} \colon \tilde{P}X \longrightarrow X$$

is a simplicial map. If X is fibrant, then $\tilde{p}_X$ is a fibration with the fiber

$$\tilde{\Omega}X = \tilde{p}_X^*.$$

The explicit construction of $\tilde{\Omega}X$ is given by

$$(\tilde{\Omega}X)_n = \{x \in X_{n+1} \mid d_0^n x = d_0 d_0 \cdots d_0 x = * \text{ and } d_{n+1} x = *\}.$$

with faces $d_i^\Omega = d_i^X$ and degeneracies $s_i^\Omega = s_i^X$ for $0 \leq i \leq n$.

By considering simplicial objects as contravariant functors from $\mathcal{O}$, one can see some insights for the dual construction of simplicial objects. Consider the category $\mathcal{O}$ consisting of finite ordered sets as objects and order preserving functions as morphisms. Let

$$\chi_n \colon [n] = \{0, 1, \ldots, n\} \longrightarrow [n] = \{0, 1, \ldots, n\}$$

be the function given by $\chi_n(j) = n - j$ for $0 \leq j \leq n$. Then there is a commutative diagram

$$
\begin{array}{ccc}
[n-1] & \xrightarrow{\chi_{n-1}} & [n-1] \\
\downarrow{\scriptstyle d^i} & & \downarrow{\scriptstyle d^{n-i}} \\
[n] & \xrightarrow{\chi_n} & [n] \\
\uparrow{\scriptstyle s^i} & \quad {\scriptstyle s^{n-i}} & \uparrow \\
[n+1] & \xrightarrow{\chi_{n+1}} & [n+1],
\end{array}
$$

which can be seen from the following tables

$$
d^i = \begin{pmatrix} 0 & \cdots & i-1 & i & \cdots & n-1 \\ 0 & \cdots & i-1 & i+1 & \cdots & n \end{pmatrix}
$$

$$
\Bigg\downarrow \binom{\chi_{n-1}}{\chi_n}
$$

$$
d^{n-i} = \begin{pmatrix} n-1 & \cdots & n-i & n-i-1 & \cdots & 0 \\ n & \cdots & n-i+1 & n-i-1 & \cdots & 0 \end{pmatrix}
$$

$$
s^i = \begin{pmatrix} 0 & \cdots & i-1 & i & i+1 & i+2 & \cdots & n+1 \\ 0 & \cdots & i-1 & i & i & i+1 & \cdots & n \end{pmatrix}
$$

$$
\Bigg\downarrow \binom{\chi_{n-1}}{\chi_n}
$$

$$
s^{n-i} = \begin{pmatrix} n+1 & \cdots & n-i+2 & n-i+1 & n-i & n-i-1 & \cdots & 0 \\ n & \cdots & n-i+1 & n-i & n-i & n-i-1 & \cdots & 0 \end{pmatrix}.
$$

In categorical language, we have a functor

$$
\chi \colon \mathcal{O} \longrightarrow \mathcal{O}
$$

which sends the object $[n]$ to itself, the morphism $d^i \colon [n-1] \to [n]$ to the morphism $d^{n-i} \colon [n-1] \to [n]$, and the morphism $s^i \colon [n+1] \to [n]$ to the morphism $s^{n-i} \colon [n+1] \to [n]$.

Now let $\mathcal{C}$ be any category and let

$$
X \colon \mathcal{O} \longrightarrow \mathcal{C}
$$

be a contravariant functor, that is, X is a simplicial object over $\mathcal{C}$. Then the composite

$$X \circ \chi \colon \mathcal{O} \longrightarrow \mathcal{C}$$

is also a contravariant functor and so $X \circ \chi$ is also a simplicial object over $\mathcal{C}$. Let $X^\chi = \{X \circ \chi([n])\}_{n \geq 0}$. Then

$$(X \circ \chi)_n = X \circ \chi([n]) = X([n]) = X_n$$

with faces

$$d_i^{X^\chi} = X \circ \chi(d^i) = X(d^{n-i}) = d_{n-i}^X \colon X_n \to X_{n-1}$$

and degeneracies

$$s_i^{X^\chi} = X \circ \chi(s^i) = X(s^{n-i}) = s_{n-i}^X \colon X_n \to X_{n+1}$$

for $0 \leq i \leq n$. The simplicial object X^χ is called a *mirror reflection* of X. Then we have the functor $\chi \colon X \mapsto X^\chi$ from the category of simplicial objects over $\mathcal{C}$ to itself with $\chi \circ \chi = \mathrm{id}$.

By using this terminology, we have

$$\tilde{P}(X) = (P(X^\chi))^\chi$$

for any simplicial set X. In fact, for any functor F from the category of simplicial objects over $\mathcal{C}$ to the category of simplicial objects over $\mathcal{D}$, we always have the dual construction given by

$$\tilde{F}(X) = (F(X^\chi))^\chi$$

for any simplicial object X over $\mathcal{C}$.

2.6.4. *Moore-Postnikov system*

It was also due to J. C. Moore to give a musical way to construct Postnikov systems in simplicial setting. The construction is just to put some canonical equivalence relations on simplicial sets to produce a tower of simplicial sets.

Let X be a simplicial set. For each $n \geq 0$, define the equivalence relation $\sim_n$ on X as follows:

> *For $x, y \in X_q$, we call $x \sim_n y$ if each iterated face of x of dimension $\leq n$ agrees with the correspondent iterated face of y.*

Let $f_x, f_y \colon \Delta[q] \to X$ be the representing maps of x and y, respectively. Then the above definition gives

$$x \sim_n y \quad \Longleftrightarrow \quad f_x|_{\mathrm{sk}_n \Delta[q]} = f_y|_{\mathrm{sk}_n \Delta[q]}. \tag{2.23}$$

Clearly $\sim_n$ is an equivalence relation. Define

$$P_n X = X/\sim_n$$

for each $n \geq 0$. Then we have the tower

$$X \longrightarrow \cdots \longrightarrow P_n X \longrightarrow P_{n-1}X \longrightarrow \cdots \longrightarrow P_0 X \qquad (2.24)$$

called *Moore-Postnikov system* of X or *coskeleton filtration* of X. By the construction, any simplicial map $f \colon X \to Y$ induces a unique simplicial map

$$P_n f \colon P_n X \to P_n Y$$

such that the diagram

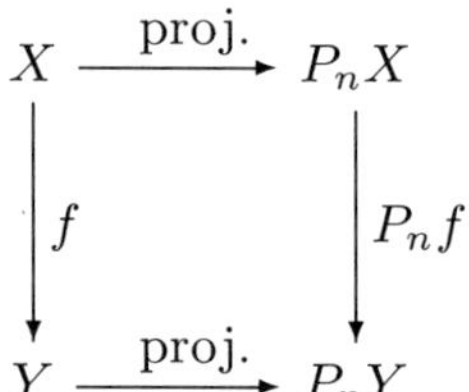

and $P_n \colon X \mapsto P_n X$, $f \mapsto P_n f$ is functor from the category of simplicial sets to the category of simplicial sets. Thus the Moore-Postnikov system is a functorial resolution of the identity functor of the category of simplicial sets.

Note. If $q \leq n$, then $x \sim_n y$ if and only if $x = y$. Some the equivalence relation $\sim_n$ only identifies the elements with dimension greater than n. Thus $(P_n X)_q = X_q$ for $q \leq n$.

Proposition 2.98. *Let X be a fibrant simplicial set. Then the following hold:*

(1). Each projection $X \to P_n X$ is a fibration.
(2). Each $P_n X$ is fibrant.
(3). For each $n \geq m$, the projection $P_n X \to P_m X$ is a fibration.

Proof. (1). Let $p_n \colon X \to P_n X$ be the projection. Let

$$x_0, x_1, \ldots, x_{k-1}, x_{k+1}, \ldots, x_q$$

be in X_{q-1} matching faces and let $b \in (P_n X)_q$ such that

$$d_j b = p_n(x_j), \text{ that is, } d_j b \sim_n x_j$$

for $j \neq k$.

If $q \le n$, then $p_n \colon X_q \to (P_n X)_q$ is the identity and so let $y = b \in X_q$. Then $d_j y = x_j$ for $j \ne k$ and $p_n(y) = b$.

If $q = n + 1$, let $z \in X_{n+1}$ such that $p_n(z) = b$. (Since $P_n X$ is the quotient of X by the equivalence relation $\sim_n$, z is just a representative element of the equivalence class b.) Then

$$d_j z = x_j$$

for $j \ne k$ because $p_n(d_j z) = d_j p_n(z) = d_j b = x_j$ and $p_n \colon X_n \to (P_n X)_n$ is the identity.

For $q > n + 1$, we need the extension condition on X. Since X is fibrant, there exists $y \in X_q$ such that

$$d_j y = x_j$$

for $j \ne k$. Let $z \in X_q$ such that $p_n(z) = b$. Then

$$d_j z \sim_n x_j$$

for $j \ne k$. We show that $y \sim_n z$, which will give that $p_n(y) = b$. Let

$$f_y, f_z \colon \Delta[q] \to X$$

be the representing maps of y and z, respectively. Then

$$p_n \circ f_y|_{\Lambda^k[q]} = p_n \circ f_z|_{\Lambda^k[q]} \tag{2.25}$$

because $d_j y \sim_n d_j z$ for $j \ne k$. Observe that

$$\mathrm{sk}_{q-2}(\Lambda^k[q]) = \mathrm{sk}_{q-2}(\Delta[q]) \tag{2.26}$$

because the nondegenerate elements in $\mathrm{sk}_{q-2}(\Delta[q])$ are given by the sequences

$$(j_0, j_1, \ldots, j_t)$$

with $t \le q - 2$ and $0 \le j_0 < j_1 < \cdots < j_t \le q$, and the $(q-1)$-dimensional nondegenerate elements in $\Lambda^k[q]$ are given by the sequences

$$(0, 1, \ldots, i - 1, i + 1, \ldots, q)$$

for $0 \le i \le q$ with $i \ne k$. Since $q \ge n + 2$, we have

$$\mathrm{sk}_n(\Lambda^k[q]) = \mathrm{sk}_n(\Lambda^k[q]).$$

By restricting Equation (2.25) to sk_n, we obtain

$$\begin{aligned}
p_n \circ f_y|_{\mathrm{sk}_n(\Delta[q])} &= p_n \circ f_y|_{\mathrm{sk}_n(\Lambda^k[q])} \\
&= p_n \circ f_z|_{\mathrm{sk}_n(\Lambda^k[q])} \\
&= p_n \circ f_z|_{\mathrm{sk}_n(\Delta[q])}
\end{aligned}$$

and so

$$f_y|_{\mathrm{sk}_n(\Delta[q])} = f_z|_{\mathrm{sk}_n(\Delta[q])}$$

because $p_n\colon X_n \to (P_nX)_n$ is the identity. Thus $y \sim_n z$ and we finish the proof of (1).

(2). Since $p_n\colon X \to P_nX$ is a fibration and p_n is onto, this assertion directly follows from Proposition 2.92.

(3). By (2), P_nX is fibrant. Observe that $P_mX = P_m(P_nX)$ by the equivalence relation $\sim_m$. Thus $P_nX \to P_mX$ is a fibration by (1). $\square$

Exercise 2.18. Prove assertion (3) of the above proposition.

Proposition 2.99. *Let X be a pointed fibrant simplicial set and let*

$$p_n\colon X \to P_n(X)$$

be the projection. Then the following hold:

(1). *$p_{n*}\colon \pi_q(X) \longrightarrow \pi_q(P_nX)$ is an isomorphism for $q \le n$.*
(2). *Let $q > n$ and let x be a spherical element in $(P_nX)_q$. Then $x = *$. In particular, $\pi_q(P_nX) = 0$ for $q > n$.*

Proof. (1). If $q < n$, this is obvious because $p_n\colon X_q \to (P_nX)_q$ is the identity map for $q \le n$. We show that

$$p_{n*}\colon \pi_n(X) \to \pi_n(P_nX)$$

is an isomorphism. Let x be a spherical element in $(P_nX)_n$. Since

$$p_n\colon X_n \to (P_nX)_n$$

is the identity map, x itself gives a spherical element in X_n and so

$$p_{n*}\colon \pi_n(X) \to \pi_n(P_nX)$$

is onto. Now let y be a spherical element in X_n such that $p_n([y]) = 0$ in $\pi_n(P_nX)$. Let $F = p_n^{-1}(*)$ be the fiber of $p_n\colon X \to P_nX$ and let $i\colon F \to X$ be the inclusion. Then, by the long exact sequence for the homotopy groups for the fibration p_n,

$$[y] \in \mathrm{Im}(i_*\colon \pi_n(F) \to \pi_n(X)).$$

Since $p_n\colon X_n \to (P_nX)_n$ is the identity map, $F_n = \{*\}$ and so $\pi_n(F) = 0$. Thus $[y] = 0$ and so $p_{n*}\colon \pi_n(X) \to \pi_n(P_nX)$ is an isomorphism.

(2). Let $x \in (P_n X)_q$ be a spherical element with $q > n$. Since $p_n \colon X \to P_n X$ is a fibration, there exists $y \in X_q$ such that $p_n(y) = x$ and $d_j y = *$ for $j > 0$.

If $q = n + 1$, then $p_n(d_0 y) = d_0 p_n(y) = d_0 x = *$ and so $d_0 y = *$ because $p_n \colon X_n \to (P_n X)_n$ is the identity map. Thus $d_j y = *$ for all $0 \leq j \leq n$. It follows that

$$y \sim_n *$$

because $d_j y = d_j *$ for all $0 \leq j \leq n$. Thus

$$x = p_n(y) = p_n(*) = *.$$

If $q > n + 1$, since $d_0 y$ is spherical, $d_i d_j y = *$ for all $0 \leq i, j \leq n$. It follows that $y \sim_n *$ because all iterated faces of both y and $*$ of dimension $\leq n$ are trivial. Thus

$$x = p_n(y) = p_n(*) = *. \qquad \square$$

A topological space X is called an *Eilenberg-MacLane space of $K(\pi, n)$* if X is homotopy equivalent to a *CW*-complex with

$$\pi_j(X) = \begin{cases} 0 & \text{if } j \neq n \\ \pi & \text{if } j = n. \end{cases}$$

A simplicial set X is called an *Eilenberg-MacLane complex of $K(\pi, n)$* if its geometric realization $|X|$ is an *Eilenberg-MacLane space of $K(\pi, n)$*.

Corollary 2.100. *Let X be a fibrant simplicial set and let $E_n X$ be the fiber of the projection $P_n X \to P_{n-1} X$. Then $E_n X$ is an Eilenberg-MacLane complex of $K(\pi_n(X), n)$.*

Proof. Since X is fibrant, $P_n X$ is fibrant. Note that $P_{n-1} X = P_{n-1}(P_n X)$. The assertion follows from the long exact sequence for the homotopy groups of the fibration

$$E_n X \longrightarrow P_n X \longrightarrow P_{n-1} X,$$

using $\pi_q(P_n X) \cong \pi_q(P_{n-1} X)$ for $q \leq n - 1$ and $\pi_q(P_m X) = 0$ for $q > m$. $\square$

Denote by $\mathrm{Hom}_S(X, Y)$ the set of simplicial maps from X to Y. Let X be a simplicial set. Let $\overline{\mathrm{sk}}_n X = \{X_q\}_{0 \leq q \leq n}$. For any simplicial set Z, a map $f = \{f_q\} \colon \overline{\mathrm{sk}}_n X \to Z$ with $f_q \colon X_q \to Z_q$, $0 \leq q \leq n$, is called *partially simplicial* if $d_j f_q = f_{q-1} d_j$ for $0 \leq j \leq q \leq n$ and $s_j f_q = f_{q+1} s_j$ for $0 \leq j \leq q \leq n - 1$.

Proposition 2.101. *Let X and Y be simplicial sets. Suppose that Y is fibrant. Then*

(1). *The inclusion $\mathrm{sk}_n X \to X$ induces a one-to-one function*

$$\mathrm{Hom}_S(X, P_n X) \longrightarrow \mathrm{Hom}_S(\mathrm{sk}_n X, P_n Y)$$

for each $n \geq 0$.

(2). *The inclusion $\mathrm{sk}_{n+1} X \to X$ induces an isomorphism*

$$\mathrm{Hom}_S(X, P_n Y) \xrightarrow{\;\cong\;} \mathrm{Hom}_S(\mathrm{sk}_{n+1} X, P_n Y)$$

for each $n \geq 0$.

Proof. (1). Let $f, g \colon X \to P_n Y$ be simplicial maps such that

$$f|_{\mathrm{sk}_n X} = g|_{\mathrm{sk}_n Y}.$$

Let $x \in X_q$ and let $f_x \colon \Delta[q] \to X$ be its representing map. Then $f \circ f_x$ and $g \circ f_x$ are the representing maps of $f(x)$ and $g(x)$, respectively. Since

$$(f \circ f_x)|_{\mathrm{sk}_n \Delta[q]} = (g \circ f_x)|_{\mathrm{sk}_n \Delta[q]}.$$

Thus $f(x) = g(x)$ by the equivalence relation $\sim_n$.

(2). It suffices to show that any partially simplicial map $f \colon \overline{\mathrm{sk}}_{n+1} X \to P_n Y$ can be extended to a simplicial map $f \colon X \to P_n Y$. Suppose that f can be extended to be a partially simplicial map

$$f = \{f_j\}_{0 \leq j \leq q-1} \colon \overline{\mathrm{sk}}_{q-1} X \to P_n Y$$

with $q > n + 1$. Let $x \in X_q$. Since Y is fibrant, $P_n Y$ is fibrant. Thus there exists $y \in (P_n Y)_q$ such that

$$d_j y = f_{q-1}(d_j x)$$

for $1 \leq j \leq q$ because $f_{q-1}(d_1 x), \ldots, f_{q-1}(d_q x)$ are matching faces with respect to 0. Now, for any $0 \leq k \leq q - 1$,

$$\begin{aligned}
d_k(d_0 y) &= d_0 d_{k+1} y \\
&= d_0 f_{q-1}(d_{k+1} x) \\
&= f_{q-2}(d_0 d_{k+1} x) \\
&= f_{q-2}(d_k d_0 x) \\
&= d_k f_{q-1}(d_0 x)
\end{aligned}$$

and so $d_0 y = f_{q-1}(d_0 x)$ (because, by choosing representative elements a, b for $d_0 y$ and $f_{q-1}(d_0 x)$, respectively, we have $a \sim_n b$ as $q \geq n + 2$). If y' is another element in $(P_n Y)_q$ such that

$$d_j y' = f_{q-1}(d_j x) = d_j y$$

for $1 \leq j \leq q$. Then $y' = y$ because $d_0 y' = f_{q-1}(d_0 x) = d_0 y$. This shows that there exists a unique element $y \in (P_n X)_q$ such that $d_j y = f_{q-1}(d_j x)$ for $0 \leq j \leq q$. Define $f_q \colon X_q \to Y_q$ by setting

$$f_q(x) = y$$

for $x \in X_q$. Then $d_j f_q = f_{q-1} d_j$ for $0 \leq j \leq q$. We check that $s_j f_{q-1} = f_q s_j$ for $0 \leq j \leq q-1$. Let $x \in X_{q-1}$. Then

$$d_j f_q(s_i x) = f_{q-1}(d_j s_i x)$$

$$= \begin{cases} \begin{aligned} f_{q-1}(s_{i-1} d_j x) &= s_{i-1} f_{q-2}(d_j x) \\ &= s_{i-1} d_j f_{q-1}(x) \\ &= d_j s_i f_{q-1}(x) \end{aligned} & \text{if } j < i \\[2ex] f_{q-1}(x) = d_j s_i f_{q-1}(x) & \text{if } j = i, i+1 \\[2ex] \begin{aligned} f_{q-1}(s_i d_{j-1} x) &= s_i f_{q-2}(d_{j-1} x) \\ &= s_i d_{j-1} f_{q-1}(x) \\ &= d_j s_i f_{q-1}(x) \end{aligned} & \text{if } j > i+1 \end{cases}$$

$$= d_j s_i f_{q-1}(x).$$

Thus $f_q(s_i x) = s_i f_{q-1}(x)$. The induction is finished and hence result. $\square$

Lemma 2.102. *If X is a minimal simplicial set, then each $P_n X$ is also minimal.*

Proof. We use Proposition 2.76 to check that $P_n X$ is minimal. Let $v, w \in (P_n X)_q$ such that $d_i v = d_i w$ for $i \neq k$. If $q \leq n$, then $d_k v = d_k w$ because X is minimal and $X_q = (P_n X)_q$. For $q > n$, let $\tilde{v}, \tilde{w} \in X_q$ be the representatives of the equivalence class v and w. Then $d_i \tilde{v} \sim_n d_i \tilde{w}$ for $i \neq k$. If $q = n+1$, then $d_i \tilde{v} = d_i \tilde{w}$ for $i \neq k$ and so $d_k \tilde{v} = d_k \tilde{w}$ as X is minimal. It follows that $d_k v = d_k w$. If $q > n+1$, $\tilde{v} \sim_n \tilde{w}$ and so $v = w$. In particular, $d_k v = d_k w$. This proves that $P_n X$ is minimal. $\square$

Theorem 2.103 (Uniqueness of $K(\pi, n)$). *Let X and Y be fibrant simplicial sets. Suppose that X and Y are Eilenberg-MacLane complexes $K(\pi, n)$ with $n \geq 1$. Then $X \simeq Y$.*

Proof. We may assume that both X and Y are minimal simplicial sets. Since X is minimal, each $P_q X$ is minimal. Since $\pi_q(X) = 0$ for $q \neq n$, we have $\pi_*(P_q X) = 0$ for $q < n$. It forces that $P_q X = *$ for $q < n$. Since the projection $X \to P_n X$ induces an isomorphisms on homotopy groups, $X \to P_n X$ is an isomorphism because they are minimal. Similarly, $P_q Y = *$ for $q < n$ and $Y = P_n Y$.

Now we construct a simplicial map $f \colon X \to Y$. From the above, $X_q = Y_q = *$ for $q < n$ and so there are unique maps $f_q \colon X_q \to Y_q$ for $q < n$. Since X and Y are minimal, $X_n = \pi_n(X)$ and $Y_n = \pi_n(Y)$. Let $f_n \colon X_n \to Y_n$ be the identity map. This gives a partial simplicial map

$$f = \{f_q\}_{0 \leq q \leq n} \colon \overline{\mathrm{sk}}_n X \longrightarrow Y = P_n Y.$$

Let $x \in X_{n+1}$. Then $d_j x$ is spherical for each $0 \leq j \leq n+1$. By applying Homotopy Addition Theorem, we have

$$\sum_{j=0}^{n+1} (-1)^j f_n(d_j x) = \sum_{j=0}^{n} (-1)^j d_j x = 0$$

in $\pi_n(Y) = \pi_n(X)$ and there exists an element $y \in Y_{n+1}$ such that

$$d_j y = f_n(d_j x)$$

for $0 \leq j \leq n+1$. (**Note.** The Homotopy Addition Theorem also works for $n = 1$ by the definition of the product in fundamental groups.) If y' is another element in Y_{n+1} such that

$$d_j y' = f_n(d_j x) = d_j y$$

for $0 \leq j \leq n+1$. Then $y = y'$ because $Y = P_n Y$. Define

$$f_{n+1}(x) = y$$

for $x \in X_{n+1}$. Then $d_j f_{n+1} = f_n d_j$ for $0 \leq j \leq n+1$. Since

$$d_j f_{n+1}(s_i x) = f_n(d_j s_i x) = d_j s_i f_n(x)$$

for $0 \leq j \leq n+1$ (as f_n is the identity map), we have $s_i f_n = f_{n+1} s_i$ for $0 \leq i \leq n+1$. From the proof of Proposition 2.101, the partially simplicial map $f = \{f_q\}_{0 \leq q \leq n+1}$ can be extended uniquely to a simplicial map $f \colon X \to Y$.

By switching X and Y, there is a unique simplicial map $g \colon Y \to X$ such that $g_n \colon Y_n \to X_n$ is the identity map. From the uniqueness property of f and g, we have $g = f^{-1}$ and hence the result. $\qquad\square$

The existence of $K(\pi, n)$ (where π is abelian if $n > 1$) can be proved by constructing examples. The space $K(\pi, 1)$ can be given by the classifying space $\bar{W}\pi$ of π. (See Section 1.)

Exercise 2.19. Let π be any group. Show that $\bar{W}\pi$ is a minimal simplicial set.

There is a simple construction for $K(\pi, n)$ for $n > 1$, where π is abelian. Start with simplicial n-sphere S^n. Define the simplicial group $\mathbb{Z}[S^n]$ by setting

$$\mathbb{Z}[S^n]_q = \mathbb{Z}(S^n_q)/\langle * \rangle,$$

the free abelian group generated by S^n_q modulo the single relation that the basepoint $* = 0$, with faces $d_i^{\mathbb{Z}[S^n]}$ and degeneracies $s_i^{\mathbb{Z}[S^n]}$ are the group homomorphisms induced by the faces $d_i^{S^n}$ and the degeneracies $s_i^{S^n}$, respectively. For any abelian group G, Define $G[S^n]$ by

$$G[S^n] = \mathbb{Z}[S^n] \otimes_{\mathbb{Z}} G = \{\mathbb{Z}[S^n]_q \otimes_{\mathbb{Z}} G\}_{q \geq 0}$$

with induces faces and degeneracies. We will see that

$$\pi_*(G[S^n]) \cong \bar{H}_*(S^n; G)$$

and so $G[S^n]$ is $K(G, n)$.

Exercise 2.20. Let G be any abelian group. Show that the simplicial abelian group $G[S^n]$ is a minimal simplicial set.

3. Simplicial Group Theory

As a combinatorial tool for studying homotopy theory, simplicial groups were first studied by J. C. Moore [27]. The classical Moore theorem states that

$$\pi_*(|\mathcal{G}|) \cong H_*(N\mathcal{G}),$$

where $|\mathcal{G}|$ is the geometric realization of $\mathcal{G}$ and $N\mathcal{G}$ is the Moore chain complex of $\mathcal{G}$ described in next section. Milnor [25] then proved that any loop space is (weakly) homotopy equivalent to a geometric realization of a simplicial group, and so, theoretically speaking, the homotopy groups of any space can be determined as the homology of a Moore chain complex.

It is possible that two simplicial groups with the same homotopy type have sharply different group structures. Simplicial group models for loop spaces have been studied by many people, see for instance [1, 7, 8, 20, 26, 27, 30, 33, 38]. Different simplicial group models for the same loop space may give different homotopy information. For example, the classical Adams spectral sequence arises as the associated graded by taking the mod p descending central series of Kan's G-construction on reduced simplicial sets, [4, 5, 11].

On the other hand, one could have a perfect simplicial group model (that is, the abelianization is the trivial group) for certain loop spaces by using Carlsson's construction [38]. For this model, the descending central series will not give any information as the groups are perfect, but word filtration provides different information.

By using Milnor's $F[K]$-construction on the simplicial circle, a combinatorial description of the general homotopy groups $\pi_n(S^2)$ was obtained in [42], where it was proved that the general homotopy group $\pi_n(S^2)$ is isomorphic to the center of a combinatorially given group G_n with n generators and certain systematic (infinitely) relations [42, Theorem 1.4]. It was then asked by many people whether there is a finitely presented group whose center is given by $\pi_n(S^2)$. A positive answer to this question is given in [22, Theorem 1].

A connection between the braid groups and the general homotopy groups of S^2 was found in [40], where it was proved that the Artin braid group B_n acts on the group G_n and the homotopy group $\pi_n(S^2)$ is given by the fixed set of the pure braid group P_n action on G_n [40, Theorem 1.2]. Moreover it was proved in [2, Theorem 1.2] that $\pi_n(S^2)$ are given by the $(n+1)$-strand Brunnian braids over the sphere modulo the $(n+1)$-strand Brunnian braids over the disk for $n \geq 4$. Some relations between the homotopy groups and Vassiliev invariants have been studied in [9, 10]. Theorems 1 and 2 in [22] then give further connections between the braid groups and the homotopy groups for addressing the conjugation problem and the mirror reflection problem on the braids.

3.1. *Moore chains and homotopy groups*

Let X be a pointed fibrant simplicial set. Recall that the product in $\pi_n(X)$, $n \geq 1$, is defined as follows: Let x and y be spherical elements in X_n. By the extension conditions, there exists $z \in X_{n+1}$ such that $d_2z = x$, $d_0z = y$ and $d_jz = *$ for $j \geq 2$. Then xy is defined to the homotopy class represented by d_1z.

A pointed simplicial set X is called a *simplicial H-set* if there exists a simplicial multiplication $\mu\colon X \times X \to X$ such that $\mu(x,*) = \mu(*,x) = x$ for any $x \in X$. Namely, the multiplication has the basepoint as the identity element. (At moment we do not require that the multiplication is associative.) Observe that for spherical elements $x, y \in X_n$, the product $\mu(x,y)$ is also a spherical element. The following proposition states that $\mu(x,y)$ is the same as xy in $\pi_n(X)$.

Lemma 3.1. *Let X be a fibrant simplicial H-set with a multiplication*

$$\mu \colon X \times X \to X.$$

Let x, y be spherical elements in X_n with $n \geq 1$. Then $\mu(x, y) \sim xy \sim yx$.

Proof. We write 1 for the basepoint $*$ in X and also for the identity in $\pi_n(X)$. Let $[x]$ denote the homotopy class of spherical element x. The simplicial map

$$\mu \colon X \times X \to X$$

induces a group homomorphism

$$\mu_* \colon \pi_n(X \times X) = \pi_n(X) \times \pi_n(X) \longrightarrow \pi_n(X).$$

We have

$$\begin{aligned}
\mu_*([x], [y]) &= \mu_*(([x], 1)(1, [y])) \\
&= \mu_*([x], 1)\mu_*(1, [y]) \\
&= [x][y] \qquad\qquad \text{because } \mu(x, *) = \mu(*, x) = x
\end{aligned}$$

$$\begin{aligned}
\mu_*([x], [y]) &= \mu_*((1, [y])([x], 1)) \\
&= \mu_*(1, [y])\mu_*([x], 1) \\
&= [y][x]
\end{aligned}$$

and hence the result. $\qquad\qquad\qquad\qquad\qquad\qquad\qquad\qquad\square$

Thus, for a fibrant simplicial H-set, we can simply use the multiplication μ to define the product in $\pi_n(X)$. In addition, we also have the product $([x], [y]) \mapsto [\mu(x, y)]$ in $\pi_0(X)$. (**Note.** If μ is not associative, then the product in $\pi_0(X)$ may not be associative. However, $\pi_n(X)$ are always abelian groups for $n \geq 1$ by the above lemma.)

Now we move to simplicial groups G. By the above lemma, the product in $\pi_n(G)$ are induced by the multiplication of G. In this $\pi_0(G)$ is a group and $\pi_n(G)$ are abelian groups for $n \geq 1$. Let

$$\mathcal{Z}_n G = \{x \in G_n \mid x \text{ is spherical}\}.$$

Then $\mathcal{Z}_n G$ is a subgroup of G_n and $\pi_n(G)$ is a quotient group of $\mathcal{Z}_n G$. Let $\mathcal{B}_n G$ be the kernel of $\mathcal{Z}_n G \to \pi_n(G)$. Then

$$\mathcal{B}_n G = \{x \in G_n \mid x \text{ is spherical and } x \simeq 1\}.$$

Lemma 3.2. *Let G be a simplicial group and let x be a spherical element in G_n. Then $x \in \mathcal{B}_n G$ if and only if there exists an element $y \in G_{n+1}$ such that $d_j y = 1$ for $j > 0$ and $d_0 y = x$.*

Proof. Let $x_0 = x$ and $x_j = 1$ for $1 \leq j \leq n + 1$. Then the assertion a direct consequence of the Homotopy Addition Theorem. $\qquad\square$

This lemma indicates that we can construct a chain complex such that $\pi_n(G)$ can be obtained from the homology of the chain complex. The precise construction is as follows:

Let G be a simplicial group. Define

$$N_n G = \bigcap_{j=1}^{n} \mathrm{Ker}(d_j \colon G_n \to G_{n-1}).$$

Let $x \in N_n G$, that is $x \in G_n$ with $d_j x = 1$ for $j > 0$. Then

$$d_k(d_0 x) = d_0 d_{k+1} x$$
$$= 1$$

for any $0 \leq k \leq n - 1$. In other words, $d_0 x \in N_{n-1}G$ with $d_0(d_0 x) = 1$.

Definition 3.3. A *chain complex* (C, ∂) is a sequence of (possibly non-commutative) groups and homomorphisms

$$\cdots \longrightarrow C_{n+1} \xrightarrow{\partial_{n+1}} C_n \xrightarrow{\partial_n} C_{n-1} \longrightarrow \cdots$$

such that $\mathrm{Im}(\partial_{n+1}) \subseteq \mathrm{Ker}(\partial_n)$, that is the composite $\partial_n \circ \partial_{n+1}$ is the trivial homomorphism. The homology $H_n(C, \partial)$ is defined to be the coset $\mathrm{Ker}(\partial_n)/\mathrm{Im}(\partial_{n+1})$.

Definition 3.4. Let G be a simplicial group. The *Moore chain complex* NG is the sequence of groups

$$\cdots \longrightarrow N_{n+1}G \xrightarrow{d_0} N_n G \xrightarrow{d_0} N_{n-1}G \longrightarrow \cdots .$$

The elements in $\mathcal{Z}_n G$, that is the spherical elements in G_n, are called *Moore cycles* and the elements in $\mathcal{B}_n G$ are called *Moore boundaries*.

By definition,

$$H_n(NG, d_0) = \mathrm{Ker}(d_0 \colon N_n G \to N_{n-1}G)/d_0(N_{n+1}G)$$
$$= \left(\bigcap_{j=0}^{n} \mathrm{Ker}(d_j \colon G_n \to G_{n-1}) \right) \Big/ \mathcal{B}_n G$$
$$= \mathcal{Z}_n G / \mathcal{B}_n G$$
$$= \pi_n(G).$$

Thus we have the following musical result of J. C. Moore.

Theorem 3.5. *Let G be a simplicial group. Then*

$$H_n(NG, d_0) \cong \pi_n(G) \cong \pi_n(|G|)$$

for each $n \geq 0$. $\qquad\qquad\square$

The significance of this theorem is that the homotopy groups (of spaces) can be combinatorially defined as the homology of chain complexes.

As one may know from the textbooks of algebraic topology, the fundamental groups act on higher homotopy groups. For simplicial groups G, the correspondent action is that $\pi_0(G)$ acts on the homotopy groups $\pi_n(G)$ given as follows:

First let G_0 act on G by conjugation, namely

$$v \cdot x = (s_0^n v)^{-1} x (s_0^n v) \tag{3.1}$$

for $v \in G_0$ and $x \in G_n$.

Proposition 3.6. *The action of G_0 on G by conjugation induces an action of $\pi_0(G)$ on $\pi_n(G)$ for each $n \geq 0$.*

Proof. Let $v, w \in G_0$ such that $v \simeq w$. Then there exists a element $a \in G_1$ such that $d_0 a = v$ and $d_1 a = w$. Note that

$$
\begin{aligned}
d_{n+1} s_0^n a &= s_0 d_n s_0^{n-1} a \\
&\;\;\cdots \\
&= s_0^n d_1 a \\
&= s_0^n w
\end{aligned}
$$

$$
\begin{aligned}
d_n s_0^n a &= s_0 d_{n-1} s_0^{n-1} a \\
&\;\;\cdots \\
&= s_0^n d_0 a \\
&= s_0^n v.
\end{aligned}
$$

Now let $x \in \mathcal{Z}_n G$ be a Moore cycle. Let $y = (s_0^n a)^{-1}(s_n x)(s_0^n a) \in G_{n+1}$. Then

$$
\begin{aligned}
d_{n+1} y &= (d_{n+1} s_0^n)^{-1}(d_{n+1} s_n x)(d_{n+1} s_0^n a) \\
&= (s_0^n w)^{-1} x (s_0^n w) \\
&= w \cdot x
\end{aligned}
$$

$$
\begin{aligned}
d_n y &= (d_n s_0^n)^{-1}(d_n s_n x)(d_n s_0^n a) \\
&= (s_0^n v)^{-1} x (s_0^n v) \\
&= v \cdot x
\end{aligned}
$$

and, for $j < n$,

$$\begin{aligned}
d_j y &= (d_j s_0^n a)^{-1}(d_j s_n x)(d_j s_0^n a) \\
&= (d_j s_0^n a)^{-1}(s_{n-1} d_j x)(d_j s_0^n a) \\
&= (d_j s_0^n a)^{-1} \cdot 1 \cdot (d_j s_0^n a) \\
&= 1.
\end{aligned}$$

By the Homotopy Addition Theorem, we have $v \cdot x \simeq w \cdot x$ and hence the result. $\qquad\square$

Definition 3.7. We call a simplicial group G *n-simple* if $\pi_0(G)$ acts trivially on $\pi_n(G)$. A simplicial group G is called *simple* if it is n-simple for every n.

Note. According to the definition of the fundamental group actions on homotopy groups (see for instance [35, Section 3 of Chapter 4]), $\pi_0(G)$-action on $\pi_n(G)$ is the same as $\pi_1(\bar{W}G) \cong \pi_0(G)$-action on $\pi_{n+1}(\bar{W}G) \cong \pi_n(G)$, where $\bar{W}G$ is the classifying space of G. Thus a simplicial group G is n-simple if and only if its classifying space $\bar{W}G$ is $(n+1)$-simple.

The following results are from [42] for describing the homotopy groups as the centers of certain groups.

Theorem 3.8. *Let G be a simplicial group and let $n \geq 0$. Suppose that G is n-simple. Then the homotopy group $\pi_n(G)$ is contained in the center of $G_n/\mathcal{B}_n G$.*

Proof. If $n = 0$, then $\pi_0(G) \cong G_0/\mathcal{B}_0 G$, which is abelian if $\pi_0(G)$ acts trivially on itself. Thus we may assume that $n \geq 1$. Note $\pi_n(G) \cong \mathcal{Z}G_n/\mathcal{B}_n G$. It suffices to show that the commutator $[x, y] \in \mathcal{B}G_n$ for any $x \in \mathcal{Z}_n G$ and $y \in G_n$.

Now let $x \in \mathcal{Z}G_n$ and let $y \in G_n$. Let z denote $s_0^n d_0^n(y^{-1}) \cdot y$. Observe that x is a cycle. There is a simplicial map $f_x \colon S^n \to G$ such that $f_x(\sigma_n) = x$, where S^n is the standard n-sphere with a nondegenerate n-simplex σ_n. Let the simplicial map $f_z \colon \Delta[n] \to G$ be the representative of z, i.e., $f_z(\tau_n) = z$ for the nondegenerate n-simplex τ_n. Consider the simplicial map

$$[f_x, f_z] \colon S^n \times \Delta[n] \to G$$

defined by $[f_x, f_z](a, b) = [f_x(a), f_z(b)] = (f_x(a))^{-1}(f_z(b))^{-1} f_x(a) f_z(b)$, the commutator of $f_x(a)$ and $f_z(b)$. Notice the equality

$$f_z(d_0^n(\tau)) = d_0^n s_0^n d_0^n(y^{-1}) \cdot d_0^n y = 1.$$

Let v be the simplicial subset of $\Delta[n]$ generated by the vertex $d_0^n \tau$ and let $*$ be the base-point of S^n. Then $[f_x, f_z]$ is trivial restricted to the simplicial subset

$$S^n \vee \Delta[n] = (S^n \times v) \cup (* \times \Delta[n]).$$

Thus $[f_x, f_z]$ factors through the quotient simplicial set $S^n \wedge \Delta[n]$. Let ϕ be the composite

$$S^n \xrightarrow{\ j\ } S^n \wedge \Delta[n] \xrightarrow{\ [f_x,\,f_z]\ } G$$

with $j(\sigma_n) = \sigma_n \wedge \tau_n$. Notice that we have $\phi(\sigma_n) = [x, z]$. Let $\{[x, z]\}$ denote the homotopy class in $\pi_n(G)$ represented by the cycle $[x, z]$. Then the simplicial map $\phi \colon S^n \to G$ is a representative of the cycle $[x, z]$ and $|\phi|_*(\iota_n) = \{[x, z]\}$, where the homomorphism $|\phi|_* \colon \pi_*(|S^n|) \to \pi_*(|G|)$ is induced by the geometric realization $|\phi|$ and ι_n is a generator for $\pi_n(|S^n|)$. Observe that the geometric realization $|S^n \wedge \Delta[n]|$ is contractible. Thus $\{[x, z]\} = 0$ in $\pi_n(G)$ and so $[x, z] \in \mathcal{B}_n G$.

Let $\alpha = d_0^n y$. Then $\{(s_0^n \alpha)^{-1} x s_0^n \alpha\} = \{x\}$ since $\pi_0(G)$ acts trivially on $\pi_n(G)$. Thus $[x, s_0^n \alpha] \in \mathcal{B}_n G$ and so $[x, y] = [x, s_0^n \alpha z] \in \mathcal{B}_n G$ by the Witt-Hall identity that $[a, bc] = [a, c][a, b][[a, b], c]$. The assertion follows. $\square$

A simplicial group G is called *connected* if $\pi_0(G)$ is trivial.

Corollary 3.9. *Let G be a simplicial group. Suppose that the classifying space $\bar{W}G$ is simple, for example, if G is connected. Then the homotopy group $\pi_n(G)$ is contained in the center of $G_n/\mathcal{B}_n G$ for each n.* $\square$

In some cases, the homotopy group $\pi_n(G)$ is the same as the center of $G_n/\mathcal{B}_n G$.

Definition 3.10. A simplicial group G is said to be *r-centerless* if the center $Z(G_n)$ is trivial for $n \geq r$.

Proposition 3.11. *Let G be a connected r-centerless simplicial group. Then*

$$\pi_n(G) \cong Z(G_n/\mathcal{B}_n G)$$

for $n \geq r + 1$.

Proof. We have proved that $\mathcal{Z}_n G/\mathcal{B}_n G \subseteq Z(G_n/\mathcal{B}_n G)$. So it suffices to show that

$$Z(G_n/\mathcal{B}_n G) \subseteq \mathcal{Z}_n G/\mathcal{B}_n G$$

for $n \geq r + 1$.

Now let $\tilde{x} \in Z(G_n/\mathcal{B}_nG)$. Choose $x \in G_n$ with $p(x) = \tilde{x}$, where $p\colon G_n \to G_n/\mathcal{B}_nG$ is the quotient homomorphism. To check $\tilde{x} \in \mathcal{Z}_nG/\mathcal{B}_nG$, it suffices to show that $x \in \mathcal{Z}_nG$ or $d_jx = 1$ for all j. Since $Z(G_{n-1}) = \{1\}$, $d_jx = 1$ if and only if $[d_jx, y] = 1$ for all $y \in G_{n-1}$. Now $[d_jx, y] = d_j[x, s_{j-1}y]$ for $j > 0$ and $[d_0x, y] = d_0[x, s_0y]$. Since $\tilde{x} \in Z(G_n/\mathcal{B}_nG)$, $[x, z] \in \mathcal{B}_nG \subseteq \mathcal{Z}_nG$ for all $z \in G_n$ and therefore $[d_jx, y] = 1$ for all $y \in G_{n-1}$. We finish the proof. $\qquad\square$

The proof also gives:

Proposition 3.12. *Let G be a connected r-centerless simplicial group. Then*

$$Z(G_n/\mathcal{Z}_nG) = \{1\}$$

for $n \geq r + 1$. $\qquad\square$

A simplicial group G is called *reduced* if $G_0 = \{1\}$.

Lemma 3.13. *Let G be a reduced simplicial group such that G_n is cyclic or centerless for each n. Let γ_G be the smallest n such that we have $G_n \neq \{1\}$. Then $Z(G_n) = \{1\}$ for $n > \gamma_G$.*

Proof. If G_n is not abelian, any G_q for $q \geq n$ is also non-abelian because it contains copies of G_n via degeneracies.

If the simplicial group G is abelian and $G_n = NG_n$ is cyclic for $n = \gamma_G$ with $\gamma_G \geq 1$, then G_{n+1} contains $n + 1$ different copies of G_n via degeneracies. So it cannot be cyclic. $\qquad\square$

Corollary 3.14. *Let G be a reduced simplicial group such that G_n is cyclic or centerless for each n. Then*

$$\pi_n(G) \cong Z(G_n/\mathcal{B}_nG)$$

for $n \neq \gamma_G + 1$, where γ_G is defined as above. $\qquad\square$

Since free group of rank $n \geq 2$ is centerless, we have the following theorem:

Theorem 3.15. *Let G be a reduced simplicial group such that G_n is a free group for each n. Then there exits a unique integer $\gamma_G > 0$ such that $G_n = \{1\}$ for $n < \gamma_G$ and $\mathrm{rank}(G_n) \geq 2$ for $n > \gamma_G$. Furthermore,*

$$\pi_n(G) \cong Z(G_n/\mathcal{B}_nG)$$

for $n \neq \gamma_G + 1$. $\qquad\square$

3.2. *Simplicial abelian groups and the Hurewicz Theorem*

3.2.1. *Two chain complexes for simplicial abelian groups*

Let G be a simplicial abelian groups. Define

$$\partial_n \colon G_n \to G_{n-1}$$

by setting

$$\partial_n(x) = \sum_{j=0}^{n} (-1)^j d_j x.$$

As we have seen for abelian Δ-groups, (G, ∂) becomes a chain complex. Thus for a simplicial abelian group G we have two chain complexes (NG, d_0) and (G, ∂). Let $x \in N_n G$. Then $\partial(x) = d_0 x$ because $d_j x = 0$ for $j > 0$, and so (NG, d_0) is a chain subcomplex of (G, ∂).

Theorem 3.16. *Let G be a simplicial abelian group. Then the inclusion*

$$(NG, d_0) \to (G, \partial)$$

induces an isomorphism on homology

$$H_*((NG, d_0) \cong H_*(G, \partial).$$

Proof. Let

$$C_n^k = \left\{ x \in G_n \mid d_j x = 0 \text{ for } \max\{1, n-k+1\} \leq j \leq n \right\}.$$

Let $C^k = \{C_n^k\}_{n \geq 0} \subseteq G$. If $k = 0$, then $C_n^k = G_n$ and so $C^0 = \{C_n^0\}_{n \geq 0} = G$. If $n \leq k$, then $C_n^k = N_n G$.

First we check that each C^k is a chain subcomplex of (G, ∂). Namely we need to show that $\partial_n(C_n^k) \subseteq (C_{n-1}^k)$. Let $x \in C_n^k$. If $n \leq k$, then $C_n^k = N_n G$ and so

$$\partial(C_n^k) = \partial(N_n G) \subseteq N_{n-1} G \subseteq C_{n-1}^k.$$

Thus we may assume that $n > k$. Then $d_i x = 0$ for $n - k + 1 \leq i \leq n$. For $0 \leq i \leq n - k$ and $n - k \leq j \leq n - 1$, we have

$$d_j(d_i x) = d_i d_{j+1} x = 0.$$

Thus $d_i x \in C_{n-1}^k$ for each $0 \le i \le n - k$ and so

$$\partial(x) = \sum_{i=0}^{n}(-1)^i d_i x = \sum_{i=0}^{n-k}(-1)^i d_i x \subseteq C_{n-1}^k.$$

Now we have a sequence of chain subcomplexes of (G, ∂)

$$NG = \bigcap_{k=0}^{\infty} \subseteq \cdots \subseteq C^{k+1} \subseteq C^k \subseteq \cdots C^0 = G.$$

We prove that the inclusion $C^{k+1} \to C^k$ induces an isomorphism

$$H_*(C^{k+1}) \cong H_*(C^k)$$

for each $k \ge 0$. Let $i^k \colon C^{k+1} \to C^k$ be the inclusion. Define

$$\phi_n^k \colon C_n^k \to C_n^{k+1}$$

by $\phi_n^k = \mathrm{id}_{N_n G}$ if $n \le k$, and for $k < n$

$$\phi_n^k(x) = x - s_{n-k-1} d_{n-k} x$$

for $x \in C_n^k$. We check that $x - s_{n-k-1} d_{n-k} x \in C_n^{k+1}$ and $\phi^k = \{\phi_n^k\}$ is a chain map, that is $\partial_n \circ \phi_n^k = \phi_{n-1}^k \circ \partial_n$. If $n \le k$, this is true because $\phi_n^k = \mathrm{id}_{N_n G}$ in this case. If $n = k+1$, then $d_j x = 0$ for $j \ge 2$ for $x \in C_{k+1}^k$. Thus $d_j s_0 d_1 x = 0$ for $j \ge 2$ and

$$\begin{aligned}
\partial_{k+1} \phi_{k+1}^k(x) &= \partial(x) - \partial(s_0 d_1 x) \\
&= d_0 x - d_1 x - (d_1 x - d_1 x) \\
&= \partial_{k+1}(x) \\
&= \phi_k^k \partial_{k+1}(x).
\end{aligned}$$

For $n > k + 1$, we have

$$d_j(s_{n-k-1} d_{n-k} x) = \begin{cases}
s_{n-k-1} d_{j-1} d_{n-k} x = 0 & \text{if } j > n - k \\
d_{n-k} x & \text{if } j = n - k \\
d_{n-k} x & \text{if } j = n - k - 1 \\
s_{n-k-2} d_j d_{n-k} x = s_{n-2} d_{n-k-1} d_j x & \text{if } j \le n - k - 2.
\end{cases}$$

Thus $\phi_n^k(x) = x - s_{n-k-1}d_{n-k}x \in C_n^{k+1}$ and

$$\partial_n \phi_n^k(x) = \sum_{j=0}^{n}(-1)^j d_j x - \sum_{j=0}^{n}(-1)^j d_j(s_{n-k-1}d_{n-k}x)$$

$$= \sum_{j=0}^{n-k}(-1)^j d_j x - \sum_{j=0}^{n-k-2}(-1)^j s_{n-k-2}d_j d_{n-k}x$$

$$= \sum_{j=0}^{n-k}(-1)^j d_j x - \sum_{j=0}^{n-k-2}(-1)^j s_{n-k-2}d_{n-k-1}d_j x$$

$$= \sum_{j=0}^{n-k}(-1)^j d_j x - \sum_{j=0}^{n-k}(-1)^j s_{n-k-2}d_{n-k-1}d_j x$$

$$\text{because } d_{n-k-1}d_{n-k-1} = d_{n-k-1}d_{n-k}$$

$$= \sum_{j=0}^{n-k}(-1)^j(d_j x - s_{n-k-2}d_{n-k-1}d_j x)$$

$$= \phi_{n-1}^k \partial_n(x)$$

Thus $\phi^k \colon C^k \to C^{k+1}$ is a chain map with

$$\phi^k \circ i^k = \mathrm{id}_{C^{k+1}} \colon C^{k+1} \longrightarrow C^{k+1}.$$

Now we are going to construct a chain homotopy. Define

$$\Phi_n^k \colon C_n^k \to C_{n+1}^k$$

by setting

$$\Phi_n^k(x) = \begin{cases} 0 & \text{for } n < k \\ (-1)^{n-k}s_{n-k}x & \text{for } n \geq k \end{cases}$$

for $x \in C_n^k$. Then, for $n > k$ and $x \in C_n^k$,

$$\partial_{n+1}\Phi_n(x) = (-1)^{n-k}\sum_{j=0}^{n+1}(-1)^j d_j s_{n-k}x$$

$$= (-1)^{n-k}\sum_{j=0}^{n+1-k}(-1)^j d_j s_{n-k}x$$

$$= (-1)^{n-k} \sum_{j=0}^{n-k-1} (-1)^j s_{n-k-1} d_j x$$

$$= -s_{n-k-1} d_{n-k} x - \left((-1)^{n-k-1} \sum_{j=0}^{n-k} (-1)^j s_{n-k-1} d_j x \right)$$

$$= -s_{n-k-1} d_{n-k} x - \Phi_{n-1} \partial_n(x)$$

$$= i^k \phi^k(x) - \mathrm{id}(x) - \Phi_{n-1} \partial_n(x).$$

For $n = k$ and $x \in C_n^k = N_n G$,

$$\partial_{n+1} \Phi_n(x) = \partial_{n+1}(s_0 x) = d_0 s_0 x - d_1 s_0 x = 0 = i^k \phi^k(x) - \mathrm{id}(x) - \Phi_{n-1} \partial_n(x).$$

For $n < k$, clearly $\partial_{n+1} \Phi_n(x) = i^k \phi^k(x) - \mathrm{id}(x) - \Phi_{n-1} \partial_n(x)$. Thus there is a chain homotopy

$$i^k \phi^k \simeq \mathrm{id} \colon C^k \longrightarrow C^k$$

and so

$$H_n(C^{k+1}) \cong H_n(C^k)$$

for all $n, k \geq 0$.

Finally since $C_n^k = N_n G$ for each $n \leq k$, we have

$$H_q(NG) = H_q(C^{q+1}) \cong H_q(C^q) \cong H_q(C^{q-1}) \cong \cdots \cong H_q(C^0) = H_q(G, \partial)$$

for each q and hence the result. $\qquad\square$

3.2.2. *Hurewicz Theorem*

Let X be a simplicial set. Let $\mathbb{Z}(X) = \{\mathbb{Z}(X_n)\}_{n \geq 0}$ be the sequence of the free abelian group generated by X_n. Note that the faces $d_i \colon X_n \to X_{n-1}$ and degeneracies $s_i \colon X_n \to X_{n+1}$ extend linearly to the faces $d_i^{\mathbb{Z}(X)} \colon \mathbb{Z}(X_n) \to \mathbb{Z}(X_{n-1})$ and degeneracies $s_i^{\mathbb{Z}(X)} \colon \mathbb{Z}(X_n) \to \mathbb{Z}(X_{n+1})$. Thus $\mathbb{Z}(X)$ is a simplicial abelian group. The integral homology of X is defined by

$$H_*(X) = \pi_*(\mathbb{Z}(X)) \cong H_*(\mathbb{Z}(X), \partial). \tag{3.2}$$

Observe that any simplicial map $f \colon X \to Y$ induces a simplicial homomorphism $Z(f) \colon \mathbb{Z}(X) \to \mathbb{Z}(Y)$. Thus $X \mapsto \mathbb{Z}(X)$, $f \mapsto \mathbb{Z}(f)$ gives a functor from the category of simplicial sets to the category of simplicial abelian groups. If $f \simeq g \colon X \to Y$ (assuming that Y is fibrant), then $\mathbb{Z}(f) \simeq \mathbb{Z}(g) \colon \mathbb{Z}(X) \to \mathbb{Z}(Y)$. Thus if $X \simeq Y$ with X and Y fibrant, then $\mathbb{Z}(X) \simeq$

$\mathbb{Z}(Y)$ and so $H_*(X) \cong H_*(Y)$. Since the geometric realization of a simplicial set is a Δ-complex, from the definition, the homology $H_*(X) = H_*(|X|)$ is the simplicial homology of the Δ-complex $|X|$. Thus, if $|X| \simeq |Y|$, then $H_*(X) \cong H_*(Y)$.

Let G be abelian group. Then the homology $H_*(X; G)$ with coefficients in G is defined by

$$H_*(X; G) = \pi_*(\mathbb{Z}(X) \otimes_{\mathbb{Z}} G) \cong H_*(\mathbb{Z}(X) \otimes_{\mathbb{Z}} G, \partial), \qquad (3.3)$$

where $\mathbb{Z}(X) \otimes_{\mathbb{Z}} G = \{\mathbb{Z}(X_n) \otimes_{\mathbb{Z}} G\}_{n \geq 0}$.

For having reduced homology, we can add a single relation on $\mathbb{Z}(X)$. Let $\mathbb{Z}[X]$ be the quotient simplicial group of $\mathbb{Z}(X)$ modulo the simplicial subgroup generated by the basepoint $*$. Then the reduced integral homology $\bar{H}_*(X)$ is defined by

$$\bar{H}_*(X) = \pi_*(\mathbb{Z}[X]) \cong H_*(\mathbb{Z}[X], \partial). \qquad (3.4)$$

Similarly the reduced homology with coefficients in G is defined by

$$\bar{H}_*(X; G) = \pi_*(\mathbb{Z}[X] \otimes_{\mathbb{Z}} G) \cong H_*(\mathbb{Z}[X] \otimes_{\mathbb{Z}} G, \partial). \qquad (3.5)$$

The inclusion $j \colon X \to \mathbb{Z}(X)$ is a simplicial map and the composite

$$\bar{j} \colon X \overset{j}{\hookrightarrow} \mathbb{Z}(X) \longrightarrow \mathbb{Z}[X]$$

is pointed simplicial map. (**Note.** The basepoint in X is $*$ and the basepoint in $\mathbb{Z}(X)$ is 0.) The map $\bar{j}$ induces a group homomorphism

$$h_n = \bar{j}_* \colon \pi_n(X) \longrightarrow \pi_n(\mathbb{Z}[X]) = \bar{H}_n(X) \qquad (3.6)$$

for any fibrant simplicial set X and $n \geq 1$, called *Hurewicz homomorphism*.

Theorem 3.17 (Hurewicz Theorem). *Let X be a fibrant simplicial set with $\pi_i(X) = 0$ for $i < n$ with $n \geq 2$. Then $\bar{H}_i(X) = 0$ for $i < n$ and*

$$h_n \colon \pi_n(X) \to \bar{H}_n(X)$$

is an isomorphism.

Proof. We may assume that X is a minimal simplicial set. Consider

$$\bar{j} \colon X \to \mathbb{Z}[X].$$

Since X is minimal, we have $X_q = *$ for $q < n$ and $X_n = \pi_n(X)$. Thus $\mathbb{Z}[X]_q = \{0\}$ for $q < n$ and $\mathbb{Z}[X]_n = \mathbb{Z}[X_n]$.

(1). h_n *is onto*: From the commutative diagram

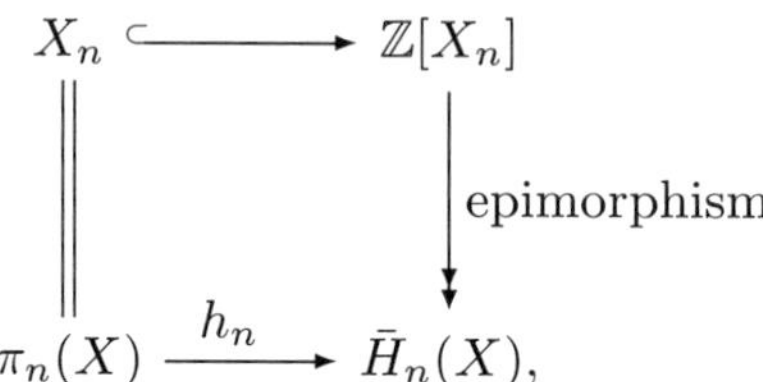

$\bar{H}_n(X)$ is generated by X_n as a group. Since h_n is a group homomorphism, it must be onto because every generator of $\bar{H}_n(X)$ lies in its image.

(2). $\mathrm{Ker}(h_n) = \{0\}$: Let $x \in \mathrm{Ker}(h_n)$. Since x is 0 in

$$\bar{H}_n(X) = H_n(\mathbb{Z}[X], \partial),$$

there exists an element $c \in \mathbb{Z}[X]_{n+1}$ such that $\partial(y) = x$ in $\mathbb{Z}[X]_n$. Let

$$y = \sum_{j=1}^{t} n_j y_j$$

with $n_j \in \mathbb{Z}$ and $y_j \in X_{n+1}$. Let $\phi \colon \mathbb{Z}[X_n] \to \pi_n(X)$ be the group homomorphism such that $\phi|_{X_n} \colon X_n \to \pi_n(X) = X_n$ is the identity map. Consider the commutative diagram

$$
\begin{array}{ccc}
\mathbb{Z}[X_{n+1}] & \xrightarrow{\ \partial\ } & \mathbb{Z}[X_n] \\
\Big\uparrow{\scriptstyle \bar{j}_{n+1}} & & \Big\uparrow{\scriptstyle \bar{j}_n}\ \ \searrow{\scriptstyle \phi} \\
X_{n+1} & & X_n = \!=\!= \pi_n(X).
\end{array}
\tag{3.7}
$$

For each $y_j \in X_{n+1}$, we have

$$
\begin{aligned}
\phi \circ \partial(y_j) &= \phi\left(\sum_{i=0}^{n+1}(-1)^i d_i y_j\right) \\
&= \sum_{i=0}^{n+1}(-1)^i \phi(d_i y_j) \\
&= \sum_{i=0}^{n+1}(-1)^i d_i y_j \qquad \text{because each } d_i y_j \in X_n
\end{aligned}
$$

in $\pi_n(X)$. By Homotopy Addition Theorem, $\phi(\partial(y_j)) = 0$ for each j. It follows that

$$x = \phi \circ \bar{j}_n(x) = \phi(\partial y) = \sum_{j=1}^{t} n_j \phi(\partial y_j) = 0.$$

in $\pi_n(X)$. The proof is finished. $\qquad\square$

A simplicial set X is called *connected* if $\pi_0(|X|) = 0$.

Theorem 3.18 (Poincaré Theorem). *Let X be a connected fibrant simplicial set. Then there is an isomorphism*

$$h' \colon \pi_1(X)/[\pi_1(X), \pi_1(X)] \xrightarrow{\ \cong\ } \bar{H}_1(X)$$

induced by $h_1 \colon \pi_1(X) \to \bar{H}_1(X)$.

Proof. The proof is similar to the proof of Hurewicz Theorem. We can choose X to be a minimal simplicial set. Then, following the same lines of the above theorem, h' is onto. For showing that h' is one-to-one, let

$$\phi \colon \mathbb{Z}[X_1] \longrightarrow \pi_1(X)/[\pi_1 X, \pi_1 X]$$

be the group homomorphism such that $\phi|_{X_1} \colon X_1 = \pi_1(X) \to \pi_1(X)/[\pi_1 X, \pi_1 X]$ is the quotient map. Change Diagram (3.7) to be

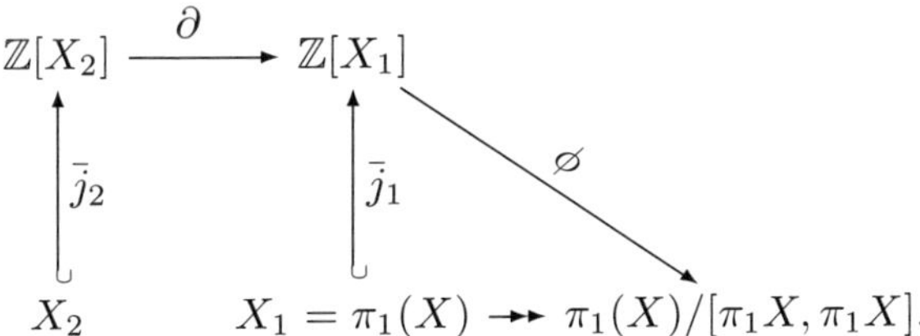

$$\mathbb{Z}[X_2] \xrightarrow{\ \partial\ } \mathbb{Z}[X_1]$$

with vertical maps $\bar{j}_2$ and $\bar{j}_1$, diagonal map ϕ,

$$X_2 \qquad X_1 = \pi_1(X) \twoheadrightarrow \pi_1(X)/[\pi_1 X, \pi_1 X].$$

Then the arguments follow the same lines in the proof of the Hurewicz Theorem. $\qquad\square$

3.2.3. *Group homology*

Let G be a monoid. Recall that WG is the simplicial set given by

$$(WG)_n = \{(g_0, g_1, \ldots, g_n) \mid g_i \in G\} = G^{n+1}$$

with faces and degeneracies given by

$$d_i(g_0, g_1, \ldots, g_n) = \begin{cases} (g_0, g_1, \ldots, g_{i-1}, g_i g_{i+1}, g_{i+2} \ldots, g_n) & \text{if } i < n \\ (g_0, \ldots, g_{n-1}) & \text{if } i = n \end{cases}$$

$$s_i(g_0, g_1, \ldots, g_n) = (g_0, g_1, \ldots, g_i, e, g_{i+1} \ldots, g_n).$$

Let R be a commutative ring and let M be an $R(G)$-module. Then we have the simplicial abelian group

$$R(WG) \otimes_{R(G)} M.$$

The homology of G with coefficients in M is then defined by

$$H_*(G; M) = \pi_*(R(WG) \otimes_{R(G)} M) \cong H_*(R(WG) \otimes_{R(G)} M, \partial).$$

3.3. *Free group constructions*

Given a set S, let $F(S)$ denote the free group generated by S. Namely every element in $F(S)$ is given by the words (as the formal products on the generators given by the set S):

$$x_1^{\epsilon_1} x_2^{\epsilon_2} \cdots x_n^{\epsilon_n}$$

for $n \geq 0$, $\epsilon_i = \pm 1$ and $x_i \in S$, with the equivalence relation given by

$$x_1^{\epsilon_1} x_2^{\epsilon_2} \cdots x_{i-1}^{\epsilon_{i-1}} x_i^{\epsilon_i} x_i^{-\epsilon_i} x_{i+1}^{\epsilon_{i+1}} \cdots x_n^{\epsilon_n} \sim x_1^{\epsilon_1} x_2^{\epsilon_2} \cdots x_{i-1}^{\epsilon_{i-1}} x_{i+1}^{\epsilon_{i+1}} \cdots x_n^{\epsilon_n}.$$

Namely the equivalence relation is obtained from $x_i x_i^{-1} = 1$ and $x_i^{-1} x_i = 1$. See the book of Magnus-Karrass-Solitar [23] for combinatorial group theory.

A pointed set S means a set S with a basepoint $*$. Denote by $F[S]$ the free group generated by S subject to the relation $* = 1$. A relation between $F(S)$ and $F[S]$ is that $F[S]$ is a quotient group of $F(S)$, and, as a group, $F[S] = F(S \smallsetminus \{*\})$.

3.3.1. *Kan's construction*

A *simplicial free group* means a simplicial group G such that each G_n is a free group. Kan's construction [21] is given as follows. Recall that a simplicial set X is called *reduced* if $X_0 = *$, namely X has only one vertex.

Definition 3.19 (Kan's construction). For a reduced simplicial set X, let GX be the simplicial group defined by

(1). $(GX)_n$ is the quotient group of X_{n+1} subject to the relations:

$$s_0 x = 1$$

for every $x \in X_n$. Thus, as a group, $(GX)_n$ is the free group generated by $X_{n+1} \smallsetminus s_0(X_n)$; or $(GX)_n = F[X_{n+1}/s_0(X_n)]$.

(2). The face and degeneracy operators are the group homomorphisms such that

$$d_0^{GX} x = (d_1 x)(d_0 x)^{-1},$$
$$d_i^{GX} x = d_{i+1} x \qquad \text{for } i > 0,$$
$$s_i^{GX} x = s_{i+1} x$$

for $x \in X_{n+1}$.

Lemma 3.20. *Let X be a reduced simplicial set. Then GX is a simplicial group.*

Proof. By definition, $(GX)_n$ is the quotient group of $F(X_{n+1})$ subject to the relation $s_0 x = 1$ for $x \in X_n$. First one needs to check that

$$d_i^{GX} : (GX)_n \to (GX)_{n-1} \quad s_i^{GX} : (GX)_n \to (GX)_{n+1}$$

are well-defined group homomorphisms for $0 \le i \le n$. This can be seen from the fact that, for $x \in X_n$, $0 < i \le n$ and $0 \le j \le n$,

$$d_0^{GX}(s_0 x) = (d_1 s_0 x)(d_0 s_0 x)^{-1} = 1$$
$$d_i^{GX}(s_0 x) = d_{i+1} s_0 x = s_0 d_i x$$
$$s_i^{GX}(s_0 x) = s_{i+1} s_0 x = s_0 s_i x.$$

Next we leave as an exercise for checking that the simplicial identities hold. $\qquad\square$

The main result of Kan's construction is the following theorem.

Theorem 3.21 (Kan). *Let X be a reduced simplicial group. Then the geometric realization*

$$|GX| \simeq \Omega|X|.$$

Proof. A proof can be found in [11, 24]. $\qquad\square$

The significance of this theorem is as follows: Given any path-connected CW-complex Y, we can choose a connected simplicial set X' such that $X' \simeq Y$. Let X be the path-connected component of X' of the basepoint. (This can be obtained by taking $X = P_0 X'$ using the Moore-Postnikov system.) Then X is a reduced simplicial set with $|X| \simeq |X'| \simeq Y$ because Y is connected. Apply Kan's construction to X. Then we have $|GX| \simeq \Omega Y$ and so GX is a simplicial group model for the loop space ΩY. In other words, any loop space of a path-connected CW-complex admits a simplicial group model.

3.3.2. *Milnor's construction*

Let X be a pointed simplicial set. Recall that the *reduced cone* CX is defined by setting

$$(CX)_n = \{(x,q) \mid x \in X_{n-q}, 0 \le q \le n\} \text{ with } (*,q) \text{ all identified to } *,$$

$$
\begin{aligned}
d_i(x,q) &= \begin{cases} (x, q-1) & \text{for } 0 \le i < q, \\ (d_{i-q}x, q) & \text{for } q \le i \le n, \end{cases} \\
s_i(x,q) &= \begin{cases} (x, q+1) & \text{for } 0 \le i < q, \\ (s_{i-q}x, q) & \text{for } q \le i \le n, \end{cases}
\end{aligned}
\tag{3.8}
$$

where for $x \in X_0$, $d_1(x,1) = *$. By identifying x with $(x,0)$, X is a simplicial subset of CX. The *reduced suspension* ΣX is defined by $\Sigma X = CX/X$.

Consider Kan's construction $G\Sigma X$ on the (reduced) suspension of X. From the definition, $(G\Sigma X)_n = F[(\Sigma X)_{n+1}/s_0((\Sigma X)_n)]$. We first work out the group $(G\Sigma X)_n$. From the definition of the reduced suspension above, $(x,0) = *$ in ΣX for any $x \in X$ and $s_0(x,q) = (x, q+1)$ for $q > 0$. Thus

$$(\Sigma X)_{n+1}/s_0((\Sigma X)_n) = \{(x,1) \mid x \in X_n\}$$

with $(*,1)$ as the basepoint. It follows that

$$(G\Sigma X)_n = F[(\Sigma X)_{n+1}/s_0((\Sigma X)_n)] = F[\{(x,1) \mid x \in X_n\}] = F[X_n].$$

Next we work out the face and degeneracy operators. From the definition of Kan's construction,

$$d_0^{G\Sigma X}(x,1) = d_1(x,1)d_0(x,1)^{-1} = (d_0x, 1)(x,0)^{-1} = (d_0x, 1).$$

For $i > 0$

$$d_i^{G\Sigma X}(x,1) = d_{i+1}(x,1) = (d_ix, 1)$$

and for $j \ge 0$

$$s_j^{G\Sigma X}(x,1) = s_{j+1}(x,1) = (s_jx, 1).$$

Let $F[X]$ denote $G\Sigma X$, called *Milnor's construction*, for any pointed simplicial set X. Then

(1). The group $F[X]_n = (G\Sigma X)_n = F[X_n]$ is the free group generated by X_n subject to the single relation that the basepoint $* = 1$. (Strictly speaking, the basepoint of X_n is $s_0^n *$.)

(2). The faces $d_i\colon F[X_n] \to F[X_{n-1}]$ is the group homomorphism induced by $d_i\colon X_n \to X_{n-1}$ and the degeneracies $s_i\colon F[X_n] \to F[X_{n+1}]$ is the group homomorphism induced by $s_i\colon X_n \to X_{n+1}$ for each $0 \le i \le n$.

Roughly speaking Milnor's construction is the free group generated by X with the basepoint to be 1. Since the geometric realization $|\Sigma X| \cong \Sigma|X|$, we obtain the following theorem.

Theorem 3.22. *Let X be any pointed simplicial set. Then*

$$|F[X]| \simeq \Omega\Sigma|X|.$$

Example 3.23. Let $X = S^n$ be the simplicial n-sphere. Then $|F[S^n]| \simeq \Omega S^{n+1}$. In particular,

$$\pi_q(F[S^n]) = \pi_q(\Omega S^{n+1}) = \pi_{q+1}(S^{n+1}).$$

3.3.3. *James' construction*

Replacing free groups by free monoids, Milnor's construction then becomes James' construction. In other words, the James' construction is the free monoid generated by a pointed simplicial set subject to the single relation that the basepoint is the identity. More precisely, let X be pointed simplicial set. Then the *James' construction* $J(X)$ is the simplicial monoid with

(1). The group $J(X)_n$ is the free monoid generated by X_n subject to the single relation that the basepoint $* = 1$. (Strictly speaking, the basepoint of X_n is $s_0^n *$.)

(2). The faces $d_i\colon J(X_n) \to J(X_{n-1})$ is the homomorphism of monoids induced by $d_i\colon X_n \to X_{n-1}$ and the degeneracies $s_i\colon J(X_n) \to J(X_{n+1})$ is the homomorphism of monoids induced by $s_i\colon X_n \to X_{n+1}$ for each $0 \leq i \leq n$.

The elements in $J(X_n)$ are given by the formal products

$$x_1 x_2 \cdots x_q$$

with $x_i \in X_n$ and $* = 1$. Observe that $J(X_n)$ is the sub monoid of $F[X_n]$ consisting of *positive words* on X_n. (A *positive word* on X_n means a word $x_1^{q_1} \cdots x_t^{q_t}$ with $x_i \in X_n$ and $q_i \geq 0$ for $1 \leq i \leq t$.) Thus the James construction $J(X)$ is a simplicial sub monoid of Milnor's construction $F[X]$ with the canonical inclusion

$$j\colon J(X) \hookrightarrow F[X].$$

Theorem 3.24. *Let X be a connected pointed simplicial set. Then the inclusion $j\colon J(X) \to F[X]$ induces a homotopy equivalence*

$$|j|\colon |J(X)| \longrightarrow |F[X]|.$$

Proof. A proof of this theorem can be found in [26], where Milnor denotes $F^+[X]$ for $J(X)$.

Another proof can be given sketched as follows. From the definition of construction, the geometric realization $|J(X)|$ is the geometric James' construction given as in [35, Section 2 of Chapter VII]. Then from [35, Theorem 2.6, p. 330], $|J(X)| \simeq \Omega\Sigma|X|$ under the assumption that X is connected.

Also one can directly show that the inclusion $j\colon J(X) \to F(X)$ induces an isomorphism on homology. It follows that $|j|_*\colon |J(X)| \to |F(X)|$ induces an isomorphism on homology. Under the assumption that X is connected, $|J(X)|$ and $|F(X)|$ are path-connected H-spaces. Then $|J(X)|$ and $|F(X)|$ are simple spaces. Since both $|J(X)|$ and $|F(X)|$ are CW-complexes, the map $|j|$ is a homotopy equivalence by the generalized Whitehead Theorem in [12, 15]. $\qquad\square$

Observe that a simplicial monoid may not be fibrant in general. The James' construction cannot be directly used for giving combinatorial computation for the homotopy groups. However the James' construction gives a good combinatorial model for the space $\Omega\Sigma|X|$ when X is connected. One of the important property is that $J(X)$ has the word-length filtration, called the *James' filtration*. More precisely, Let

$$J_k(X)_n = \{x_1 x_2 \cdots x_t \in J(X)_n \mid x_i \in X_n, \ t \leq k\}$$

be the subset of $J(X)_n$ consisting of words of product length at most k. Then

$$J_k(X) = \{J_k(X)_n\}_{n \geq 0}$$

forms a simplicial subset of $J(X)$ for each $k \geq 1$, with

$$J_1(X) = X \subseteq J_2(X) \subseteq \cdots \subseteq J_k(X) \subseteq J_{k+1}(X) \subseteq \cdots$$

and

$$J(X) = \bigcup_{k=1}^{\infty} J_k(X).$$

Let

$$X^{\wedge q} = X \wedge \cdots \wedge X$$

denote the q-fold self smash product of X.

Theorem 3.25. *Let X be any pointed simplicial set. Then there is an isomorphism of simplicial sets*

$$J_q(X)/J_{q-1}(X) \cong X^{\wedge q}$$

for each $q \geq 1$.

Proof. From the construction, $J_q(X)$ is the quotient of the q-fold Cartesian product

$$X^q = X \times \cdots \times X$$

by the equivalence relation

$$(x_1, \ldots, x_{i-1}, *, x_i, \ldots, x_{q-1}) \sim (x_1, \ldots, x_{j-1}, *, x_j, \ldots, x_{q-1})$$

for $1 \leq i, j \leq q$. Let $\pi \colon X^q \to J_q(X)$ be the quotient map. Then the composite

$$X^q \xrightarrow{\ \pi\ } J_q(X) \longrightarrow J_q(X)/J_{q-1}(X)$$

induces a simplicial isomorphism

$$X^{\wedge q} \xrightarrow{\ \cong\ } J_q(X)/J_{q-1}(X). \qquad \square$$

As a combinatorial model for $\Omega\Sigma Y$ for connected CW-complexes Y, the James' construction induces *James-Hopf maps* or *James-Hopf invariants* that can be combinatorially constructed by

$$H_k \colon J(X) \longrightarrow J(X^{\wedge k})$$

with

$$H_k(x_1 \cdots x_q) = \prod_{1 \leq i_1 < \cdots < i_k \leq q} (x_{i_1} \wedge x_{i_2} \wedge \cdots \wedge x_{i_k})$$

for $x_1, \ldots, x_q \in X$, where $(x_{i_1} \wedge x_{i_2} \wedge \cdots \wedge x_{i_k}) \in X^{\wedge k}$ and the order of the product is given by the right-lexicographic order. (See [35, Section 2 of Chapter VII, p. 334] for the James-Hopf maps. In simplicial case, we only need to check that the functions

$$H_k \colon J(X)_n \longrightarrow J(X^{\wedge k})_n$$

are well-defined with $d_i \circ H_k = H_k \circ d_i$ and $s_i \circ H_k = H_k \circ s_i$, that is, H_k is a well-defined simplicial map. Then its geometric realization is automatically continuous.)

By using the James-Hopf maps, one gets the following suspension splitting theorem.

Theorem 3.26 (James). *There is a homotopy equivalence*

$$\Sigma|J(X)| \simeq \bigvee_{k=1}^{\infty} \Sigma|X|^{\wedge k}$$

for any pointed simplicial set X. Thus

$$\Sigma\Omega\Sigma Y \simeq \bigvee_{k=1}^{\infty} \Sigma Y^{\wedge k}$$

for any path-connected CW-complex Y. $\qquad\qquad\qquad\qquad\qquad\square$

Further combinatorial information on the James construction can be found in [38, 39, 43, 44, 45, 22].

3.3.4. *The Moore chains of simplicial free groups*

Lemma 3.27 (Decomposition Theorem). *Let G be a simplicial group such that (1) each G_n is a free group and (2) $\pi_0(G)$ acts trivially on each $\pi_n(G)$. Then there is a decomposition*

$$N_n G/\mathcal{B}_n G \cong \pi_n(G) \times \mathcal{B}_{n-1} G$$

for each n.

Proof. From the definition, $\mathcal{B}_{n-1}G = d_0(N_n G)$ and there is a group extension

$$\mathcal{Z}_n G \ \longhookrightarrow\ N_n G \ \xrightarrow{\ d_0\ }\ \mathcal{B}_{n-1} G$$

and so a group extension

$$\pi_n(G) = \mathcal{Z}_n G/\mathcal{B}_n G \ \longhookrightarrow\ N_n G/\mathcal{B}_n G \ \xrightarrow{\ \bar{d}_0\ }\ \mathcal{B}_{n-1} G.$$

Since G_{n-1} is a free group, the subgroup $\mathcal{B}_{n-1}G$ is also a free group. Thus the epimorphism $\bar{d}_0\colon N_n G/\mathcal{B}_n G \to \mathcal{B}_{n-1} G$ admits a cross-section, that is, there is a group homomorphism $s\colon \mathcal{B}_{n-1} G \to N_n/\mathcal{B}_n G$ such tat the composite $\bar{d}_0 \circ s = \mathrm{id}_{\mathcal{B}_{n-1} G}$. Let

$$\phi\colon \mathcal{Z}_n G/\mathcal{B}_n G \times \mathcal{B}_{n-1} G \ \longrightarrow\ N_n G/\mathcal{B}_n G$$

be the function defined by

$$\phi(x, y) = x \cdot s(y).$$

Then ϕ is one-to-one and onto. By the central extension theorem, $\mathcal{Z}_n G/\mathcal{B}_n G$ is contained in the center of $G_n/\mathcal{B}_n G$. Thus $\mathcal{Z}_n G/\mathcal{B}_n G$ is contained in the center of $N_nG/\mathcal{B}_nG$. It follows that the function ϕ is a group homomorphism and hence the result. $\qquad\square$

Let G be a group. Write $G^{\mathrm{ab}} = G/[G, G]$ for the *abelianization* of the group G. Now given a simplicial group G, we have a chain complex of (non-abelian in general) groups (NG, d_0). Take the abelianization $(N_nG)^{\mathrm{ab}}$ of N_nG for each n, the differential

$$d_0 \colon N_nG \longrightarrow N_{n-1}G$$

induces a homomorphism

$$d_0^{\mathrm{ab}} \colon (N_nG)^{\mathrm{ab}} \longrightarrow (N_{n-1}G)^{\mathrm{ab}}$$

and so a chain complex of **abelian** groups $((NG)^{\mathrm{ab}}, d_0^{\mathrm{ab}})$. Since abelian groups are much easier than non-abelian groups, we may ask whether there is a connection between the homotopy groups $\pi_*(G)$ and the abelian chain complex $((NG)^{\mathrm{ab}}, d_0^{\mathrm{ab}})$. The following theorem gives an answer to this question.

Theorem 3.28 (Abelianization of the Moore chains). *Let G be a simplicial group such that (1) each G_n is a free group and (2) $\pi_0(G)$ acts trivially on each $\pi_n(G)$. Then there is a decomposition*

$$H_n((NG)^{\mathrm{ab}}) \cong \pi_n(G) \oplus A_n,$$

where A_n is a free abelian group.

This theorem is a new result. The significance of this theorem is that it is possible to systematically construct abelian complexes whose homology give the homotopy groups. This arises the chance to determine the homotopy groups of spaces. It should be pointed out that there are abelian chain complex model for rational homotopy groups such as Sullian model and Quillen model. But the most difficult and most challenging problem in algebraic topology is to determine the torsion component of the homotopy groups. As we see from the theorem that, the torsion component of $\pi_n(G)$ is actually the same as the torsion component of $H_n((NG)^{\mathrm{ab}})$ because A_n is torsion free. As we mentioned, this theorem is a new result. It is expected that the subsequent development of this result may produce new theory in algebraic topology towards to a possible solution to the homotopy groups of spaces, one of the most challenging problems in mathematics.

Proof. From the Decomposition Theorem, we have

$$N_nG/\mathcal{B}_nG \cong \mathcal{Z}_nG/\mathcal{B}_nG \times \mathcal{B}_{n-1}G = \pi_n(G) \times \mathcal{B}_{n-1}G. \qquad (3.9)$$

By taking the abelianization, we have the decomposition of abelian groups

$$(N_nG/\mathcal{B}_nG)^{\mathrm{ab}} \cong \mathcal{Z}_nG/\mathcal{B}_nG \oplus (\mathcal{B}_{n-1}G)^{\mathrm{ab}} = \pi_n(G) \oplus (\mathcal{B}_{n-1}G)^{\mathrm{ab}}, \quad (3.10)$$

where $(\mathcal{B}_{n-1}G)^{\mathrm{ab}}$ is a free abelian group because $\mathcal{B}_{n-1}G$ is a free group. From the short exact sequence

$$\mathcal{B}_nG \hookrightarrow N_nG \twoheadrightarrow N_nG/\mathcal{B}_nG,$$

there is a right exact sequence

$$(\mathcal{B}_nG)^{\mathrm{ab}} \longrightarrow (N_nG)^{\mathrm{ab}} \xrightarrow{\ q\ } (N_nG/\mathcal{B}_nG)^{\mathrm{ab}}.$$

Notice that the differential

$$d_0^{\mathrm{ab}} \colon (N_{n+1}G)^{\mathrm{ab}} \to (N_nG)^{\mathrm{ab}}$$

is induced by $d_0 \colon N_{n+1}G \to N_nG$. Thus $d_0^{\mathrm{ab}} \colon (N_{n+1}G)^{\mathrm{ab}} \to (N_nG)^{\mathrm{ab}}$ factors through $(\mathcal{B}_nG)^{\mathrm{ab}}$ and so the quotient map

$$q \colon (N_nG)^{\mathrm{ab}} \longrightarrow (N_nG/\mathcal{B}_nG)^{\mathrm{ab}}$$

factors through the cokernel

$$\mathrm{Coker}(d_0^{\mathrm{ab}}) = \mathrm{Coker}(d_0^{\mathrm{ab}} \colon (N_{n+1}G)^{\mathrm{ab}} \to (N_nG)^{\mathrm{ab}}).$$

Let

$$p \colon (N_nG)^{\mathrm{ab}} \longrightarrow \mathrm{Coker}(d_0^{\mathrm{ab}} \colon (N_{n+1}G)^{\mathrm{ab}} \to (N_nG)^{\mathrm{ab}})$$

be the quotient. Then the composite

$$N_nG \twoheadrightarrow (N_nG)^{\mathrm{ab}} \twoheadrightarrow \mathrm{Coker}(d_0^{\mathrm{ab}})$$

restricted to the subgroup

$$\mathcal{B}_nG = \mathrm{Im}(d_0 \colon N_{n+1}G \to N_nG)$$

is trivial. Thus the quotient map p factors through $(N_nG/\mathcal{B}_nG)^{\mathrm{ab}}$. It follows that

$$\mathrm{Coker}(d_0^{\mathrm{ab}} \colon (N_{n+1}G)^{\mathrm{ab}} \to (N_nG)^{\mathrm{ab}}) \cong (N_nG/\mathcal{B}_nG)^{\mathrm{ab}}.$$

From Decomposition 3.10, the composite

$$\pi_n(G) \longrightarrow H_n((NG)^{\mathrm{ab}}) \hookrightarrow \mathrm{Coker}(d_0^{\mathrm{ab}} \colon (N_{n+1}G)^{\mathrm{ab}} \to (N_nG)^{\mathrm{ab}})$$

has a retraction. Thus there is a decomposition

$$H_n((NG)^{\mathrm{ab}}) \cong \pi_n(G) \oplus A_n,$$

where A_n is a subgroup of the free abelian group $(\mathcal{B}_{n-1}G)^{\mathrm{ab}}$. The assertion follows. $\qquad\square$

3.3.5. *The simplicial group $F[S^1] \simeq \Omega S^2$*

Let S^1 be the simplicial 1-sphere. The elements in S^1_n can be listed as follows.

$$
\begin{aligned}
S^1_0 &= \{*\}, \\
S^1_1 &= \{s_0*, \sigma\}, \\
S^1_2 &= \{s_0^2*, s_0\sigma, s_1\sigma\}, \\
S^1_3 &= \{s_0^3*, s_2 s_1\sigma, s_2 s_0\sigma, s_1 s_0\sigma\},
\end{aligned}
$$

and in general $S^1_{n+1} = \{s_0^{n+1}*, x_0, \ldots, x_n\}$, where $x_j = s_n \cdots \hat{s}_j \cdots s_0\sigma$. The face

$$
d_i \colon S^1_{n+1} = \{*, x_0, \ldots, x_n\} \longrightarrow S^1_n = \{*, x_0, \ldots, x_{n-1}\}
$$

is given by $d_i s_0^{n+1}* = s_0^n *$ and

$$
d_i x_j = d_i s_n \cdots \hat{s}_j \cdots s_0\sigma =
\begin{cases}
s_0^n * & \text{if } j = i = 0 \text{ or } i = j+1 = n+1 \\
x_j & \text{if } j < i \\
x_{j-1} & \text{if } j \geq i.
\end{cases}
$$

Similarly,

$$
s_i x_j = s_i s_n \cdots \hat{s}_j \cdots s_0\sigma =
\begin{cases}
x_j & \text{if } j < i \\
x_{j+1} & \text{if } j \geq i.
\end{cases}
$$

Consider the special case of Milnor's free group construction for S^1, $F[S^1]$ is a simplicial group model for ΩS^2. As a sequence of groups, the group $F[S^1]_n$ is the free group of rank n generated by $x_0, x_1, \ldots, x_{n-1}$ with faces as above. Tietze transformations may be used to change the basis of the free group $F[S^1]_{n+1}$, so as to reformulate the faces d_i in a canonical way.

Let $y_0 = x_0 x_1^{-1}, \ldots, y_{n-1} = x_{n-1} x_n^{-1}$ and $y_n = x_n$ in $F[S^1]_{n+1}$. Clearly $\{y_0, y_1, \ldots, y_n\}$ is a set of free generators for $F[S^1]_{n+1}$ with

$$
d_i y_j \ (0 \leq i \leq n+1, \ -1 \leq j \leq n)
$$

given by

$$
d_i y_j = d_i(x_j x_{j+1}^{-1}) =
\begin{cases}
y_j & \text{if } j < i-1 \\
1 & \text{if } j = i-1 \\
y_{j-1} & \text{if } j \geq i,
\end{cases}
\qquad
s_i y_j =
\begin{cases}
y_j & \text{if } j < i-1 \\
y_j y_{j+1} & \text{if } j = i-1 \\
y_{j+1} & \text{if } j \geq i,
\end{cases}
$$

$$
\tag{3.11}
$$

where $y_{-1} = (y_0 y_1 \cdots y_{n-1})^{-1}$ and in this formula $x_{n+1} = 1$. Under the generating system of y_j's, the faces d_i with $i > 0$ are projection maps in the sense that d_i sends y_{i-1} to 1 and other generators to the generators for $F[S^1]_n$ so as to retain the order. The first face d_0 differs from the others as d_0 sends y_0 to the product element $(y_0 y_1 \cdots y_{n-1})^{-1}$ and each other generator y_j to y_{j-1} for $F[S^1]_n$.

To describe all faces d_i systematically in terms of projections, consider the free group of rank n in the following way. Let $\hat{F}_{n+1}$ be the quotient of the free group $F(z_0, z_1, \ldots, z_n)$ subject to the single relation

$$z_0 z_1 \cdots z_n = 1.$$

Let $\hat{z}_j$ be the image of z_j in $\hat{F}_{n+1}$. The group $\hat{F}_{n+1}$ is written $\hat{F}(\hat{z}_0, \hat{z}_1, \ldots, \hat{z}_n)$ in case the generators $\hat{z}_j$ are used. Clearly

$$\hat{F}_{n+1} \cong F(\hat{z}_0, \hat{z}_1, \ldots, \hat{z}_{n-1})$$

is a free group of rank n. Define the faces d_i and degeneracies s_i on $\{\hat{F}_{n+1}\}_{n \geq 0}$ as follows:

$$d_i \hat{z}_j = \begin{cases} \hat{z}_j & \text{if } j < i \\ 1 & \text{if } j = i \\ \hat{z}_{j-1} & \text{if } j > i, \end{cases} \qquad s_i \hat{z}_j = \begin{cases} \hat{z}_j & \text{if } j < i \\ \hat{z}_i \hat{z}_{i+1} & \text{if } j = i \\ \hat{z}_{j+1} & \text{if } j > i. \end{cases} \qquad (3.12)$$

It is straightforward to check that the sequence of groups $\hat{F} = \{\hat{F}_{n+1}\}_{n \geq 0}$ is a simplicial group under d_i and s_i defined as above. Let

$$\phi_n \colon \hat{F}_{n+1} \to F[S^1]_n$$

be the group homomorphism given by $\phi_n(\hat{z}_0) = (y_0 y_1 \cdots y_{n-1})^{-1}$ and $\phi_n(\hat{z}_j) = y_{j-1}$ for $1 \leq j \leq n$. It is routine to check part (1) of the following result. Part (2) follows from Milnor's Theorem that $|F[X]| \simeq \Omega \Sigma^{n+1} |X|$.

Proposition 3.29. *Let $\hat{F}$ be the simplicial group defined above. Then*

(1). *$\phi = \{\phi_n\} \colon \hat{F} \to F[S^1]$ is an isomorphism of simplicial groups; and*
(2). *the geometric realization of $\hat{F}$ is homotopy equivalent to ΩS^2.* $\square$

3.3.6. A combinatorial description of the general homotopy groups of S^2

The materials in this subsection are from [41]. So one could read [41] for details. The key point in this section is to determine the Moore chain complex $NF[S^1]$ using the generating system given in the previous subsection.

Definition 3.30. Let S be a set and let $T \subseteq S$ be a subset. The *projection homomorphism*

$$\pi \colon F(S) \to F(T)$$

is defined by

$$\pi(x) = \begin{cases} x & \text{if } x \in T, \\ 1 & \text{if } x \notin T. \end{cases}$$

Now let $\pi \colon F(S) \to F(T)$ be the projection homomorphism and let R be the kernel of π. Define subsets of the free group $F(S)$ as follows.

For $x \in S - T$ and y a reduced word in $F(T)$ define $\mu(x, y)$ by induction on the length of y:

$\mu(x, y) = x$ if y is the empty word;
$\mu(x, y) = [\mu(x, y'), z^\epsilon]$ if $y = y'z^\epsilon$ with $z \in T$ and $\epsilon = \pm 1$.

Let $\mathcal{A}_T$ be the set of $\mu(x, y)$. Let $\mathcal{B}_T$ be the set of $y^{-1}xy$ for $x \in S - T$ and $y \in F(T)$. By the classical Kurosch-Schreier theorem (see [23, Theorem 18.1]), we have the following.

Proposition 3.31. *The subgroup R is a free group freely generated by $\mathcal{B}_T$.*

In fact $\mathcal{A}_T$ is also a set of free generators for R.

Proposition 3.32. *The subgroup R is a free group freely generated by $\mathcal{A}_T$.*

Proof. The proof follows from the following two steps:

(a) Any finite part of $\mathcal{B}_T$ can be replaced, using Tietze moves (see [23]) by a finite part of $\mathcal{A}_T$. Thus $\mathcal{A}_T$ generates the subgroup R.

(b) In the same way any finite part of $\mathcal{A}_T$ can be replaced by a finite part of $\mathcal{B}_T$. Thus $\mathcal{A}_T$ freely generates R. $\qquad\square$

Now let us consider the intersection of kernels of projection homomorphisms. Let S be a set and let T_j be a subset of S for $1 \leq j \leq k$. Let $\pi_j \colon F(S) \to F(T_j)$ be the projection homomorphism for $1 \leq j \leq k$. We construct a subset $\mathcal{A}(T_1, \ldots, T_k)$ of $F(S)$ by induction on k as follows.

$$\mathcal{A}(T_1) = \mathcal{A}_{T_1},$$

where $\mathcal{A}_T$ has been defined above. Let

$$T_2^{(2)} = \{w \in \mathcal{A}(T_1) \mid w = [[x, y_1^{\epsilon_1}], \ldots], y_t^{\epsilon_t}] \text{ with } x, y_j \in T_2 \text{ for all } j\}$$

and define

$$\mathcal{A}(T_1, T_2) = \mathcal{A}(T_1)_{T_2^{(2)}}.$$

Suppose that $\mathcal{A}(T_1, T_2, \ldots, T_{k-1})$ is well-defined such that all of the elements in $\mathcal{A}(T_1, T_2, \ldots, T_{k-1})$ are written down as certain commutators in $F(S)$ in terms of elements in S. Let $T_k^{(k)}$ be the subset of $\mathcal{A}(T_1, T_2, \ldots, T_{k-1})$ defined by

$$T_k^{(k)} = \{w \in \mathcal{A}(T_1, T_2, \ldots, T_{k-1}) \mid w = [x_1^{\epsilon_1}, \ldots, x_l^{\epsilon_l}] \text{ with } x_j \in T_k \text{ for all } j\},$$

where $[x_1^{\epsilon_1}, \ldots, x_l^{\epsilon_l}]$ are the elements in $\mathcal{A}(T_1, T_2, \ldots, T_{k-1})$ that are written down as commutators. Then let $\mathcal{A}(T_1, T_2, \ldots, T_k)$ be defined by

$$\mathcal{A}(T_1, T_2, \ldots, T_k) = \mathcal{A}(T_1, T_2, \ldots, T_{k-1})_{T_k^{(k)}}.$$

Theorem 3.33. *Let S be a set and let T_j be a subset of S for $1 \leq j \leq k$. Let $\pi_j \colon F(S) \to F(T_j)$ be the projection homomorphism for $1 \leq j \leq k$. Then the intersection $\bigcap_{j=1}^{k} \mathrm{Ker}(\pi_j)$ is a free group freely generated by $\mathcal{A}(T_1, T_2, \ldots, T_k)$.*

Proof. The proof is given by induction on k. If $k = 1$, the assertion follows from Proposition 3.32. Suppose that $\bigcap_{j=1}^{k-1} \mathrm{Ker}(\pi_j) = F(\mathcal{A}(T_1, T_2, \ldots, T_{k-1}))$, and consider $\pi_k \colon F(S) \to F(T_k)$. Then

$$\bigcap_{j=1}^{k} \mathrm{Ker}(\pi_j) = \mathrm{Ker}(\bar{\pi}_k \colon F(\mathcal{A}(T_1, T_2, \ldots, T_{k-1})) \to F(T_k)),$$

where $\bar{\pi}_k$ is π_k restricted to the subgroup $F(\mathcal{A}(T_1, T_2, \ldots, T_{k-1}))$. Let $w = [x_1^{\epsilon_1}, \ldots, x_l^{\epsilon_l}] \in \mathcal{A}(T_1, T_2, \ldots, T_{k-1})$. If $w \notin T_k^{(k)}$, then $x_j \notin T_k$ for some j and $\bar{\pi}_k(w) = 1$. Thus $\bar{\pi}_k$ factors through $F(T_k^{(k)})$, that is, there is a homomorphism $j \colon F(T_k^{(k)}) \to F(T_k)$ such that $\bar{\pi}_k = j \circ \pi$, where $\pi \colon F(\mathcal{A}(T_1, T_2, \ldots, T_{k-1})) \to F(T_k^{(k)})$ is the projection homomorphism. We claim that $j \colon F(T_k^{(k)}) \to F(T_k)$ is a monomorphism. Consider the commutative diagram

$$
\begin{array}{ccccc}
F(\mathcal{A}(T_1, T_2, \ldots, T_{k-1})) & \xrightarrow{\;\pi_k\;} & F(T_k) & \hookrightarrow & F(S) \\
\uparrow & & \uparrow{\scriptstyle j} & & \uparrow \\
F(T_k^{(k)}) & \xrightarrow{\;=\;} & F(T_k^{(k)}) & \hookrightarrow & F(\mathcal{A}(T_1, T_2, \ldots, T_{k-1})),
\end{array}
$$

where $F(T_k^{(k)}) \to F(\mathcal{A}(T_1, T_2, \ldots, T_{k-1}))$ and $F(\mathcal{A}(T_1, T_2, \ldots, T_{k-1})) \to F(S)$ are inclusions of subgroups. Thus $j \colon F(T_k^{(k)}) \to F(T_k)$ is a monomorphism and

$$\mathrm{Ker}(\bar{\pi}_k) = \mathrm{Ker}(F(\mathcal{A}(T_1, T_2, \ldots, T_{k-1})) \to F(T_k^{(k)}) = F(\mathcal{A}(T_1, T_2, \ldots, T_k)).$$

The assertion follows. $\qquad\qquad\square$

We list these group-theoretical descriptions as follows. Let G be a group and let $[x, y] = x^{-1} y^{-1} xy$ in G.

Definition 3.34. A *bracket arrangement* of weight n in a group G is a map $\beta^n \colon G^n \to G$ which is defined inductively as follows:

$$\beta^1 = \mathrm{id}_G, \quad \beta^2(a_1, a_2) = [a_1, a_2]$$

for any $a_1, a_2 \in G$. Suppose that the bracket arrangements of weight k are defined for $1 \le k < n$ with $n \ge 3$. A map $\beta^n \colon G^n \to G$ is called a bracket arrangement of weight n if β^n is the composite

$$G^n = G^k \times G^{n-k} \xrightarrow{\beta^k \times \beta^{n-k}} G \times G \xrightarrow{\beta^2} G$$

for some bracket arrangements β^k and β^{n-k} of weight k and $n - k$, respectively, with $1 \le k < n$.

For instance, if $n = 3$, there are two bracket arrangements given by $[[a_1, a_2], a_3]$ and $[a_1, [a_2, a_3]]$.

Definition 3.35. Let G be a group and let S be a subset of G. Let $\langle S \rangle$ denote the normal subgroup generated by S. Let H_j be a sequence of subgroups of G for $1 \le j \le k$. Let $[[H_1, \ldots, H_k]]$ denote the subgroup of G generated by all of the commutators $\beta^t(h_{i_1}^{(1)}, \ldots, h_{i_t}^{(t)})$, where

(1). $1 \le i_s \le k$;
(2). all integers in $\{1, 2, \ldots, k\}$ appear as at least one of the integers i_s;
(3). $h_j^{(s)} \in H_j$;
(4). for each $t \ge k$, β^t runs over all of the bracket arrangements of weight t.

Corollary 3.36. *Let π_j be the projection homomorphisms as in Theorem 3.33. If S is the union of the sets T_j, then the intersection subgroup $\bigcap_{j=1}^k \mathrm{Ker}(\pi_j)$ equals the commutator subgroup $[[\langle T_1 \rangle, \ldots, \langle T_k \rangle]]$.*

Now according to the previous section, $F[S^1]_{n+1} = F(y_0, y_1, \ldots, y_n)$ with

$$d_i y_j \ (0 \le i \le n + 1, \ -1 \le j \le n)$$

given by

$$d_i y_j = d_i(x_j x_{j+1}^{-1}) = \begin{cases} y_j & \text{if } j < i-1 \\ 1 & \text{if } j = i-1 \\ y_{j-1} & \text{if } j \geq i, \end{cases} \qquad s_i y_j = \begin{cases} y_j & \text{if } j < i-1 \\ y_j y_{j+1} & \text{if } j = i-1 \\ y_{j+1} & \text{if } j \geq i, \end{cases}$$

$$(3.13)$$

where $y_{-1} = (y_0 y_1 \cdots y_{n-1})^{-1}$ and in this formula $x_{n+1} = 1$. Under the generating system of y_j's, the faces d_i with $i > 0$ are projection maps in the sense that d_i sends y_{i-1} to 1 and other generators to the generators for $F[S^1]_n$ so as to retain the order. Thus a set of generators for $N_{n+1}F[S^1]$ can be determined according to Theorem 3.33.

Corollary 3.37. *In* $F[S^1]$,

$$N_{n+1}F[S^1] = [[\langle y_0 \rangle, \langle y_1 \rangle, \ldots, \langle y_n \rangle]]$$

for each n. □

Since $\mathcal{B}_n F[S^1] = d_0(N_{n+1}F[S^1])$, a set of generator for $\mathcal{B}_n F[S^1]$ can be given by the image of d_0 of the elements in a set of generators for $N_{n+1}F[S^1]$. Notice that $d_0 y_j = y_{j-1}$. We have the following.

Corollary 3.38. *In* $F[S^1]$,

$$\mathcal{B}_n F[S^1] = [[\langle y_{-1} \rangle, \langle y_0 \rangle, \ldots, \langle y_{n-1} \rangle]]$$

for each n. □

Together with the central extension theorems of the first section of this chapter, we have the following theorem.

Theorem 3.39. *The homotopy group* $\pi_{n+2}(S^2)$ *is isomorphic to the center of*

$$F(y_0, y_1, \ldots, y_n)/[[\langle y_{-1} \rangle, \langle y_0 \rangle, \ldots, \langle y_n \rangle]]$$

for $n \geq 1$. □

Note that the subgroup $[[\langle y_{-1} \rangle, \langle y_0 \rangle, \ldots, \langle y_n \rangle]]$ is generated by the commutators

$$\beta^t(a_1^{\epsilon_1}, a_2^{\epsilon_2}, \ldots, a_t^{\epsilon_t}),$$

where

(1). $\epsilon_j = \pm 1$;
(2). $a_s \in \{y_{-1}, y_0, y_1, \ldots, y_n\}$;

(3). all elements in $\{y_{-1}, y_0, y_1, \ldots, y_n\}$ appear as at least one of the a_j;

(4). for each $t \geq n+2$, β^t runs over all of the commutator bracket arrangements of weight t.

Thus the group

$$F(y_0, y_1, \ldots, y_n)/[[\langle y_{-1}\rangle, \langle y_0\rangle, \ldots, \langle y_n\rangle]]$$

can be combinatorially given in the sense that the generators are $y_0, \ldots, y_n$ with specific defining relations given as above. So the significance of the above theorem is that the general homotopy groups of S^2 can be given as the centers of certain combinatorially defined groups with specific regenerators and systematic relations.

References

1. M. G. Barratt and P. J. Eccles, Γ^+-*structures-II: Recognition principle for infinite loop spaces*, Topology **13** (1974), 25–45.

2. J. A. Berrick, F. R. Cohen, Y. L. Wong and J. Wu, *Braids, configurations and homotopy groups*, J. Amer. Math. Soc. **19** (2006), 265–326.

3. J. S. Birman, *Braids, Links, and Mapping Class Groups*, Annals of Math. Studies **82** Princeton University Press (1975).

4. A. K. Bousfield and E. B. Curtis, *A spectral sequence for the homotopy of nice spaces*, Trans. Amer. Math. Soc. **151** (1970), 457–479.

5. A. K. Bousfield, E. B. Curtis, D. M. Kan, D. G. Quillen, D. L. Rector and J. W. Schlesinger, *The mod-p lower central series and the Adams spectral sequence*, Topology **5** (1966), 331–342.

6. G. E. Bredon, *Topology and Geometry*, Graduate Texts Math. **139** Springer (1993).

7. G. Carlsson, *A simplicial group construction for balanced products*, Topology **23** (1985), 85–89.

8. F. R. Cohen, *On combinatorial group theory in homotopy*, Contemp. Math. **188** (1995), 57–63.

9. F. R. Cohen and J. Wu, *On braids, free groups and the loop space of the 2-sphere. Categorical decomposition techniques in algebraic topology (Isle of Skye, 2001)*, Prog. Math. **215** Birkhäuser, Basel (2004), 93–105.

10. F. R. Cohen and J. Wu, *On braid groups and homotopy groups*, Geometry and Topology, to appear.

11. E. B. Curtis, *Simplicial homotopy theory*, Advances in Math. (1971), 107–209.

12. E. Dror, *A Generalization of the Whitehead Theorem*, Lecture Notes in Math. **249** Springer-Verlag, Berlin and New York (1971), 13–22.

13. S. Eilenberg and N. Steenrod, *Foundations of Algebraic Topology*, Princeton University Press (1952).

14. Z. Feidorowicz and J.-L. Loday, *Crossed simplicial groups and their associated homology*, Trans. Amer. Math. Soc. **326** (1991), 57–87.

15. S. M. Gersten, *Shorter notes: The Whitehead Theorem for nilpotent spaces*, Proc. AMS **47** (1975), 259–260.

16. P. G. Goerss and J. F. Jardine, *Simplicial Homotopy Theory*, Progress in Mathematics **174** Birkhäuser-Verlag (1999).

17. A. Hatcher, *Algebraic Topology*, Cambridge University Press (2002).

18. P. J. Hilton and U. Stammbach, *A Course in Homological Algebra*, GTM **4** Springer-Verlag (1971).

19. M. W. Hirsch, *Differential Topology*, Graduate Texts Math. **33** Springer (1976).

20. D. Kan, *A combinatorial definition of homotopy groups*, Ann. of Math. **67** (1958), 288–312.

21. D. Kan, *On homotopy theory and c.s.s. groups*, Ann. of Math. **68** (1958), 38–53.

22. J. Li and J. Wu, *Artin Braid groups and homotopy groups*, Proc. London Math. Soc., to appear.

23. W. Magnus, A. Karrass and D. Solitar, *Combinatorial group theory*, Pure and Applied Mathematics **XIII** (1966).

24. J. P. May, *Simplicial Objects in Algebraic Topology*, Van Nostrand (1967) (reprinted by University of Chicago Press).

25. J. W. Milnor, *The geometric realization of a semi-simplicial complex*, Ann. of Math. (2) **65** (1957), 357–362.

26. J. Milnor, *On the construction $F[K]$*, Algebraic Topology — A Student Guide, by J. F. Adams, Cambridge Univ. Press (1972), 119–136.

27. J. C. Moore, *Homotopie des complexes monöideaux*, Seminaire Henri Cartan (1954-55).

28. J. R. Munkres, *Elements of Algebraic Topology*, Addison-Wesley Publishing Company (1984).

29. D. G. Quillen, *The geometric realization of a Kan fibration is a Serre fibration*, Proc. Amer. Math. Soc. **19** (1968), 1499–1500.

30. D. G. Quillen, *Homotopical Algebra*, Lect. Notes in Math. **43** Springer-Verlag (1967).

31. B. J. Sanderson, *The simplicial extension theorem*, Math. Proc. Camb. Phil. Soc. **77** (1975), 497–498.

32. P. S. Selick and J. Wu, *On natural decompositions of loop suspensions and natural coalgebra decompositions of tensor algebras*, Memoirs AMS **148** (2000), No. 701.

33. J. H. Smith, *Simplicial group models for $\Omega^n \Sigma^n X$*, Israel J. Math. **66** (1989), 330–350.

34. E. H. Spanier, *Algebraic Topology*, McGraw-Hill (1966).

35. G. W. Whitehead, *Elements of Homotopy Theory*, Graduate Texts in Mathematics **61** Springer-Verlag (1978).

36. J. H. C. Whitehead, *On the asphericity of regions in a 3-sphere*, Fund. Math. **32** (1939), 149–166.

37. J. Wu, *On Combinatorial Descriptions of Homotopy Groups and the Homotopy Theory of mod 2 Moore Spaces*, Ph. D. Thesis of the University of Rochester (1995).

38. J. Wu, *On fiberwise simplicial monoids and Milnor-Carlsson's constructions*, Topology **37** (1998), 1113–1134.

39. J. Wu, *On combinatorial calculations for the James-Hopf maps*, Topology **37** (1998), 1011–1023.

40. J. Wu, *On products on minimal simplicial sets*, J. Pure and Appl. Algebra **148** (2000), 89–111.

41. J. Wu, *Combinatorial descriptions of the homotopy groups of certain spaces*, Math. Proc. Camb. Philos. Soc. **130** (2001), 489–513.

42. J. Wu, *A braided simplicial group*, Proc. London Math. Soc. **84** (2002), 645–662.

43. J. Wu, *Homotopy theory of the suspensions of the projective plane*, Memoirs AMS **162** (2003), No. 769.

44. J. Wu, *On maps from loop suspensions to loop spaces and the shuffle relations on the Cohen groups*, Memoirs AMS **180** (2006), No. 851.

45. J. Wu, *The EHP sequences for $(p-1)$-cell complexes and the functor $A^{\min}$*, to appear.

46. C. S. Ding, H. Niederreiter and C. P. Xing, *Some new codes from algebraic curves*, IEEE Trans. Inform. Theory, **46** (2000), 2638–2642.

47. H. Niederreiter and C. P. Xing, *Rational Points on Curves over Finite Fields: Theory and Applications*, Cambridge University Press, Cambridge (2001).

48. H. Stichtenoth, *Algebraic Function Fields and Codes*, Springer, Berlin (1993).

49. M. A. Tsfasman and S. G. Vlădut, *Algebraic-Geometric Codes*, Kluwer, Dordrecht (1991).

50. C. P. Xing and S. Ling, *A class of linear codes with good parameters from algebraic curves*, IEEE Trans. Inform. Theory **46** (2000), 1527–1532.

INTRODUCTION TO CONFIGURATION SPACES AND THEIR APPLICATIONS

Frederick R. Cohen

Department of Mathematics
University of Rochester
Rochester, NY 14627, USA
E-mail: cohf@math.rochester.edu

1. Introduction

This chapter, an introduction to the subject, develop basic, classical properties of configuration spaces as well as pointing out several natural connections between these spaces and other subjects. The main topics here arise from classical fibrations, homogeneous spaces, configuration spaces of surfaces, mapping class groups and loop spaces of configuration spaces, together with the relationships of these objects to simplicial groups and homotopy groups. Properties of the simplicial setting of homotopy groups are analogous to features of the Borromean rings or "Brunnian" links and braids.

The confluence of structures encountered here is within low-dimensional topology, as well as homotopy theory. These structures appear in a variety of contexts given by knots, links, homotopy groups and simplicial groups. Thus some homological consequences are developed, together with a description of how these results fit with linking phenomena in Sec. 20.

The structure of a simplicial group and Δ-group, which arise in the context of configuration spaces, are also described below. Connections to homotopy groups show how classical congruence subgroups arise in this context and coincide with certain natural subgroups of braid groups occurring in geometric group theory. These structures date back to the 1800's [48].

This chapter is intended as a short introduction to a few basic properties and applications of configuration spaces. Much excellent as well as beautiful work of many people on this subject has been deliberately omitted because of space and time restrictions. The author apologizes to many friends and colleagues for these omissions.

A more thorough development of configuration spaces is a book in preparation with Sam Gitler and Larry Taylor. One final remark: Secs. 16 through 21 give a revised version of notes in [21].

2. Basic Definitions

The first definition is that of the configuration space.

Definition 2.1. Let M denote a topological space. Define the *configuration space of ordered k-tuples of distinct points in M* as the subspace of M^k given by

$$\mathrm{Conf}(M, k) = \{(m_1, m_2, \ldots, m_k) \mid m_i \neq m_j \text{ for all } i \neq j\}.$$

The symmetric group on k-letters, Σ_k, acts on $\mathrm{Conf}(M, k)$ from the left by

$$\sigma(m_1, \ldots, m_k) = (m_{\sigma(1)}, \ldots, m_{\sigma(k)}).$$

One basic example is given next.

Example 2.2. This example gives classical properties of the configuration space of points in the plane $\mathbb{R}^2$ which is also regarded as the complex numbers $\mathbb{C}$. In this case, Artin's braid group with k strands, B_k, as well as Artin's pure braid group with k strands, P_k, defined in Sec. 8 below, arise naturally.

If $M = \mathbb{R}^2$, then $\mathrm{Conf}(\mathbb{R}^2, k)$ is a $K(P_k, 1)$ and $\mathrm{Conf}(\mathbb{R}^2, k)/\Sigma_k$ is a $K(B_k, 1)$ with proof first given in [32, 37] as well as sketched as Theorem 12.2 below. This example arises in the context of classical polynomials in one complex variable. Consider the space of unordered k-tuples of points in the complex numbers $\mathbb{C}^k/\Sigma_k$, a space well-known as the k-fold symmetric product. A point in $\mathbb{C}^k/\Sigma_k$ may be regarded as the set of roots $\{r_1, \ldots, r_k\}$, possibly repeated, of any monic, complex polynomial of degree k in one indeterminate z over $\mathbb{C}$.

There is a homeomorphism

$$\mathrm{Root} : \mathbb{C}^k/\Sigma_k \to \mathbb{C}^k$$

for which

$$\mathrm{Root}(\{r_1, \ldots, r_k\}) = p(z)$$

where

$$p(z) = \prod_{1 \leq i \leq k} (z - r_i).$$

Thus the space of complex polynomials $p(z) = z^k + a_{k-1}z^{k-1} + \cdots + a_1z + a_0$ is identified by this homeomorphism which sends the roots of $p(z)$, $\{r_1, \ldots, r_k\}$, to the point in $\mathbb{C}^k$ with coordinates $(a_{k-1}, \ldots, a_0)$, the coefficients of $p(z)$, where the a_j are given, up to sign, by the elementary symmetric functions in the r_i.

The subspace $\mathrm{Conf}(\mathbb{C}, k)/\Sigma_k$ of $\mathbb{C}^k/\Sigma_k$ is homeomorphic to the space of monic, complex polynomials $p(z) = z^k + a_{k-1}z^{k-1} + \cdots + a_1z + a_0$ for which $p(z)$ has exactly k distinct roots. The classical homeomorphism sends an equivalence class $[r_1, \ldots, r_k] \in \mathrm{Conf}(\mathbb{C}, k)/\Sigma_k$ to the polynomial

$$p(z) = \prod_{1 \leq i \leq k} (z - r_i).$$

Features of the inverse of this homeomorphism are one of the main topics in classical Galois theory. Further features of $\mathrm{Conf}(\mathbb{C}, k)/\Sigma_k$ concerning the homotopy groups of the 2-sphere will be addressed in Sec. 20.

Example 2.3. The configuration space $\mathrm{Conf}(\mathbb{R}^n, k)$ is homeomorphic to

$$\mathbb{R}^n \times \mathrm{Conf}(\mathbb{R}^n - Q_1, k - 1)$$

where Q_1 is the set with a single point given by the origin in $\mathbb{R}^n$. An extension of this fact is given in Example 2.6 below.

Notice that $\mathrm{Conf}(\mathbb{R}^n, 2)$ has S^{n-1} as a strong deformation retract with one choice of equivalence given by

$$A : S^{n-1} \to \mathrm{Conf}(\mathbb{R}^n, 2),$$

the map defined on points by the formula $A(z) = (z, -z)$ for z in S^{n-1}, where S^{n-1} is regarded as the points of unit norm in $\mathbb{R}^n$. A map

$$B : \mathrm{Conf}(\mathbb{R}^n, 2) \to S^{n-1}$$

is defined by the formula

$$B((x, y)) = \frac{(x - y)}{|x - y|}.$$

Observe that $B \circ A$ is the identity, and $A \circ B$ is homotopic to the identity via a homotopy leaving S^{n-1} point-wise fixed.

Example 2.4. Some features of configuration spaces for a sphere are listed next.

(1) The configuration space $\mathrm{Conf}(S^n, 2)$ is homotopy equivalent to S^n with one choice of equivalence given by the map

$$g : S^n \to \mathrm{Conf}(S^n, 2)$$

for which $g(z) = (z, -z)$.

(2) Let $\tau(S^n)$ denote the unit sphere bundle in the tangent bundle for S^n. There is a homotopy equivalence $E : \tau(S^n) \to \mathrm{Conf}(S^n, 3)$ defined on points (z, v) by

$$E(z, v) = (z, \exp(v), \exp(-v)).$$

Furthermore, this map is a fiber homotopy equivalence as implied by Theorem 3.2 below [32, 31].

(3) Properties of the tangent bundle and normal bundle for a smooth manifold M arise repeatedly in [20] where cofiber sequences are developed for configuration spaces given in terms of Thom spaces of associated normal bundles. Similar features arise in Totaro's spectral sequence [77].

Example 2.5. A classical fact is that the configuration space $\mathrm{Conf}(S^2, 3)$ is homeomorphic to $PGL(2, \mathbb{C})$, as for example, stated as Lemma 9.3 below. It follows that the configuration space $\mathrm{Conf}(S^2, k+3)$ is homeomorphic to the product $PGL(2, \mathbb{C}) \times \mathrm{Conf}(S^2 - Q_3, k)$ for all $k \geq 0$, for which Q_3 denotes a set of three distinct points in S^2. The group $PGL(2, \mathbb{C})$ has $SO(3)$ as a maximal compact subgroup and is thus homotopy equivalent to the real projective space $\mathbb{RP}^3$. This case was basic in [5, 11].

Example 2.6. If G is a topological group, then there is a homeomorphism

$$h : \mathrm{Conf}(G, k) \to G \times \mathrm{Conf}(G - \{1_G\}, k - 1)$$

where $h(g_1, \ldots, g_k) = (g_1, (g_1^{-1} g_2, \ldots, g_1^{-1} g_k))$. Thus if $k \geq 2$, there are homeomorphisms

$$\mathrm{Conf}(\mathbb{R}^2, k) \to \mathbb{R}^2 \times (\mathbb{R}^2 - \{0\}) \times \mathrm{Conf}(\mathbb{R}^2 - Q_2, k - 2)$$

where $Q_2 = \{0, 1\} \subset \mathbb{R}^2$ [32, 31, 15].

Natural variations are listed next. The first, a fiber-wise analogue of $\mathrm{Conf}(M, k)$ has been used to give certain natural $K(\pi, 1)$'s and to provide an application of a classical "incidence bundle" [11] to produce computations of the cohomology of certain discrete groups.

Definition 2.7. Let $\alpha\colon M \to B$ be any continuous map. The *fiber-wise configuration space*

$$\mathrm{Conf}_\alpha(M, k)$$

is the subspace of $\mathrm{Conf}(M, k)$ given by

$$\mathrm{Conf}_\alpha(M, k) = \{(m_1, \ldots, m_k) \mid m_i \neq m_j \text{ for all } i \neq j \text{ and}$$
$$\alpha(m_i) = \alpha(m_1) \text{ for all } i, j\}\,.$$

Given the projection map for a fiber bundle $\beta\colon E \to B$ with fiber X, define the *incidence bundle of k points in E* to be

$$\mathrm{Conf}_\beta(E, k).$$

There is a natural projection

$$\pi : \mathrm{Conf}_\beta(E, k) \to B$$

which, with mild restrictions given in the next example, is fibration with fiber $\mathrm{Conf}(X, k)$.

One example is listed next.

Example 2.8. Let G be a topological group which acts on the left of a space M and thus diagonally on $\mathrm{Conf}(M, k)$. Consider the Borel construction for $\mathrm{Conf}(M, k)$ given by

$$EG \times_G \mathrm{Conf}(M, k)$$

together with the natural projection maps

$$\gamma : EG \times_G M \to BG$$

and

$$\gamma_k : EG \times_G \mathrm{Conf}(M, k) \to BG.$$

Then the natural map

$$EG \times_G \mathrm{Conf}(M, k) \to \mathrm{Conf}_\gamma(EG \times_G M, k)$$

is a homeomorphism. If G is a compact Lie group, then the natural projection map

$$EG \times_G \mathrm{Conf}(M, k) \to BG$$

is a fibration with fiber $\mathrm{Conf}(M, k)$, as implied by [64]. A more general setting arises with the proof of Theorem 3.2 in Sec. 6.

In Secs. 8 through 11, we will use this example in the special case of

$$\eta : BSO(2) = ESO(3) \times_{SO(3)} S^2 \to BSO(3)$$

to obtain $K(\pi, 1)$'s closely connected to mapping class groups. In this case, the group

$$\pi = \pi_1(\mathrm{Conf}_\eta(BSO(2), k)/\Sigma_k)$$

is isomorphic to the group of path-components of the orientation preserving group of diffeomorphisms of S^2 which preserve a given set of k points [11], i.e., the mapping class group for a punctured 2-sphere.

A small modification to principal $U(2)$-bundles gives a $K(\pi, 1)$ for which π is the mapping class for genus two surfaces, see Example 9.10 below.

Example 2.9. Consider the natural action of $O(n)$ on S^{n-1}. Regard $O(k)$ as the subgroup of $O(n + k)$ given by $O(k) \times 1^n$, and $O(n)$ as the subgroup of $O(n + k)$ given by $1^k \times O(n)$. Let $V(n + k, k)$ denote the Stiefel manifold $O(n + k)/O(k)$ and $Gr(n + k, k)$ denote the Grassmann manifold $O(n + k)/O(k) \times O(n)$.

Consider

$$V(n + k, k) \times_{O(n)} S^{n-1}$$

as the total space of a fiber bundle with projection

$$\gamma : V(n + k, k) \times_{O(n)} S^{n-1} \to Gr(n + k, k)$$

and with fiber S^{n-1}. The associated incidence bundle is

$$\gamma_q : V(n + k, k) \times_{O(n)} \mathrm{Conf}(S^{n-1}, q) \to Gr(n + k, k)$$

with fiber $\mathrm{Conf}(S^{n-1}, q)$.

One result from [11] is that the "incidence bundle"

$$\bigcup_{k \to \infty} V(3 + k, k) \times_{O(3)} \mathrm{Conf}(S^2, q)/\Sigma_q$$

is a $K(\pi, 1)$ where π is a $\mathbb{Z}/2\mathbb{Z}$ extension of the mapping class group for a 2-sphere which has been punctured q times, for $q \geq 3$. The mapping class group for the punctured 2-sphere is the fundamental group of an analogous bundle where $Gr(3 + k, k)$ is replaced by $O(3 + k)/(O(k) \times SO(3))$, the subject of Sec. 9 here.

A third construction is given next.

Definition 2.10. Let Γ be a group which acts freely and properly discontinuously on the space M so that the projection map

$$M \to M/\Gamma$$

is the projection map in a principal Γ-bundle. Define the *orbit configuration space*

$$\mathrm{Conf}^{\Gamma}(M,k) = \{(m_1,\ldots,m_k) \mid m_i\Gamma \cap m_j\Gamma = \emptyset \text{ for all } i \neq j\}.$$

The group Γ^k acts on $\mathrm{Conf}^{\Gamma}(M,k)$ (from the left) by the formula

$$(\gamma_1,\ldots,\gamma_k)(m_1,\ldots,m_k) = (\gamma_1 \cdot m_1,\ldots,\gamma_k \cdot m_k).$$

Example 2.11. If S is a surface homeomorphic to $S^1 \times S^1$, then

$$S = \mathbb{R}^2/\Gamma$$

where $\Gamma = \mathbb{Z} \oplus \mathbb{Z}$ or any parametrized lattice in $\mathbb{C}$. If S is an orientable surface of genus g with $g > 1$, then S is the quotient of the upper half-plane $\mathbb{H}^2$ by the fundamental group of the surface. The spaces $\mathrm{Conf}^{\mathbb{Z}\oplus\mathbb{Z}}(\mathbb{R}^2,k)$ and $\mathrm{Conf}^{\Gamma}(\mathbb{H}^2,k)$ were studied in [85, 63, 14] where Γ is a Fuchsian group. In addition, further useful properties of $\mathrm{Conf}^{\Gamma}(M,k)$ were developed in [85].

3. Fibrations

Throughout this section M denotes a topological manifold with $\mathrm{Top}(M)$ the group of homeomorphisms of M, topologized with the compact-open topology. The purpose of this section is to introduce two theorems first proven in [32] with a small modification below in terms of homogeneous spaces as well as classifying spaces of certain homeomorphism groups.

In the special case for which M is a closed orientable surface, the theorems in this section are used in Secs. 8 through 11 below to construct certain natural $K(\pi,1)$'s associated to the mapping class group of a surface.

Definition 3.1. The group $\mathrm{Top}(M)$ acts diagonally on the configuration space $\mathrm{Conf}(M,k)$. Thus if $f : M \to M$ is an element in $\mathrm{Top}(M)$, then the action

$$\theta : \mathrm{Top}(M) \times \mathrm{Conf}(M,k) \to \mathrm{Conf}(M,k)$$

is defined by the formula

$$\theta((f,(m_1,\ldots,m_k))) = (f(m_1),f(m_2),\ldots,f(m_k)).$$

Let $Q_k = \{q_1, \ldots, q_k\}$ denote the underlying set of points obtained from a fixed point $\vec{q} = (q_1, \ldots, q_k)$ in $\mathrm{Conf}(M, k)$. The subgroup of elements in $\mathrm{Top}(M)$ which point-wise fixes the set Q_k is denoted $\mathrm{Top}(M, k)$ here. There is an induced map

$$\rho_{\vec{q}} : \mathrm{Top}(M)/\mathrm{Top}(M, k) \to \mathrm{Conf}(M, k)$$

defined by

$$\rho_{\vec{q}}(f) = (f(q_1), \ldots, f(q_k)).$$

The action specified by θ gives rise to natural fiber bundles as given in the next two theorems which are proven in Secs. 7 and 6, with original sources [32, 31].

Theorem 3.2. *Assume that M is a topological manifold without boundary.*

(1) *The group $\mathrm{Top}(M, k)$ acts on $\mathrm{Top}(M)$ by composition (from the left). The natural quotient map*

$$\mathrm{Top}(M) \to \mathrm{Top}(M)/\mathrm{Top}(M, k)$$

is the projection map for the principal $\mathrm{Top}(M, k)$-bundle

$$\mathrm{Top}(M, k) \to \mathrm{Top}(M) \to \mathrm{Top}(M)/\mathrm{Top}(M, k).$$

(2) *The induced map $\rho_{\vec{q}} : \mathrm{Top}(M)/\mathrm{Top}(M, k) \to \mathrm{Conf}(M, k)$ is a homeomorphism.*

(3) *The homotopy theoretic fiber of the natural map*

$$B\mathrm{Top}(M, k) \to B\mathrm{Top}(M)$$

is $\mathrm{Conf}(M, k)$, and, if $G = \mathrm{Top}(M)$, then $EG \times_G \mathrm{Conf}(M, k)$ is homeomorphic to $B\mathrm{Top}(M, k)$.

Observe that the natural projection maps

$$p_i : M^k \to M^{k-1}$$

which delete the i-th coordinate restrict to maps on the level of configuration spaces

$$p_i : \mathrm{Conf}(M, k) \to \mathrm{Conf}(M, k-1).$$

The second theorem is as follows [32, 31].

Theorem 3.3. *If M is a manifold without boundary, the natural projection map*

$$p_i : \mathrm{Conf}(M, k) \to \mathrm{Conf}(M, k-1)$$

is a fibration with fiber

$$M - Q_{k-1}.$$

Corollary 3.4. *Assume that M is a manifold without boundary, and $I = (i_1, \ldots, i_r)$ is a sequence of integers with $1 \leq i_1 < i_2 < \cdots < i_r \leq k$. Then the natural composite of projection maps $p_I = p_{i_1} \circ \cdots \circ p_{i_r}$ is a fibration*

$$p_I : \mathrm{Conf}(M, k) \to \mathrm{Conf}(M, k - r)$$

with fiber

$$\mathrm{Conf}(M - Q_{k-r}, r).$$

Remark 3.5. Several related remarks concerning features of Theorems 3.3 and 3.2 and their proofs are given next.

(1) Theorem 3.3 was stated and proven in a classical paper by Fadell-Neuwirth [32]. The result also follows from earlier work of R. Palais who was addressing a different question analogous to Theorem 3.2 [62]. Elegant further developments are in the book [31].

(2) The proofs below are those of [32, 31] with a small addition concerning principal fibrations.

(3) Additional hypothesis on M, such as $M = \mathbb{R}^n \times N$, imply that the collection of groups $\pi_1(\mathrm{Conf}(\mathbb{R}^n \times N, k))$, $k \geq 1$, form a simplicial group (see Sec. 16 for the concept of simplicial group). The group $\pi_1(\mathrm{Conf}(\mathbb{R}^n \times N, k))$ is regarded as the $(k - 1)$-st group in the simplicial group, see [5] and Sec. 16 below.

One basic example is given by the collection $\pi_1(\mathrm{Conf}(\mathbb{R}^2, k)) = P_k$, $k \geq 1$, a simplicial group denoted $AP_{\bullet}$ in Sec. 19. This simplicial group is closely tied to the homotopy groups of the 2-sphere [21].

For general M, the collection of fundamental groups $\pi_1(\mathrm{Conf}(M, k))$, $k \geq 1$, admit the structure of a Δ-group, with the Δ-structure induced by the projection maps $p_i : \mathrm{Conf}(M, k) \to \mathrm{Conf}(M, k - 1)$. This idea is developed in [5]; in the special case of $M = S^2$ there is a connection with the homotopy groups of the 2-sphere. These structures are also described in Sec. 16 below, as well as [83].

(4) In the case of the simplicial group $\{\pi_1(\mathrm{Conf}(\mathbb{R}^n \times N, k))\}$ for $M = \mathbb{R}^n \times N$, consider the kernel of the induced map on fundamental groups

$$\pi_1(p_k) : \pi_1(\mathrm{Conf}(M, k)) \to \pi_1(\mathrm{Conf}(M, k - 1)).$$

These kernels inherit the structure of a simplicial group isomorphic to Moore's simplicial loop space construction [61] applied to the simplicial group $\{\pi_1(\mathrm{Conf}(\mathbb{R}^n \times N, k))\}$. Thus, the natural projection maps in Theorem 3.3 also naturally give Moore's simplicial loop space for these simplicial groups, a point described in Secs. 19, 20, and 21 below.

The main feature here is that the fibers of the projection maps

$$p_i : \mathrm{Conf}(M, k) \to \mathrm{Conf}(M, k - 1)$$

give a precise topological analogue for the group-theoretic process of forming a simplicial loop space from a simplicial group. Thus these projection maps are informative for other subjects.

4. On Cross-sections for Configuration Spaces

The purpose of this section is to describe certain cross-sections for the projection maps $p_i : \mathrm{Conf}(M, k) \to \mathrm{Conf}(M, k - 1)$ when they exist for direct reasons. These maps are useful in what follows below.

Example 4.1. The first natural case is given by cross-sections for the projection maps

$$p_k : \mathrm{Conf}(M \times \mathbb{R}^n, k) \to \mathrm{Conf}(M \times \mathbb{R}^n, k - 1).$$

A section is specified by

$$\sigma_k : \mathrm{Conf}(M \times \mathbb{R}^n, k - 1) \to \mathrm{Conf}(M \times \mathbb{R}^n, k)$$

with

$$\sigma_k((m_1, r_1), \ldots, (m_{k-1}, r_{k-1})) = ((m_1, r_1), \ldots, (m_{k-1}, r_{k-1}), (m_1, L\vec{e}_1))$$

where

$$L = 1 + \max_i \|r_i\|$$

and $\vec{e}_1 = (1, 0, \ldots, 0)$.

Notice that a direct variation of this map applies to give sections for the projection maps

$$p_k : \mathrm{Conf}(\zeta, k) \to \mathrm{Conf}(\zeta, k - 1)$$

where ζ is an n-plane bundle over M which supports a nowhere vanishing cross-section $\sigma : M \to \zeta$ for the bundle projection $p : \zeta \to M$. See the next example.

Example 4.2. Define

$$\sigma_k : \mathrm{Conf}(\zeta, k - 1) \to \mathrm{Conf}(\zeta, k)$$

by the formula

$$\sigma_k((z_1,\ldots,z_{k-1})) = (z_1,\ldots,z_{k-1},\lambda\widehat{\sigma(z_1)})$$

where

$$\widehat{\sigma(z_1)} = \sigma(z_1)/|\sigma(z_1)|$$

and $\lambda = 1 + \max_i |\sigma(z_i)|$.

Next consider a manifold without boundary M, together with a fixed subset $Q \in M$ consisting of a single point.

Example 4.3. The projection maps

$$p_k : \mathrm{Conf}(M - Q, k) \to \mathrm{Conf}(M - Q, k - 1)$$

admit cross-sections up to homotopy [32].

Example 4.4. An example for which the projection map

$$p_3 : \mathrm{Conf}(M, 3) \to \mathrm{Conf}(M, 2)$$

does not admit a cross-section is given by $M = S^2$. Observe that $\mathrm{Conf}(S^2, 3)$ is homeomorphic to $PGL(2, \mathbb{C})$, a classical fact stated as Lemma 9.3 below. Since $PGL(2, \mathbb{C})$, $SO(3)$, and $\mathbb{RP}^3$ are homotopy equivalent, there does not exist a section for $p_3 : \mathrm{Conf}(S^2, 3) \to \mathrm{Conf}(S^2, 2)$ as $H_2(S^2) = H_2(\mathrm{Conf}(S^2, 2)) = \mathbb{Z}$, but $H_2(\mathrm{Conf}(S^2, 3)) = H_2(\mathbb{RP}^3) = \{0\}$.

Similarly, the projection maps $p_3 : \mathrm{Conf}(S^{2n}, 3) \to \mathrm{Conf}(S^{2n}, 2)$ do not admit sections for all $n > 0$ as $\mathrm{Conf}(S^{2n}, 3)$ is homotopy equivalent to the unit sphere bundle in the tangent bundle of S^{2n} with $H_{2n}(\mathrm{Conf}(S^{2n}, 3)) = \{0\}$, and $H_{2n}(\mathrm{Conf}(S^{2n}, 2)) = \mathbb{Z}$.

5. Preparation for Theorems 3.2 and 3.3

Lemma 5.1. *Assume that M is a non-empty manifold without boundary, of dimension at least 1. Then $\mathrm{Top}(M, k)$ is a closed subgroup of $\mathrm{Top}(M)$.*

Proof. Given any point f in the complement of $\mathrm{Top}(M, k)$ in $\mathrm{Top}(M)$, there is at least one point q_i in Q_k that is not fixed by f. Since M is Hausdorff, there is a non-empty open set U in M that does not contain q_i. An open set in the complement of $\mathrm{Top}(M, k)$ in $\mathrm{Top}(M)$ containing f is given by the set of continuous functions that carry the point q_i into U. The lemma follows. $\square$

The next definition is given in [76, p. 30].

Definition 5.2. Let H be a closed subgroup of a topological group G with natural quotient map $p : G \to G/H$. A *local cross-section of G in H* is a continuous function

$$f : U \to G$$

for U an open set in G/H such that

$$pf(x) = x$$

for every $x \in U$. To be more precise, such a continuous function f is also called a *local cross-section of G in H over the open set U*. A *local cross-section over a point $x \in G/H$* is a local cross-section of G in H over some open set U with $x \in U \subset G/H$.

The next theorem is given in Steenrod's book *The Topology of Fiber Bundles* [76, p. 30].

Theorem 5.3. *Let H be a closed subgroup of G and assume that the map $p : G \to G/H$ admits local cross-sections over every point $x \in G/H$. Then the projection map*

$$BH \to BG$$

is the projection map in a fiber bundle with fiber given by the space of left cosets G/H.

Remark 5.4. Steenrod's proof gives a homeomorphism

$$EG \times_G G/H \to BH$$

under the conditions of the theorem. A statement and proof of this fact is recorded next for completeness.

Theorem 5.5. *Let H be a closed subgroup of G and assume that the map $p : G \to G/H$ admits local cross-sections over every point $x \in G/H$. Then the natural map*

$$\pi : EG \times_G G/H \to BH$$

is a homeomorphism.

Proof. Let G/H denote the space of left cosets $\cup_{g \in G} gH$. Let X be any space that has a right G-action that is free and properly discontinuous, thus the projection $p : X \to X/G$ is a principal G-bundle. Notice that G acts on the product $X \times G/H$ via the formula

$$\gamma \cdot (x, g) = (x \cdot \gamma^{-1}, \gamma \cdot g).$$

Furthermore there is a natural map

$$q : X \times G/H \to X/H$$

defined by

$$q(x, gH) = [x\bar{g}],$$

the class of $x\bar{g}$ in X/H which is independent of the choice of $\bar{g} \in gH$. Since

$$q(x \cdot \gamma^{-1}, \gamma \cdot g) = q(x, g),$$

there is an induced map

$$\pi : X \times_G G/H \to X/H$$

defined by the equation $\pi([x, gH]) = q(x, gH)$.

A second map $\alpha : X \to X \times_G G/H$, defined by the equation

$$\alpha(x) = [x, 1H],$$

passes to quotients

$$\beta : X/H \to X \times_G G/H.$$

The composites $\pi \circ \beta$ and $\beta \circ \pi$ are both the identity. Thus the map

$$q : (X \times_G G/H) \to X/H$$

is a homeomorphism. Steenrod's theorem follows by setting $X = EG$ with the identification $BH = EG/H$. $\qquad\square$

Remark 5.6. Steenrod showed that the natural quotient map $\pi : G \to G/H$ is the projection in a principal fiber bundle in case H is a closed subgroup of G and the map has local sections [74]. Similarly, if the map π is the projection in a bundle, then it has local sections. Thus local sections are necessary as well as sufficient in order that $\pi : G \to G/H$ be the projection in a principal fiber bundle, in case H is a closed subgroup of G.

The proofs of Theorems 3.2 and 3.3 depend on the next lemma. Here, let D^n denote the n-disk, the points in $\mathbb{R}^n$ of Euclidean norm at most 1, with interior denoted $\overset{o}{D}{}^n$ and with the origin in D^n denoted $(0, 0, \ldots, 0)$. The map θ in the next lemma was useful in [32], the article by Fadell and Neuwirth, while the formula here is given explicitly in [85].

Lemma 5.7. (1) *There is a continuous map*

$$\theta : \overset{o}{D}{}^n \times D^n \to D^n$$

such that $\theta(x, -)$ fixes the boundary of D^n point-wise and

$$\theta(x, x) = (0, 0, \ldots, 0)$$

for every x in $\overset{o}{D}{}^n$.

(2) *If M is a topological manifold without boundary, then there exists a basis of open sets U for the topology of $\mathrm{Conf}(M, k)$ together with local sections $\phi : U \to \mathrm{Top}(M)$ such that the composite $U \to \mathrm{Top}(M) \to \mathrm{Top}(M)/\mathrm{Top}(M, k) \to \mathrm{Conf}(M, k)$ is a homeomorphism onto U.*

(3) *The natural map $\rho_{\vec{q}} : \mathrm{Top}(M)/\mathrm{Top}(M, k) \to \mathrm{Conf}(M, k)$ given by evaluation at a point $\vec{q} = (q_1, \ldots, q_k) \in \mathrm{Conf}(M, k)$ is a homeomorphism.*

Proof. Define

$$\alpha : \overset{o}{D}{}^n \to \mathbb{R}^n$$

by the formula $\alpha(x) = x/(1 - |x|)$, and so $\alpha^{-1}(z) = z/(1 + |z|)$. Let $\partial(D^n)$ denote the boundary of D^n.

For a fixed element q in $\overset{o}{D}{}^n$, define

$$\gamma_q : D^n \to D^n$$

by the formula

$$\gamma_q(y) = \begin{cases} y & \text{if } y \in \partial(D^n), \\ \alpha^{-1}\left(\dfrac{y}{1 - |y|} - \dfrac{q}{1 - |q|} \right) & \text{if } y \in \overset{o}{D}{}^n. \end{cases}$$

Define $\theta : \overset{o}{D}{}^n \times D^n \to D^n$ by the formula

$$\theta(q, y) = \gamma_q(y).$$

Notice that θ is continuous, and $\theta(q, q) = (0, 0, \ldots, 0)$. Thus part (1) of the lemma follows.

To prove part (2), consider a point $(q_1, q_2, \ldots, q_k)$ in $\mathrm{Conf}(M, k)$ together with disjoint open discs $\overset{o}{D}{}^n(q_1), \overset{o}{D}{}^n(q_2), \ldots, \overset{o}{D}{}^n(q_k)$ where $\overset{o}{D}{}^n(q_i)$ is a disc with center q_i. (There is a choice of homeomorphism in the identification of each such open disc with an open coordinate patch of M; this choice is suppressed here.) Let

$$U = \overset{o}{D}{}^n(q_1) \times \overset{o}{D}{}^n(q_2) \times \cdots \times \overset{o}{D}{}^n(q_k).$$

Observe that U is an open set in $\mathrm{Conf}(M, k)$ and that the sets U give a basis for the topology of $\mathrm{Conf}(M, k)$ as the $(q_1, q_2, \ldots, q_k)$ range over the points in $\mathrm{Conf}(M, k)$.

Define

$$\phi : U \to \mathrm{Top}(M)$$

by the formula

$$\phi((y_1, y_2, \ldots, y_k)) = H$$

for H in $\mathrm{Top}(M)$ where $(y_1, y_2, \ldots, y_k)$ is in $U = \overset{o}{D}{}^n(q_1) \times \overset{o}{D}{}^n(q_2) \times \cdots \times \overset{o}{D}{}^n(q_k)$, and H is the homeomorphism of M given as follows.

(1) $H(x) = x$ if x is in the complement in the disjoint union $\coprod_{1 \leq i \leq k} \overset{o}{D}{}^n(q_i)$, and
(2) $H(x) = \theta(q_i, x)$ if x is in $D^n(q_i)$.

Clearly H is in $\mathrm{Top}(M)$ as the two parts of the definition for H agree on the boundary of $D^n(q_i)$. To finish part (2) of the lemma, it suffices to check that ϕ is continuous. Notice that all spaces here are locally compact, and Hausdorff. Thus it follows that ϕ is continuous if and only if the adjoint

$$\mathrm{adj}(\phi) : U \times M \to M$$

defined by the formula

$$\mathrm{adj}(\phi)(u, m) = H(u, m)$$

is continuous. Then continuity of ϕ follows at once from the continuity of H. The second part of the lemma follows.

To finish the third part of the lemma, it must be checked that the natural map

$$\rho : \mathrm{Top}(M)/\mathrm{Top}(M, k) \to \mathrm{Conf}(M, k)$$

is a homeomorphism. Notice that part (2) of the lemma gives local sections

$$\phi : U \to \mathrm{Top}(M).$$

Thus consider the composite $\lambda : U \to \mathrm{Top}(M)/\mathrm{Top}(M,k)$ given by the composite $p \circ \phi$ where $p : G \to G/H$ is the natural quotient map. Notice that $\lambda : U \to \lambda(U)$ is a continuous bijection, and $\lambda(U) = \phi^{-1}(U)$. Thus $\lambda(U)$ is open, and the map ϕ is open. Thus ρ is open, and hence a homeomorphism. The lemma follows. $\qquad\qquad\square$

6. Proof of Theorem 3.2

By Lemma 5.1, $\mathrm{Top}(M,k)$ is a closed subgroup of $\mathrm{Top}(M)$. Furthermore, local sections exist for $\mathrm{Top}(M) \to \mathrm{Top}(M)/\mathrm{Top}(M,k)$ by Lemma 5.7. Thus there is a principal fibration

$$\mathrm{Top}(M,k) \to \mathrm{Top}(M) \to \mathrm{Top}(M)/\mathrm{Top}(M,k).$$

The first part of the theorem follows.

In addition, the natural evaluation map $\mathrm{Top}(M) \to \mathrm{Conf}(M,k)$ factors through the quotient map $\mathrm{Top}(M) \to \mathrm{Top}(M)/\mathrm{Top}(M,k)$. Thus, the induced map

$$\rho_{\vec{q}} : \mathrm{Top}(M)/\mathrm{Top}(M,k) \to \mathrm{Conf}(M,k)$$

is a homeomorphism by Lemma 5.7. Part 2 of the theorem follows.

The third statement in the theorem follows at once from Theorems 5.3 and 5.5.

7. Proof of Theorem 3.3

The statement to be proven is that if M is a manifold without boundary, then the natural projection map

$$p_i : \mathrm{Conf}(M,k) \to \mathrm{Conf}(M,k-1)$$

is a fibration with fiber homeomorphic to

$$M - Q_{k-1}$$

where p_i denotes the projection map which deletes the i-th coordinate. To prove that p_i is a fibration, it suffices to check that the map is locally trivial, by a theorem of Dold [73]. It suffices to check the result in case $i = k$ by applying the permutation which swaps i and k.

Consider the projection which deletes the last coordinate

$$p_k : \mathrm{Conf}(M, k) \to \mathrm{Conf}(M, k - 1).$$

The inverse image of the point $(q_1, \ldots, q_{k-1}) \in \mathrm{Conf}(M, k - 1)$, $p_k^{-1}((q_1, \ldots, q_{k-1}))$, is homeomorphic to $M - Q_{k-1}$ with $Q_{k-1} = \{q_1, \ldots, q_{k-1}\}$ because the natural inclusion

$$\iota_k : M - Q_{k-1} \to p_k^{-1}((q_1, \ldots, q_{k-1})),$$

defined by the equation $\iota_k(m) = (q_1, \ldots, q_{k-1}, m)$, is a homeomorphism.

Consider the point $(q_1, \ldots, q_{k-1})$ in $\mathrm{Conf}(M, k - 1)$ together with disjoint open discs $\overset{o}{D}{}^n(q_1), \ldots, \overset{o}{D}{}^n(q_{k-1})$ with centers q_i. Then consider the open set

$$V = \overset{o}{D}{}^n(q_1) \times \cdots \times \overset{o}{D}{}^n(q_{k-1}).$$

To show that p_k is locally trivial, we need to show that the following holds. Given any point $\vec{q} = (q_1, \ldots, q_{k-1})$ in $\mathrm{Conf}(M, k - 1)$, there is an open set V, containing $\vec{q}$, together with a homeomorphism

$$\Phi : V \times (M - Q_{k-1}) \to p_k^{-1}(V)$$

for which there is a commutative diagram

$$
\begin{array}{ccc}
V \times (M - Q_{k-1}) & \overset{\Phi}{\longrightarrow} & p_k^{-1}(V) \\
{\scriptstyle p}\downarrow & & \downarrow{\scriptstyle p_k} \\
V & \overset{1}{\longrightarrow} & V
\end{array}
$$

where p is the natural projection map.

We will now define Φ. The definition given is exactly that of [32] or [31] and is given by the formula

$$\Phi((m_1, \ldots, m_{k-1}), m_k)$$

$$= \begin{cases} (m_1, \ldots, m_{k-1}, m_k) & \text{if } m_k \notin \cup_{1 \le i \le k-1} D^n(q_i), \quad \text{and} \\ (m_1, \ldots, m_{k-1}, \theta(m_i, m_k)) & \text{if } m_k \in D^n(q_i) \end{cases}$$

where the map θ is that of Lemma 5.7. That Φ is a homeomorphism and the theorem then follows.

8. Surfaces, Braid Groups and Connections to Mapping Class Groups

The subject of this section is basic properties of braid groups of surfaces as well as their connections to mapping class groups. Throughout this section S denotes a surface, possibly open or possibly non-orientable. The definition of the braid group of a surface is given next.

Definition 8.1. Let S denote a surface.

(1) The *k-stranded braid group for S* is

$$B_k(S) = \pi_1(\mathrm{Conf}(S, k)/\Sigma_k).$$

(2) The *k-stranded pure braid group for S* is

$$P_k(S) = \pi_1(\mathrm{Conf}(S, k)).$$

(3) In case $S = \mathbb{R}^2$, let B_k, respectively P_k, denote $B_k(\mathbb{R}^2)$, respectively $P_k(\mathbb{R}^2)$.

Remark 8.2. Useful consequences of the definition of the (pure) braid groups rely heavily on the fact that S is a surface. In this case, the natural inclusion

$$i : \mathrm{Conf}(S, k) \to S^k$$

does not induce an isomorphism on the level of fundamental groups, a feature which has been proven to be quite useful. Vershinin has written an informative survey of braid groups of surfaces in [78].

However, in case N is a manifold of dimension at least 3, the natural inclusion

$$i : \mathrm{Conf}(N, k) \to N^k$$

does induce an isomorphism on fundamental groups. Thus to obtain interesting structures which are analogous to braid groups of surfaces in the case of manifolds of dimension at least 3, new constructions are required.

Constructions which provide non-trivial analogues of braid groups for any space M are defined in [19]. The definition arises by considering the structure of the space of "suitably compatible maps" $(S^1)^n \to \mathrm{Conf}(M, k)$ which provides an alternative definition of the braid group of a surface, and which extends in a natural way to give analogues of braid groups for any manifold M. These analogues of braid groups have their own version of Vassiliev invariants as well as other natural properties, and are reminiscent of Fox's torus homotopy groups [36], but with further global structure. One version is the group of pointed homotopy classes of pointed maps $[\Omega(S^2), \Omega(\mathrm{Conf}(M,k))]$, a group with tractable structure in case all spaces have been localized at the rational numbers. In the case of $[\Omega(S^2), \Omega(\mathrm{Conf}(\mathbb{R}^{2m}, k))]$ for $m > 1$, the "rationalization" of this group is isomorphic to the Malĉev completion of P_k [19].

Theorem 8.3. *If S is a surface not equal to either S^2 or $\mathbb{RP}^2$, and $Q_i = \{q_1, \ldots, q_i\}$ is a sub-set of S having cardinality i, possibly zero, then $\mathrm{Conf}(S - Q_i, k)$ and consequently $\mathrm{Conf}(S - Q_i, k)/\Sigma_k$ are $K(\pi, 1)$'s.*

Proof. Notice that S as well as $S - Q_i$ are both $K(\pi, 1)$'s. An induction on k using the fibrations in Theorem 3.3 implies that $\mathrm{Conf}(S - Q_i, k)$ is a $K(\pi, 1)$, as follows.

Since $S - Q_i$ is a surface not equal to either S^2 or $\mathbb{RP}^2$, it follows that $S - Q_i$ is a $K(\pi, 1)$. The inductive step is to observe that there is a fibration

$$\mathrm{Conf}(S - Q_i, k) \to S - Q_i$$

with fiber given by $\mathrm{Conf}(S - Q_{i+1}, k - 1)$, by Theorem 3.3. Since $\mathrm{Conf}(S - Q_{i+1}, k - 1)$ and $S - Q_i$ may be assumed to be $K(\pi, 1)$'s, it follows that $\mathrm{Conf}(S - Q_i, k)$ is also a $K(\pi, 1)$.

Furthermore, the natural quotient maps $\mathrm{Conf}(S, k) \to \mathrm{Conf}(S, k)/\Sigma_k$ are projections in a covering space. Thus, $\mathrm{Conf}(S, k)/\Sigma_k$ is also a $K(\pi, 1)$ and the theorem follows. □

Remark 8.4. In case S is the surface S^2 or $\mathbb{RP}^2$, then constructions of $K(B_k(S^2), 1)$ and $K(B_k(\mathbb{RP}^2), 1)$ are derived in Sec. 9. These spaces are given by total spaces of various natural choices of fiber bundles obtained from the natural $SO(3)$-actions on either S^2 or $\mathbb{RP}^2$.

One (classical) definition of the mapping class group for a closed, orientable Riemann surface S with fundamental group $\pi_1(S)$ is the group of outer-automorphisms $\mathrm{Out}(\pi_1(S))$ [55, p. 175]. An alternative definition is given by the group of path-components of the orientation preserving homeomorphisms of the surface. Useful variations are given next which are obtained by restricting to homeomorphisms which leave certain subspaces of S fixed.

Definition 8.5. Let S be a closed orientable surface of genus g with a given point $*$ in S.

(1) The mapping class group Γ_g is the group of path-components of the orientation preserving homeomorphisms of S, $\mathrm{Top}^+(S)$.

(2) The mapping class group Γ_g^k is the group of path-components of the orientation preserving homeomorphisms of S which leave a set of k distinct points Q_k in S invariant, and is equal to $\pi_0(\mathrm{Top}^+(S, k))$. The pure mapping class group $P\Gamma_g^k$ is the kernel of the natural homomorphism $\Gamma_g^k \to \Sigma_k$.

(3) The pointed mapping class group $\Gamma_g^{k,*}$ is the group of path-components of the orientation preserving homeomorphisms which (i) preserve the point $*$, and (ii) leave a set of k distinct points in $S - *$, invariant, $\mathrm{Top}^+(S, \{*\}, k)$. The pure pointed mapping class group $P\Gamma_g^{k,*}$ is the kernel of the natural homomorphism $\Gamma_g^{k,*} \to \Sigma_k$. We use $\mathrm{Top}^+(S, \{*\})$ to denote the group of orientation preserving homeomorphisms which leaves the point $*$ fixed.

(4) If, in addition, m disjoint disks are given in S together with k distinct points in the complement of the union of these disks, then define $\Gamma_{g,m}^k$ as the group of path-components of the orientation preserving homeomorphisms of S which leave the set of k distinct points invariant, as well as the boundaries of all m disks fixed point-wise.

Remark 8.6. Definition 8.5 provides a definition of the "pure pointed mapping class group" $P\Gamma_g^{k,*}$ and the "pointed mapping class group" $\Gamma_g^{k,*}$. The remarks here give some motivation for this variation as there are several natural applications as illustrated next.

First, consider the function spaces of all continuous maps $\mathrm{Map}(S, X)$ together with the subspace of pointed continuous maps $\mathrm{Map}_*(S, X)$ for a pointed space X. The homology of the space $\mathrm{Map}_*(S, X)$ is sometimes much more accessible than that of the "free mapping space" $\mathrm{Map}(S, X)$. One case in point is where S is S_g a closed, orientable surface of genus g, and X is S^{2L}, an even dimensional sphere. The homology of the function space $\mathrm{Map}_*(S_g, S^{2L})$ is easily accessible while that of $\mathrm{Map}(S_g, S^{2L})$ is complicated.

In these cases, the group $\mathrm{Top}^+(S_g)$ acts naturally on the function space $\mathrm{Map}(S_g, S^{2L})$. Consider the homotopy orbit space $E\mathrm{Top}^+(S_g) \times_{\mathrm{Top}^+(S_g)} \mathrm{Map}(S_g, S^{2L})$. There is a "pointed-version" given by

$$E\mathrm{Top}^+(S_g, \{*\}) \times_{\mathrm{Top}^+(S_g, \{*\})} \mathrm{Map}_*(S_g, S^{2L}).$$

One application is that the homology of all of the groups Γ_g^k are given at once by the homology of the space $E\mathrm{Top}^+(S_g) \times_{\mathrm{Top}^+(S_g)} \mathrm{Map}(S_g, S^{2L})$

with a degree shift depending on the choices of L and k [12]. The actual computations do not appear to be accessible even in the cases $g = 0, 1$.

An analogous result is satisfied for the homology of the "pointed version" given by the space $E\mathrm{Top}^+(S_g, \{*\}) \times_{\mathrm{Top}^+(S_g, \{*\})} \mathrm{Map}_*(S_g, S^{2L})$. In this case, the homology of all of the groups $\Gamma_g^{k,*}$ are given at once by the homology of this space with a degree shift depending on the choices of L and k [12]. However, these homology groups are much more accessible in the "pointed" case, as given in [12].

In the case of $g = 1$, the cohomology of this space was worked out and gives the cohomology of $\Gamma_1^{k,*}$ in terms of classical modular forms. This case is also addressed in Sec. 10.

A classical theorem concerning the homeomorphism group and diffeomorphism group of a closed orientable Riemann surface S_g of genus g is stated next [27, 28]. Let $\mathrm{Diff}^+(S_g)$ denote the group of orientation preserving diffeomorphisms. The next theorem follows from results proven in [27, 28]. That is, the group of path-components for $\mathrm{Diff}^+(S_g)$ and $\mathrm{Top}^+(S_g)$ are isomorphic, and the components of the identity have the same homotopy type. The author would like to thank Benson Farb for a late Saturday night e-mail conversation regarding this point.

Theorem 8.7. *The natural inclusion*

$$\mathrm{Diff}^+(S_g) \to \mathrm{Top}^+(S_g)$$

is a homotopy equivalence.

In what follows, the groups $\mathrm{Diff}^+(S_g)$ and $\mathrm{Top}^+(S_g)$ will be used in different ways in which these differences are stated explicitly. The groups $\mathrm{Top}(S_g)$ and $\mathrm{Top}^+(S_g)$ act on the configuration space of points in S_g, $\mathrm{Conf}(S_g, k)$, diagonally. A useful "folk theorem" gives (1) there are natural $K(\pi, 1)$'s obtained from the associated Borel construction (homotopy orbit spaces) for groups $\mathrm{Top}(M)$ acting on configuration spaces, and (2) these configuration spaces are analogous to homogeneous spaces in the sense that they are frequently homeomorphic to a quotient of a topological group by a closed subgroup.

Namely, let G be a subgroup of $\mathrm{Top}(M)$, and consider the diagonal action of G on $\mathrm{Conf}(M, k)$ together with the homotopy orbit spaces

$$EG \times_G \mathrm{Conf}(M, k)$$

and

$$EG \times_G \mathrm{Conf}(M, k)/\Sigma_k.$$

In case M is a surface, these constructions frequently give $K(\pi, 1)$'s where the group π is given by certain mapping class groups. Three different cases which depend on the genus of the surface are given in the next three sections.

A remark concerning orientations is listed next. Observe that S_g has an orientation reversing involution which leaves k points invariant. Consider the natural map

$$h : \mathrm{Top}(S_g) \to \mathrm{Aut}(H_2(S_g))$$

defined by

$$h(f) = f_* : H_2(S_g) \to H_2(S_g).$$

This map satisfies the following properties:

(1) The map h surjects to $\mathrm{Aut}(H_2(S_g)) = \mathbb{Z}/2\mathbb{Z}$.
(2) The map h restricts to a surjection $h|_{\mathrm{Top}(S_g,k)} : \mathrm{Top}(S_g, k) \to \mathrm{Aut}(H_2(S_g))$.
(3) The kernel of h is $\mathrm{Top}^+(S_g)$ while the kernel of $h|_{\mathrm{Top}(S_g,k)}$ is $\mathrm{Top}^+(S_g, k)$.
(4) Thus the natural map

$$\mathrm{Top}^+(S_g)/\mathrm{Top}^+(S_g, k) \to \mathrm{Top}(S_g)/\mathrm{Top}(S_g, k)$$

is a homeomorphism.

This point is recorded in the following lemma (which of course is a special case).

Lemma 8.8. *Assume that S_g is an orientable surface of genus g. The natural map*

$$\mathrm{Top}^+(S_g)/\mathrm{Top}^+(S_g, k) \to \mathrm{Top}(S_g)/\mathrm{Top}(S_g, k)$$

is a homeomorphism.

9. On Configurations in S^2

The natural actions of $SO(3)$ on S^2 by rotations, as well as on $\mathbb{RP}^2$ by rotation of a line through the origin in $\mathbb{R}^3$, are applied in this section to give $K(\pi, 1)$'s where the groups π are certain mapping class groups. We will also make use of the group S^3, the connected double cover of $SO(3)$, and its action on $\mathrm{Conf}(S^2, q)$ through the diagonal action of $SO(3)$ on products of S^2.

The purpose of this section is to derive properties of the configuration spaces $\mathrm{Conf}(S^2, k)$ as well as the spaces

$$ESO(3) \times_{SO(3)} \mathrm{Conf}(S^2, k)/\Sigma_k.$$

The first theorem in this direction is due to Smale who proved the following result [72].

Theorem 9.1. *The natural inclusions*

$$SO(3) \subset \mathrm{Diff}^+(S^2) \subset \mathrm{Top}^+(S^2)$$

are homotopy equivalences.

Smale's theorem has the following consequences as pointed out in [11, 6].

Theorem 9.2. *Assume that* $q \geq 3$.

(1) *The space*

$$ESO(3) \times_{SO(3)} \mathrm{Conf}(S^2, q)/\Sigma_q$$

  is a $K(\Gamma_0^q, 1)$.
(2) *The space*

$$ESO(3) \times_{SO(3)} \mathrm{Conf}(S^2, q)$$

  is a $K(P\Gamma_0^q, 1)$.

A slightly stronger version of Theorem 9.2 is obtained from the next classical lemma, proven by considering cross-ratios, for which S^2 is regarded as the space of complex lines though the origin in $\mathbb{C}^2$, namely $\mathbb{CP}^1$. Consider the evaluation map

$$e : PGL(2, \mathbb{C}) \times \mathrm{Conf}(\mathbb{CP}^1, 3) \to \mathrm{Conf}(\mathbb{CP}^1, 3)$$

defined by the equation

$$e(\alpha, (L_1, L_2, L_3)) = (\alpha(L_1), \alpha(L_2), \alpha(L_3))$$

where L_i are 3 fixed, distinct lines through the origin in $\mathbb{C}^2$ with (L_1, L_2, L_3) a point in $\mathrm{Conf}(\mathbb{CP}^1, 3)$ and α is an element in $PGL(2, \mathbb{C})$.

Lemma 9.3. *The restriction of the map*

$$e : PGL(2, \mathbb{C}) \times \mathrm{Conf}(\mathbb{CP}^1, 3) \to \mathrm{Conf}(\mathbb{CP}^1, 3),$$

to the subspace

$$PGL(2, \mathbb{C}) \times \{(L_1, L_2, L_3)\},$$

for any point (L_1, L_2, L_3) *in* $\mathrm{Conf}(\mathbb{CP}^1, 3)$, *is a homeomorphism. Thus* $PGL(2, \mathbb{C})$ *is homeomorphic to* $\mathrm{Conf}(\mathbb{CP}^1, 3)$.

A cruder version of Lemma 9.3 which exhibits a homotopy equivalence rather than a homeomorphism follows directly from Theorem 3.3 as follows:

Lemma 9.4. *Restrict the map*

$$e : SO(3) \times \mathrm{Conf}(\mathbb{CP}^1, 3) \to \mathrm{Conf}(\mathbb{CP}^1, 3)$$

defined by the equation

$$e(\alpha, (L_1, L_2, L_3)) = (\alpha(L_1), \alpha(L_2), \alpha(L_3))$$

to the subspace

$$SO(3) \times \{(L_1, L_2, L_3)\}$$

for any point (L_1, L_2, L_3) *in* $\mathrm{Conf}(\mathbb{CP}^1, 3)$. *This restriction is a homotopy equivalence.*

Proof. Since $SO(3)$ is path-connected, it suffices to check the lemma for the points in S^2 given by

(1) $L_1 = (1, 0, 0)$,
(2) $L_2 = (0, 1, 0)$, and
(3) $L_3 = (0, 0, 1)$.

The map

$$E : SO(3) \to \mathrm{Conf}(S^2, 3)$$

defined by the equation

$$E(\alpha) = (\alpha(L_1), \alpha(L_2), \alpha(L_3))$$

gives a morphism of fibrations

$$
\begin{array}{ccc}
SO(2) & \xrightarrow{\ E|SO(2)\ } & S^2 - Q_2 \\
\downarrow & & \downarrow \\
SO(3) & \xrightarrow{\ E\ } & \mathrm{Conf}(S^2, 3) \\
\downarrow & & \downarrow \\
S^2 & \xrightarrow{\ \beta\ } & \mathrm{Conf}(S^2, 2)
\end{array}
$$

where $Q_2 = \{L_1, L_2\}$ with $\beta : S^2 \to \mathrm{Conf}(S^2, 2)$ defined by

$$\beta(\alpha(L_1)) = (\alpha(L_1), \alpha(L_2)).$$

The induced maps on the fiber $E|SO(2) : SO(2) \to S^2 - Q_2$, as well as on the base, $\beta : S^2 \to \mathrm{Conf}(S^2, 2)$, are homotopy equivalences by inspection. Since the right-hand side is a fibration by Theorem 3.3, the lemma follows. $\qquad\square$

Lemma 9.5. *The spaces*

$$ESO(3) \times_{SO(3)} \mathrm{Conf}(S^2, 3),$$

and

$$EPGL(2, \mathbb{C}) \times_{PGL(2,\mathbb{C})} \mathrm{Conf}(S^2, 3)$$

are contractible.

Remark 9.6. The preceding lemma gives a special case of Theorem 9.2 for $q = 3$ in which the group π is the trivial group.

Proof of Lemma 9.5. It suffices to check that the space $ESO(3) \times_{SO(3)}$ $\mathrm{Conf}(S^2, 3)$ is contractible as $SO(3)$ is the maximal compact subgroup of $PGL(2, \mathbb{C})$ and the inclusion $SO(3) \subset PGL(2, \mathbb{C})$ is a homotopy equivalence.

Since the natural map

$$E : SO(3) \to \mathrm{Conf}(S^2, 3)$$

is a homotopy equivalence by the proof of Lemma 9.4, and is also $SO(3)$-equivariant by construction, the result follows as $ESO(3) \times_{SO(3)} SO(3)$ is contractible. $\qquad\square$

The proof of Theorem 9.2 is given next with an application to the co-homology of the associated mapping class groups [11].

Proof of Theorem 9.2. There are two steps to the proof of this theorem. The first is that if $q \geq 3$, then the resulting space $ESO(3) \times_{SO(3)} \mathrm{Conf}(S^2, q)$ is a $K(\pi, 1)$. The second step is to work out the fundamental group π of the space in question.

Assume that $q \geq 3$ and observe that the natural projection map to the first three coordinates

$$p(1, 2, 3) : \mathrm{Conf}(S^2, q) \to \mathrm{Conf}(S^2, 3)$$

is $SO(3)$-equivariant by inspection of the definitions. In addition, the map $p(1,2,3)$ is a fibration with fiber $\mathrm{Conf}(S^2 - Q_3, q - 3)$, by Theorem 3.3. Thus the induced projection map

$$ESO(3) \times_{SO(3)} \mathrm{Conf}(S^2, q) \to ESO(3) \times_{SO(3)} \mathrm{Conf}(S^2, 3)$$

is a fibration with fiber $\mathrm{Conf}(S^2 - Q_3, q - 3)$.

On the other hand, the space $ESO(3) \times_{SO(3)} \mathrm{Conf}(S^2, 3)$ is contractible by Lemma 9.5. Thus the inclusion $\mathrm{Conf}(S^2 - Q_3, q - 3) \subset ESO(3) \times_{SO(3)} \mathrm{Conf}(S^2, q)$ is a homotopy equivalence. Furthermore, the space $\mathrm{Conf}(S^2 - Q_3, q - 3)$ is a $K(\pi, 1)$ by Theorem 8.3 as $S^2 - Q_3$ is a surface which is not S^2 or $\mathbb{RP}^2$. Thus $\mathrm{Conf}(S^2 - Q_3, q - 3)$ and consequently $ESO(3) \times_{SO(3)} \mathrm{Conf}(S^2, q)$ are $K(\pi, 1)$'s.

The final step is to show that if $q \geq 3$, the spaces

$$ESO(3) \times_{SO(3)} \mathrm{Conf}(S^2, q),$$

and

$$ESO(3) \times_{SO(3)} \mathrm{Conf}(S^2, q)/\Sigma_q$$

are, respectively, $K(P\Gamma_0^q, 1)$, and $K(\Gamma_0^q, 1)$. Thus it suffices to work out their fundamental groups.

To carry out this step, it is useful to note that by Theorem 9.1, the natural inclusions

$$SO(3) \subset \mathrm{Diff}^+(S^2) \subset \mathrm{Top}^+(S^2)$$

are homotopy equivalences. Furthermore, by Theorem 3.2, there is a fibration sequence

$$\mathrm{Top}(S^2, q) \to \mathrm{Top}(S^2) \to \mathrm{Conf}(S^2, q) \to B\mathrm{Top}(S^2, q) \to B\mathrm{Top}(S^2).$$

Since the induced map

$$\mathrm{Top}^+(S^2)/\mathrm{Top}^+(S^2, q) \to \mathrm{Top}(S^2)/\mathrm{Top}(S^2, q)$$

is a homeomorphism by Lemma 8.8, it follows that there is a homotopy equivalence

$$ESO(3) \times_{SO(3)} \mathrm{Conf}(S^2, q) \to B\mathrm{Top}^+(S^2, q). \qquad \square$$

Remark 9.7. A similar argument, given in the thesis of J. Wong [80], gives the construction of $K(\pi, 1)$'s where π is the braid group of either S^2 or $\mathbb{RP}^2$. Furthermore, Wong used these spaces to work out the cohomology of the associated braid groups.

Theorem 9.8. *Assume that $q \geq 3$.*

(1) *The space $ES^3 \times_{S^3} \mathrm{Conf}(S^2, q)/\Sigma_q$ is a $K(B_q(S^2), 1)$. Furthermore, the space $ES^3 \times_{S^3} \mathrm{Conf}(S^2, q)$ is a $K(P_q(S^2), 1)$.*
(2) *The space $ES^3 \times_{S^3} \mathrm{Conf}(\mathbb{RP}^2, q)/\Sigma_q$ is a $K(B_q(\mathbb{RP}^2), 1)$. Furthermore, the space $ES^3 \times_{S^3} \mathrm{Conf}(\mathbb{RP}^2, q)$ is a $K(P_q(\mathbb{RP}^2), 1)$.*

In addition, an attractive presentation of the braid group $B_k(S^2)$, the fundamental group of $\mathrm{Conf}(S^2, k)/\Sigma_k$, is given in an elegant paper by Fadell and van Buskirk [33, 3].

Remark 9.9. The spaces $ES^3 \times_{S^3} \mathrm{Conf}(S^2, q)/\Sigma_q$ and their fundamental groups are intimately connected with mapping class groups. One connection is to a group $\Delta_{2g+2} \subset \Gamma_g^0$ called the hyper-elliptic mapping class group, which is the centralizer of a certain choice of involution of Γ_g^0. There is a central extension

$$1 \to \mathbb{Z}/2\mathbb{Z} \to \Delta_{2g+2} \to \Gamma_0^{2g+2} \to 1.$$

The group Δ_{2g+2} is the fundamental group of a bundle arising from complex 2-plane bundles and associated configuration space bundles, a point described in the next example.

Geometrically, a $K(\Gamma_2^0, 1)$ can be obtained by "twisting together" actions of $S^3 \times S^1$ to give an action of $U(2)$ on $\mathrm{Conf}(S^2, k) \times_{\Sigma_k} S^1$. This is addressed in the next example.

Example 9.10. The variation of the above construction for Δ_{2g+2} is described in this example as developed in [11]. This construction gives models for certain mapping class groups.

There are three ingredients here:

(1) The group $\pi = \mathbb{Z}/2\mathbb{Z}$ is the center of $SU(2)$ while $SO(3)$ is the quotient $SU(2)/\pi$. Furthermore $U(2)$ is a quotient of $SU(2) \times S^1$ obtained from this action as follows. Regard $\mathbb{Z}/2\mathbb{Z}$ as the central subgroup of $S^3 \times S^1$ generated by $(-1, -1)$. Then form the central quotient $SU(2) \times_{\mathbb{Z}/2\mathbb{Z}} S^1$. There is an isomorphism of groups

$$SU(2) \times_{\mathbb{Z}/2\mathbb{Z}} S^1 \to U(2),$$

a construction given in work of Atiyah, Bott, and Shapiro known as $Spin^c(3)$.
(2) Define an action of the product $S^3 \times S^1$ on $S^2 \times S^1$ by requiring that (i) S^3 act on S^2 through the natural action of rotations by $SO(3)$, and

(ii) the group S^1 acts on itself as follows:

$$(\alpha, \beta) = \alpha^2 \cdot \beta.$$

(3) The diagonal action of $S^3 \times S^1$ on $\mathrm{Conf}(S^2, k) \times_{\Sigma_k} S^1$ descends to an action of $U(2)$.

Consider the Borel construction of this action. One result of that is

$$EU(2) \times_{U(2)} \mathrm{Conf}(S^2, k) \times_{\Sigma_k} S^1$$

is a $K(\pi, 1)$. If $k = 6$, then this space is a $K(\Gamma_2, 1)$ [11]. Furthermore, if $k = 2g + 2$ for g even, then the fundamental group of $EU(2) \times_{U(2)} \mathrm{Conf}(S^2, k) \times_{\Sigma_k} S^1$ is the centralizer of the class of a hyper-elliptic involution in Γ_g which is denoted Δ_{2g+2}.

Consider the fibration

$$ESO(3) \times_{SO(3)} \mathrm{Conf}(S^2, q)/\Sigma_q \to BSO(3)$$

with fiber $\mathrm{Conf}(S^2, q)/\Sigma_q$. Then consider the long exact homotopy sequence for this fibration together with the fact that $\pi_1(BSO(3))$ is isomorphic to $\mathbb{Z}/2\mathbb{Z}$ to prove the following theorem [3].

Theorem 9.11. *If $q \geq 3$, then there is a central extension*

$$1 \to \mathbb{Z}/2\mathbb{Z} \to B_q(S^2) \to \Gamma_0^q \to 1.$$

A similar result applies to the spaces

$$EU(2) \times_{U(2)} \mathrm{Conf}(S^2, k) \times_{\Sigma_k} S^1.$$

Theorem 9.12. *If g is even with $g \geq 2$, there are central extensions*

(1) $1 \to \mathbb{Z}/2\mathbb{Z} \to \Gamma_2 \to \Gamma_0^6 \to 1$, *and*
(2) $1 \to \mathbb{Z}/2\mathbb{Z} \to \Delta_{2g+2} \to \Gamma_0^{2g+2} \to 1$.

Remark 9.13. Any such central extension is determined by a characteristic class. The characteristic class in the case of Theorem 9.12 naturally arises from a $Spin^c(3)$-structure implicit in the underlying topology of the diffeomorphism group [11].

10. On Configurations in $S^1 \times S^1$

A second special case arises with configuration spaces for surfaces of genus 1, the subject of this section. Since the upper half-plane is closely connected to the structure of this configuration space, as well as to the applications below, such as the structure of certain "Brunnian braid groups" as given in the Appendix, Sec. 23, some introductory information on it is listed next.

Let $\mathbb{H}$ denote the upper half-plane, the complex numbers with strictly positive pure imaginary part. The group $SL(2,\mathbb{Z})$ acts on $\mathbb{H}$ by fractional linear transformations where

$$M = \begin{pmatrix} a & b \\ c & d \end{pmatrix}$$

is in $SL(2,\mathbb{Z})$, and

$$M(z) = \frac{az + b}{cz + d}$$

for $z \in \mathbb{H}$.

The orbit space $\mathbb{H}/SL(2,\mathbb{Z})$, important in classical number theory and the theory of automorphic forms [71], has the feature that the projection map

$$q : \mathbb{H} \to \mathbb{H}/SL(2,\mathbb{Z})$$

has singular points and is not a covering projection. In this case, the action of $SL(2,\mathbb{Z})$ is not free.

The action by the kernel $\Gamma(2,r)$ of the mod-r reduction map $\rho_r : SL(2,\mathbb{Z}) \to SL(2,\mathbb{Z}/r\mathbb{Z})$ is free in case $r \geq 2$. Furthermore, the group $\Gamma(2,r)$ is isomorphic to a finitely generated free group in case $r \geq 2$ [48, 71]. For example, if p is an odd prime, $\Gamma(2,p)$ is a free group on $1 + p(p^2 - 1)/12$ letters while $\Gamma(2,2)$ is a free group on 2 letters [38]. Furthermore, the group $\Gamma(2,4)$ is isomorphic to the 4-stranded Brunnian braid group $\mathrm{Brun}_4(S^2)$, as described in Sec. 23.

Observe that there is a map

$$\Phi : \mathrm{Top}^+(S^1 \times S^1) \to SL(2,\mathbb{Z})$$

defined by sending an element $f \in \mathrm{Top}^+(S^1 \times S^1)$ to the isomorphism

$$f_* : H_1(S^1 \times S^1) \to H_1(S^1 \times S^1),$$

regarded as an element in $GL(2,\mathbb{Z})$ of determinant $+1$.

There is an analogous map

$$\Phi_g : \mathrm{Top}^+(S_g) \to Sp(2g,\mathbb{Z})$$

defined by the equation

$$\Phi_g(f) = f_* : H_1(S_g) \to H_1(S_g)$$

where S_g is a closed orientable surface of genus g. In this case, the value of $\Phi_g(f)$ is an element in $GL(2g, \mathbb{Z})$ which preserves the cup-product structure for the cohomology of S_g and is thus an element in $Sp(2g, \mathbb{Z})$.

For convenience, let T^2 denote the torus $S^1 \times S^1$. Recall that the group $SL(2, \mathbb{Z})$ also acts on $T^2 = S^1 \times S^1$ with action defined by the equation

$$M(u, v) = (u^a v^b, u^c v^d)$$

for $(u, v) \in T^2$. Notice that this action preserves the point $(1, 1)$ and thus restricts to an action on $T^2 - \{(1, 1)\}$, a space which is denoted $\widehat{T^2}$ here. Furthermore,

$$M((u, v) \cdot (u', v')) = M((u, v)) \cdot M((u', v'))$$

since S^1 is abelian. Furthermore, observe that this action, on the level of the first homology group $H_1(T^2) = \mathbb{Z} \oplus \mathbb{Z}$, is precisely the "tautological" action of $SL(2, \mathbb{Z})$ on $\mathbb{Z} \oplus \mathbb{Z}$.

Notice that there is an induced homomorphism

$$E : SL(2, \mathbb{Z}) \to \mathrm{Top}^+(T^2)$$

defined by the equation $E(M)(u, v) = M(u, v)$. A lemma using this information and convenient for the proof of Theorem 10.6 below, is stated next.

Lemma 10.1. *The function*

$$E : SL(2, \mathbb{Z}) \to \mathrm{Top}^+(T^2)$$

is a continuous homomorphism. Furthermore, this map splits the natural map

$$\Phi : \mathrm{Top}^+(T^2) \to SL(2, \mathbb{Z})$$

with $\Phi \circ E$ *given by the identity self-map of* $SL(2, \mathbb{Z})$.

Variations obtained by "twisting together" configuration spaces, on the one hand, give $K(\pi, 1)$'s for $\pi = \Gamma_1^k$. On the other hand, the resulting $K(\Gamma_1^k, 1)$-spaces have real cohomology groups which are given in terms of classical modular forms. This section is an exposition of the connection between these analogues of configuration spaces with cohomology given in terms of modular forms. The first theorem in this direction is developed next [12].

It is now convenient to focus on the space $\mathrm{Diff}^+(S^1 \times S^1)$ in order to appeal directly to results of Earle and Eells [27], keeping in mind that the natural inclusion

$$\mathrm{Diff}^+(S^1 \times S^1) \subset \mathrm{Top}^+(S^1 \times S^1)$$

is a homotopy equivalence by Theorem 8.7. Notice that $S^1 \times S^1$ acts by rotations on each coordinate in T^2, giving elements in $\mathrm{Diff}^+(S^1 \times S^1)$ which are isotopic to the identity. These rotations are in the kernel of

$$\Phi : \mathrm{Diff}^+(S^1 \times S^1) \to SL(2, \mathbb{Z}),$$

with this kernel denoted $\widetilde{S^1 \times S^1}$. Earle and Eells prove that the natural inclusion

$$S^1 \times S^1 \to \widetilde{S^1 \times S^1}$$

is a homotopy equivalence [27]. Thus there is a fibration

$$B\Phi : B\mathrm{Diff}^+(S^1 \times S^1) \to BSL(2, \mathbb{Z})$$

with fiber $B(\widetilde{S^1 \times S^1})$ which is homotopy equivalent to $(\mathbb{CP}^\infty)^2$. This information will be used to prove the next result.

Theorem 10.2. *Assume that $k \geq 2$. The spaces*

$$E\mathrm{Diff}^+(S^1 \times S^1) \times_{\mathrm{Diff}^+(S^1 \times S^1)} \mathrm{Conf}(T^2, k)/\Sigma_k,$$

and

$$E\mathrm{Top}^+(S^1 \times S^1) \times_{\mathrm{Top}^+(S^1 \times S^1)} \mathrm{Conf}(T^2, k)/\Sigma_k$$

are both $K(\Gamma_1^k, 1)$.

A proof is given by the following sequence of lemmas.

Lemma 10.3. *The spaces $E\mathrm{Top}^+(T^2) \times_{S^1 \times S^1} T^2$, and $E\mathrm{Top}^+(T^2) \times_{\widetilde{S^1 \times S^1}} T^2$ are contractible. Thus $E\mathrm{Diff}^+(T^2) \times_{S^1 \times S^1} T^2$, and $E\mathrm{Diff}^+(T^2) \times_{\widetilde{S^1 \times S^1}} T^2$ are contractible.*

Proof. Observe $ET^2 \times_{T^2} T^2$ is contractible and so $E\mathrm{Top}^+(T^2) \times_{S^1 \times S^1} T^2$ is also.

Next, consider the natural inclusion $\iota : S^1 \times S^1 \to \widetilde{S^1 \times S^1}$ to obtain a map of orbit spaces

$$E\mathrm{Top}^+(T^2) \times_{S^1 \times S^1} T^2 \to E\mathrm{Top}^+(T^2) \times_{\widetilde{S^1 \times S^1}} T^2$$

which is a homotopy equivalence. Since $ETop^+(T^2) \times_{S^1 \times S^1} T^2$ is contractible, so is

$$ETop^+(T^2) \times_{\widetilde{S^1 \times S^1}} T^2.$$

The lemma follows. $\qquad\square$

Lemma 10.4. *The space*

$$ETop^+(T^2) \times_{Top^+(T^2)} T^2$$

is a $K(\pi, 1)$ where π is isomorphic to $SL(2, \mathbb{Z})$.

Proof. Observe that there is a fibration

$$ETop^+(T^2) \times_{Top^+(T^2)} T^2 \to BSL(2, \mathbb{Z})$$

with fiber

$$ETop^+(T^2) \times_{\widetilde{S^1 \times S^1}} T^2.$$

Since $ETop^+(T^2) \times_{\widetilde{S^1 \times S^1}} T^2$ is contractible by Lemma 10.3, the lemma follows. $\qquad\square$

Lemma 10.5. *If $k \geq 2$ then the space*

$$ETop^+(T^2) \times_{Top^+(T^2)} Conf(T^2, k)$$

is a $K(\pi, 1)$.

Proof. Consider the natural first coordinate projection map

$$ETop^+(T^2) \times_{Top^+(T^2)} Conf(T^2, k) \to ETop^+(T^2) \times_{Top^+(T^2)} T^2$$

with fiber $Conf(\widehat{T^2}, k - 1)$.

The base of this fibration, $ETop^+(T^2) \times_{Top^+(T^2)} T^2$ is a $K(SL(2, \mathbb{Z}), 1)$ by Lemma 10.4, and the fiber $Conf(\widehat{T^2}, k - 1)$ is also a $K(\pi, 1)$ by Theorem 8.3. It follows that

$$ETop^+(T^2) \times_{Top^+(T^2)} Conf(T^2, k)$$

is a $K(\pi, 1)$ for all $k \geq 1$. $\qquad\square$

Proof of Theorem 10.2. By Lemma 10.5, the space

$$ETop^+(T^2) \times_{Top^+(T^2)} Conf(T^2, k)$$

is a $K(\pi, 1)$ for all $k \geq 2$. Thus the space $ETop^+(T^2) \times_{Top^+(T^2)}$ $Conf(T^2, k)/\Sigma_k$ is also a $K(\pi, 1)$ as the action of Σ_k on $ETop^+(T^2)$ $\times_{Top^+(T^2)} Conf(T^2, k)$ is free.

Furthermore, by Theorem 3.2 $ETop(T^2) \times_{Top(T^2)} Conf(T^2, k)/\Sigma_k$ has fundamental group given by $\pi_0(Top(T^2, k))$. In addition, $ETop^+(T^2)$ $\times_{Top^+(T^2)} Conf(T^2, k)/\Sigma_k$ has fundamental group given by $\pi_0(Top^+(T^2, k))$ by Lemma 8.8. $\qquad\square$

The next theorem gives information about the pointed mapping class group with k marked points, $\Gamma_g^{k,*}$. Let $\widehat{1}$ denote the element $(1, 1) \in T^2$ with $\widehat{T^2} = T^2 - \{\widehat{1}\}$. The above remarks imply that the natural action of $SL(2, \mathbb{Z})$ on T^2 restricts to an action on $\widehat{T^2}$. The next theorem gives information about $ESL(2, \mathbb{Z}) \times_{SL(2,\mathbb{Z})} Conf(\widehat{T^2}, k)/\Sigma_k$ which has had computational utility [12].

Theorem 10.6. *Assume that $k \geq 1$. The space*

$$ESL(2, \mathbb{Z}) \times_{SL(2,\mathbb{Z})} Conf(\widehat{T^2}, k)/\Sigma_k$$

is a $K(\Gamma_1^{k,*}, 1)$.

Proof. By Lemma 10.5, the space

$$ETop^+(T^2) \times_{Top^+(T^2)} Conf(T^2, k)$$

is a $K(\pi, 1)$ for all $k \geq 2$. Thus the space $ETop^+(T^2) \times_{Top^+(T^2)}$ $Conf(T^2, k + 1)/(1 \times \Sigma_k)$ is also a $K(\pi, 1)$ as the action of $1 \times \Sigma_k$ on $ETop^+(T^2) \times_{Top^+(T^2)} Conf(T^2, k + 1)$ is free. Furthermore, the fundamental group of $ETop^+(T^2) \times_{Top^+(T^2)} Conf(T^2, k + 1)/(1 \times \Sigma_k)$ is $\Gamma_1^{k,*}$ by Theorem 3.2.

To finish the proof of the theorem, it suffices to exhibit a homotopy equivalence

$$\Gamma(k) : ESL(2, \mathbb{Z}) \times_{SL(2,\mathbb{Z})} Conf(\widehat{T^2}, k)/\Sigma_k$$
$$\to ETop^+(T^2) \times_{Top^+(T^2)} Conf(T^2, k + 1)/(1 \times \Sigma_k).$$

To define the map $\Gamma(k)$, first consider the action of $SL(2, \mathbb{Z})$ on T^2 obtained from the homomorphism

$$E : SL(2, \mathbb{Z}) \to Top^+(T^2)$$

of Lemma 10.1. From this, we see that $Top^+(T^2)$ is a semi-direct product of $SL(2, \mathbb{Z})$ and $\widetilde{S^1 \times S^1}$.

Now define a map

$$\gamma_k : \mathrm{Conf}(\widehat{T^2}, k) \to \mathrm{Conf}(T^2, k+1)$$

by the formula

$$\gamma_k((z_1, \ldots, z_k)) = (z_1, \ldots, z_k, 1).$$

Next consider the natural quotient map

$$\rho : ESL(2, \mathbb{Z}) \times_{SL(2,\mathbb{Z})} \mathrm{Conf}(T^2, k+1) \to ETop^+(T_2) \times_{Top^+(T_2)} \mathrm{Conf}(T^2, k+1)$$

pre-composed with the map $1 \times \gamma_k$; we let

$$\mu : ESL(2, \mathbb{Z}) \times_{SL(2,\mathbb{Z})} \mathrm{Conf}(\widehat{T^2}, k) \to ETop^+(T_2) \times_{Top^+(T_2)} \mathrm{Conf}(T^2, k+1)$$

be given by

$$\mu = \rho \circ (1 \times \gamma_k).$$

Since the map μ is equivariant with respect to the action of Σ_k on the source and $1 \times \Sigma_k$ on the target, that μ is a homotopy equivalence then implies that the induced quotient map $\Gamma(k)$ is also a homotopy equivalence and the theorem follows.

That μ is an equivalence follows from a comparison of fibrations as given next. First consider the commutative diagram

$$
\begin{array}{ccccc}
E \times_{SL(2,\mathbb{Z})} \mathrm{Conf}(\widehat{T^2}, k) & \xrightarrow{\ \pi\ } & BSL(2, \mathbb{Z}) & \xrightarrow{\ 1\ } & BSL(2, \mathbb{Z}) \\
\downarrow{\scriptstyle 1 \times \gamma_k} & & \downarrow{\scriptstyle 1} & & \downarrow \\
E \times_{SL(2,\mathbb{Z})} \mathrm{Conf}(T^2, k+1) & \xrightarrow{\ \pi\ } & BSL(2, \mathbb{Z}) & \xrightarrow{\ 1\ } & BSL(2, \mathbb{Z}) \\
\downarrow{\scriptstyle \rho} & & \downarrow{\scriptstyle BE} & & \downarrow{\scriptstyle 1} \\
E \times_{Top^+(T_2)} \mathrm{Conf}(T^2, k+1) & \xrightarrow{\ \pi\ } & BTop^+(T_2) & \xrightarrow{\ B\Phi\ } & BSL(2, \mathbb{Z})
\end{array}
$$

for which $\pi : EG \times_G X \to BG$ denotes the natural projection map.

Thus there is a morphism of fibrations

$$
\begin{array}{ccccc}
\mathrm{Conf}(\widehat{T^2}, k) & \xrightarrow{\ 1\ } & E \times_{SL(2,\mathbb{Z})} \mathrm{Conf}(\widehat{T^2}, k) & \xrightarrow{\ \pi\ } & BSL(2, \mathbb{Z}) \\
\downarrow{\scriptstyle 1 \times \gamma_k} & & \downarrow{\scriptstyle \rho \circ (1 \times \gamma_k)} & & \downarrow{\scriptstyle 1} \\
E \times_{\widetilde{S^1 \times S^1}} \mathrm{Conf}(T^2, k+1) & \xrightarrow{\ 1\ } & E \times_{Top^+(T_2)} \mathrm{Conf}(T^2, k+1) & \xrightarrow{\ B\Phi \circ \pi\ } & BSL(2, \mathbb{Z})
\end{array}
$$

for which the induced map

$$1 \times \gamma_k : \mathrm{Conf}(\widehat{T^2}, k) \to E \times_{\widetilde{S^1 \times S^1}} \mathrm{Conf}(T^2, k+1)$$

is an equivalence by observing that:

(1) the projection

$$E \times_{\widetilde{S^1 \times S^1}} \mathrm{Conf}(T^2, k+1) \to E \times_{\widetilde{S^1 \times S^1}} T^2$$

is a bundle projection with fiber $\mathrm{Conf}(\widehat{T^2}, k)$,
(2) the space

$$E \times_{\widetilde{S^1 \times S^1}} T^2$$

is contractible (by Lemma 10.3), and
(3) the induced self-map of $\mathrm{Conf}(\widehat{T^2}, k)$ is the identity, by inspection.

The theorem follows. $\qquad\square$

Remark 10.7. (1) The bundle projection $B\mathrm{Diff}^+(S^1 \times S^1) \to BSL(2, \mathbb{Z})$ with fiber $B(S^1 \times S^1) = (\mathbb{CP}^\infty)^2$ was first exploited in a beautiful paper by Furusawa, Tezuka, and Yagita who showed that the real cohomology of $B\mathrm{Diff}^+(S^1 \times S^1)$ was given in terms of classical modular forms [40]. They also determined the torsion in the cohomology of $B\mathrm{Diff}^+(S^1 \times S^1)$.
(2) The naturally associated orbifold $\mathbb{H} \times_{SL(2,\mathbb{Z})} \mathrm{Conf}(S^1 \times S^1, q)$ has cohomology which is that of $ESL(2, \mathbb{Z}) \times_{SL(2,\mathbb{Z})} \mathrm{Conf}(S^1 \times S^1, q)$, as long as the primes 2 and 3 are units.
(3) The real cohomology of $ESL(2, \mathbb{Z}) \times_{SL(2,\mathbb{Z})} \mathrm{Conf}(\widehat{T^2}, k)/\Sigma_k$, and thus the cohomology of $\Gamma_1^{k,*}$, was worked out in [12] where the answer is given in terms of ranks of certain modular forms. In addition, fixing the dimension of the cohomology group while letting k increase gives a "stable answer" which is equal to the ranks of certain Jacobi forms computed by Eichler and Zagier [30], by a direct comparison. This particular interpretation also gave the cohomology groups (more easily) with coefficients in the sign representation.

11. On Configurations in a Surface of Genus Greater than 1

The purpose of this section is to focus on the configuration space of surfaces S_g of genus g greater than 1. One distinguishing feature in this case is the result of Earle and Eells which proves that

$$\mathrm{BDiff}^+(S_g)$$

is a $K(\pi, 1)$, namely, each path-component of $\mathrm{Diff}^+(S_g)$ is contractible [27].

Theorem 11.1. *Assume that S_g is an orientable surface of genus $g \geq 2$. Then*

(1) $E\mathrm{Top}^+(S_g) \times_{\mathrm{Top}^+(S_g)} \mathrm{Conf}(S_g, k)/\Sigma_k$ *is a* $K(\Gamma_g^k, 1)$, *and*

(2) $E\mathrm{Top}(S_g)^+ \times_{\mathrm{Top}^+(S_g)} \mathrm{Conf}(S_g, k+1)/\{1 \times \Sigma_k\}$ *is a* $K(\Gamma_g^{k,*}, 1)$.

Proof. First consider $E\mathrm{Top}^+(S_g) \times_{\mathrm{Top}^+(S_g)} \mathrm{Conf}(S_g, k)$, the total space of a bundle over $B\mathrm{Top}^+(S_g)$ with fiber $\mathrm{Conf}(S_g, k)$. Since $B\mathrm{Top}^+(S_g)$, and $\mathrm{Conf}(S_g, k)$ are $K(\pi, 1)$'s, so is $E\mathrm{Top}^+(S_g) \times_{\mathrm{Top}^+(S_g)} \mathrm{Conf}(S_g, k)$ as well as $E\mathrm{Top}^+(S_g) \times_{\mathrm{Top}^+(S_g)} \mathrm{Conf}(S_g, k)/H$ where H is any subgroup of the symmetric group on k letters. Thus both

$$E\mathrm{Top}^+(S_g) \times_{\mathrm{Top}^+(S_g)} \mathrm{Conf}(S_g, k)/\Sigma_k,$$

and

$$E\mathrm{Top}^+(S_g) \times_{\mathrm{Top}^+(S_g)} \mathrm{Conf}(S_g, k+1)/\{1 \times \Sigma_k\}$$

are $K(\pi, 1)$'s.

To finish, it suffices to identify the fundamental groups of these spaces. Notice that by Theorem 3.2, (i) the fundamental group of $E\mathrm{Top}^+(S_g) \times_{\mathrm{Top}^+(S_g)} \mathrm{Conf}(S_g, k)/\Sigma_k$ is Γ_g^k, and (ii) the fundamental group of $E\mathrm{Top}(S_g)^+ \times_{\mathrm{Top}^+(S_g)} \mathrm{Conf}(S_g, k+1)/\{1 \times \Sigma_k\}$ is $\Gamma_g^{k,*}$. $\square$

Remark 11.2. Many of these results appeared in a slightly different form in [42, 69, 70], with the exception of the $K(\pi, 1)$ properties.

12. Loop Spaces of Configuration Spaces

The subject of this section, as well as Secs. 14 through 15, is basic properties of loop spaces of configuration spaces. A subsequent section, Sec. 21, lists overlapping features describing connections of these structures to low dimensional topology, homotopy theory and number theory.

First recall the definitions of free loop spaces, and pointed loop spaces.

Definition 12.1. Let X denote a topological space.

(1) Define the free loop space of X to be

$$L(X) = \{f : S^1 \to X \mid f \text{ is continuous}\},$$

topologized with the compact-open topology.

(2) If X has a base-point $*$, define the pointed loop space of X to be

$$\Omega(X) = \{f : S^1 \to X \mid f \text{ is continuous and } f(1) = *\}$$

for $1 \in S^1 \subset \mathbb{R}^2$ with $\Omega(X)$ topologized as a subspace of $L(X)$.

(3) The topology on both $L(X)$ and $\Omega(X)$ is the compact-open topology. A more convenient, as well as more general, choice of topology is given by the associated compactly-generated topology. However, this generalization will not be emphasized here.

One way to view $\Omega\mathrm{Conf}(M, k)$ is through the graph of a function

$$f : [0, 1] \to \mathrm{Conf}(M, k)$$

given by

$$\mathrm{graph}(f) : [0, 1] \to [0, 1] \times \mathrm{Conf}(M, k)$$

with

$$\mathrm{graph}(f)(t) = (t, f(t)).$$

Notice that in case $M = \mathbb{R}^2$, then $\mathrm{graph}(f)$ is an embedding whose image is exactly a braid which starts at $(0, f(0))$ and quits at $(1, f(1))$. These graphs represent the precise physical meaning of a braid, see Fig. 1.

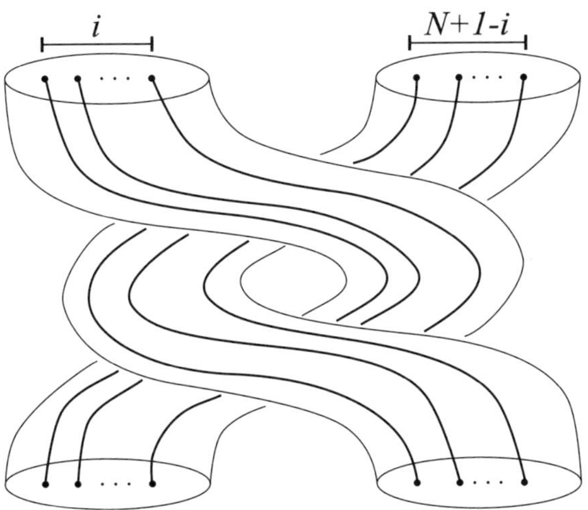

Fig. 1.   Picture of a braid in P_{N+1}.

Notice that by definition, the group of path-components satisfies the properties

$$\pi_0(\Omega(\mathrm{Conf}(\mathbb{R}^2, k))) = P_k$$

and

$$\pi_0(\Omega(\mathrm{Conf}(\mathbb{R}^2, k)/\Sigma_k)) = B_k.$$

Furthermore, the spaces $\mathrm{Conf}(\mathbb{R}^2, k)$, and $\mathrm{Conf}(\mathbb{R}^2, k)/\Sigma_k$ are $K(\pi, 1)$'s by Theorem 8.3. The next classical result follows at once [32, 37].

Theorem 12.2. *The unordered configuration space* $\mathrm{Conf}(\mathbb{R}^2, k)/\Sigma_k$ *is a* $K(B_k, 1)$ *and the ordered configuration space* $\mathrm{Conf}(\mathbb{R}^2, k)$ *is a* $K(P_k, 1)$.

One result proven here gives the structure of the homology of the pointed loop space of $\mathrm{Conf}(\mathbb{R}^m, k)$, a result which reflects elementary properties of linking invariants for pairs of linked spheres. This setting of linking is developed in two ways below. One way is by looking at the way in which these constructions extend to invariants of classical links. The second way is by looking at how the algebras associated to these linking invariants correspond to certain spectral sequences in homotopy theory, as elucidated in the section on simplicial groups, Sec. 16. This section will start with motivating examples and then continue with the proofs of some basic theorems. A summary of how and where these results fit in is given in the section "Other Connections", Sec. 21.

Recall the following facts. The projection maps

$$p_k : \mathrm{Conf}(\mathbb{R}^m, k) \to \mathrm{Conf}(\mathbb{R}^m, k - 1)$$

admit cross-sections σ defined by the formula

$$\sigma_k(x_1, \ldots, x_{k-1}) = (x_1, \ldots, x_{k-1}, w)$$

where

$$w = M(\vec{e}_1)$$

where e_1 is the unit vector $(1, 0, \ldots, 0)$ and $M = 1 + \max_{1 \leq i \leq k-1} ||x_i||$. Furthermore, the fiber of p_k is $\mathbb{R}^n - Q_{k-1}$ which is homotopy equivalent to $\vee_{k-1} S^{m-1}$.

Next recall the following classical lemma with proof given in [66].

Lemma 12.3. *Let*

$$p : E \to B$$

be a fibration with fiber F and $\iota : F \to E$ the inclusion of the fiber in the total space, for which E, B, and F are path-connected spaces. If p admits a cross-section (up to homotopy), then $\Omega(E)$ is homotopy equivalent to

$$\Omega(B) \times \Omega(F).$$

One consequence of the existence of the cross-sections σ_k above, and Lemma 12.3, is stated next.

Proposition 12.4. *Assume that M is a manifold without boundary, of dimension $m \geq 2$ such that the natural first coordinate projection*

$$p_1 : \mathrm{Conf}(M, k) \to M$$

admits a section. Then there is a homotopy equivalence

$$\prod_{0 \leq i \leq k-1} \Omega(M - Q_i) \to \Omega\mathrm{Conf}(M, k).$$

If $m \geq 3$, then there is a homotopy equivalence

$$\prod_{1 \leq i \leq k-1} \Omega(\vee_i S^{m-1}) \to \Omega\mathrm{Conf}(\mathbb{R}^m, k).$$

Proof. Since the projection p_1 admits a cross-section by hypotheses, it follows from Lemma 12.3 that there is a homotopy equivalence

$$\Omega(M) \times \Omega\mathrm{Conf}(M - Q_1, k - 1) \to \Omega\mathrm{Conf}(M, k). \qquad \square$$

The homology of the loop space of certain configuration spaces is worked out next. One reason for including this computation is that it appears in several different natural mathematical contexts. A second reason is that the homology of $\Omega\mathrm{Conf}(\mathbb{R}^m, k)$ arises in terms of Vassiliev invariants of pure braids as developed by T. Kohno [50]. It then turns out that these structures are intimately tied to the homotopy groups of spheres, as described below, and then to certain natural structures involving derivations of free Lie algebras, as described in several sections below. Since the Lie algebras encountered here are free as modules over the integers, the definitions given next will be restricted to free modules over the integers.

Certain graded Lie algebras are basic here. The first one is the free Lie algebra generated by a graded free abelian group V. First recall that any graded, associative algebra A inherits the structure of a Lie algebra with the bracket

$$[-, -] : A \otimes A \to A$$

defined by the formula

$$[a, b] = a \cdot b + (-1)^{|a||b|} b \cdot a$$

for elements a and b of degree $|a|$ and $|b|$ respectively.

Definition 12.5. Let V denote a graded free abelian group with $T[V]$ the tensor algebra generated by V. Then

$$L[V]$$

is the smallest sub-Lie algebra of $T[V]$ generated by V.

A related Lie algebra arises which is universal for the (graded) "infinitesimal braid relations", also known as the "horizontal $4T$-relations", or the "Yang-Baxter Lie algebra relations", as in [49, 52, 13, 31]. That is the largest Lie algebra over a fixed commutative ring R for which the "infinitesimal braid relations" are satisfied.

Definition 12.6. Fix a strictly positive integer q. Define

$$\mathcal{L}_k(q)$$

to be the free (graded) Lie algebra over the integers $\mathbb{Z}$ generated by elements $B_{i,j}$ of degree q, $k \geq i > j \geq 1$, modulo the graded infinitesimal braid relations:

(i) $[B_{i,j}, B_{s,t}] = 0$ if $\{i, j\} \cap \{s, t\} = \emptyset$,
(ii) $[B_{i,j}, B_{i,t} + (-1)^q B_{t,j}] = 0$ if $1 \leq j < t < i \leq k$, and
(iii) $[B_{t,j}, B_{i,j} + B_{i,t}] = 0$ if $1 \leq j < t < i \leq k$.

Kohno [49] gives a slightly different description of these relations as follows: Introduce new generators $B_{j,i}$ of degree q, $k \geq i > j \geq 1$ with the relations $B_{i,j} = (-1)^q B_{j,i}$. Then Kohno's description of the above relations simplifies to (i), and (ii) above with distinct i, j, and t.

These relations appear as special cases in the Vassiliev invariants of braids [52, 13]. They also arise in the study of the KZ (Knishnik-Zamolodchikov) equations as integrability conditions for certain flat bundles [10], as well as in work of Kohno [49, 52], and Drinfel'd [25, 26] on the Kohno-Drinfel'd monodromy theorem [10]. The next theorem, an algebraic reflection, was proven in [13, 31] where the notation Prim $H_*(\Omega\mathrm{Conf}(\mathbb{R}^m, k); \mathbb{Z})$ denotes the module of primitive elements (in a torsion free Hopf algebra).

Theorem 12.7. *If $m \geq 3$, the homology of $\Omega\mathrm{Conf}(\mathbb{R}^m, k)$ is torsion free and there is an isomorphism of Lie algebras on the level of the module of primitives:*

$$\mathcal{L}_k(m-2) \to \mathrm{Prim}\, H_*(\Omega\mathrm{Conf}(\mathbb{R}^m, k); \mathbb{Z}).$$

Furthermore, the universal enveloping algebra of $\mathcal{L}_k(m-2)$,

$$U[\mathcal{L}_k(m-2)]$$

is isomorphic to

$$H_*(\Omega\mathrm{Conf}(\mathbb{R}^m, k); \mathbb{Z})$$

as a Hopf algebra.

There is more topology behind this theorem. The loop space $\Omega\mathrm{Conf}(\mathbb{R}^m, k)$ is homotopy equivalent to a product of loop spaces

$$\prod_{1 \leq i \leq k-1} \Omega(\vee_i S^{m-1}),$$

thus it is natural to construct representative cycles. It is these cycles, in dimension $m-2$, which represent the elements $B_{i,j}$. In addition, this geometric decomposition has the following algebraic consequence. There are embeddings of Lie algebras

$$g_j : \mathcal{L}_j(m-2) \to H_*(\Omega\mathrm{Conf}(\mathbb{R}^m, k))$$

such that the natural additive extension

$$\bigoplus_{1 \leq j \leq k-1} \mathcal{L}_j(m-2) \to \mathrm{Prim}\, H_*(\Omega\mathrm{Conf}(\mathbb{R}^m, k))$$

is an isomorphism, but does not preserve the structure as Lie algebras. The failure to preserve the Lie algebra structure is important in applications.

13. Planetary Motion in Configuration Spaces

The purpose of this section is to give pairs of naively linked spheres in $\mathbb{R}^m$ which correspond to "planetary motion" and which reflect properties of the loop space of a configuration space. To illustrate, start with $\mathrm{Conf}(\mathbb{R}^m, 3)$, and regard S^{m-1} as the standard locus of points in $\mathbb{R}^m$ of norm equal to one. There is a map

$$\gamma : S^{m-1} \times S^{m-1} \to \mathrm{Conf}(\mathbb{R}^m, 3)$$

defined by

$$\gamma(v, w) = (0, v, v + w/4).$$

Observe that one may regard 0 as the coordinates of a sun S_0 with v the coordinates of a planet P_v in orbit about the sun S_0, and with $v + w/4$ the coordinates of a moon in orbit around the planet P_v.

These maps, as well as analogous maps obtained by permuting the coordinates

$$(0, v, v + w/4),$$

induce relations in the homology of the loop space of the configuration space by considering

$$\Omega(\gamma) : \Omega(S^{m-1} \times S^{m-1}) \to \Omega(\mathrm{Conf}(\mathbb{R}^m, 3)).$$

These relations give precisely the "horizontal 4T relations" or "infinitesimal braid relations" as stated in Definition 12.6.

14. Homological Calculations for $\mathbb{R}^m$

The purpose of this section is to give the computation stated in Theorem 12.7. Recall that $\mathrm{Conf}(\mathbb{R}^m, k)$ is $(m-2)$-connected and that the algebra

$$H^*(\mathrm{Conf}(\mathbb{R}^m, k); \mathbb{Z})$$

is generated by classes $A_{i,j}$, $k \geq i > j \geq 1$, of degree $m - 1$ [15, 20]. Thus the homology suspension induces an isomorphism

$$\sigma_* : H_{m-2}(\Omega\mathrm{Conf}(\mathbb{R}^m, k); \mathbb{Z}) \to H_{m-1}(\mathrm{Conf}(\mathbb{R}^m, k); \mathbb{Z})$$

for $m > 2$.

Definition 14.1. Define the homology class $B_{i,j}$ to be the unique class specified by

$$\sigma_*(B_{i,j}) = A_{i,j*},$$

the dual basis element dual to $A_{i,j}$ with $k \geq i > j \geq 1$.

Alternatively, the classes $B_{i,j}$ are represented by maps of spheres, as described in the next section. Furthermore, commutation relations for the $B_{i,j}$ are obtained by (1) exhibiting maps

$$\gamma : S^{m-1} \times S^{m-1} \to \mathrm{Conf}(\mathbb{R}^m, 3),$$

(2) looping the map γ, and (3) using the commutativity of the two fundamental cycles in

$$H_*((\Omega S^{m-1})^2; \mathbb{Z}).$$

The maps $\gamma : S^{m-1} \times S^{m-1} \to \mathrm{Conf}(\mathbb{R}^m, 3)$ correspond to naive planetary motion as given in the previous section. These relations are analyzed in the next section.

15. On Spheres Embedded in the Configuration Space

The purpose of this section is to give pairs of spheres in $\mathbb{R}^m$ which will reflect homological properties of the loop space of a configuration space. Fix integers s, t, and ℓ such that $k \geq s > t \geq 1$, $k \geq \ell \geq 1$, with $\ell \notin \{s, t\}$.

Definition 15.1. Define a map

$$\gamma(s, t, \ell) : S^{m-1} \times S^{m-1} \to \mathrm{Conf}(\mathbb{R}^m, k)$$

by the formula

$$\gamma(s, t, \ell)(u, v) = (x_1, \ldots, x_k)$$

with $\|u\| = \|v\| = 1$ such that

(1) $z_i = (4i, 0, 0, \ldots, 0)$ for $k \geq i \geq 1$,
(2) $x_i = z_i$ if $i \neq \{s, t\}$, and
(3) $x_s = z_\ell + 2v$, $x_t = z_\ell + u$.

Notice that $(x_1, \ldots, x_k)$ is indeed in $\mathrm{Conf}(\mathbb{R}^m, k)$.

Next, recall that the class $A_{i,j}$ is defined by the equation

$$A_{i,j} = \pi_{i,j}^*(\iota)$$

where $\pi_{i,j} : \mathrm{Conf}(\mathbb{R}^m, k) \to \mathrm{Conf}(\mathbb{R}^m, 2)$ denotes projection on the (i, j) coordinates and ι is a fixed fundamental cycle for $H^{m-1}(S^{m-1})$ [15, 20]. Furthermore,

$$A_{i,j} = (-1)^m A_{j,i}$$

by inspection. The notation of Definition 15.1 is used in the proof of the next lemma.

Lemma 15.2. (1) *If $\{i,j\} \cap \{s,t,\ell\}$ has cardinality 0 or 1, then*

$$\gamma(s,t,\ell)^*(A_{i,j}) = 0.$$

(2) *If $\{i,j\} \cap \{s,t,\ell\}$ has cardinality 2, then*

    (a) *for $\ell < t < s$,*

$$\gamma(s,t,\ell)^*(A_{i,j}) = \begin{cases} \iota \otimes 1 & \text{if } j = \ell \text{ and } i = t, \\ 1 \otimes \iota & \text{if } j = \ell \text{ and } i = s, \\ 1 \otimes \iota & \text{if } j = t \text{ and } i = s, \end{cases}$$

    (b) *for $t < \ell < s$,*

$$\gamma(s,t,\ell)^*(A_{i,j}) = \begin{cases} (-1)^m \iota \otimes 1 & \text{if } j = t \text{ and } i = \ell, \\ 1 \otimes \iota & \text{if } j = t \text{ and } i = s, \\ 1 \otimes \iota & \text{if } j = \ell \text{ and } i = s, \end{cases}$$

    (c) *for $t < s < \ell$,*

$$\gamma(s,t,\ell)^*(A_{i,j}) = \begin{cases} (-1)^m \iota \otimes 1 & \text{if } j = t \text{ and } i = \ell, \\ 1 \otimes \iota & \text{if } j = t \text{ and } i = s, \\ (-1)^m 1 \otimes \iota & \text{if } j = s \text{ and } i = \ell. \end{cases}$$

Proof. Notice that if $\{i,j\} \cap \{s,t,\ell\} = \emptyset$, then $\pi_{i,j} \circ \gamma(s,t,\ell)$ is constant. If $\{i,j\} \cap \{s,t,\ell\}$ has cardinality 1, then all but one of the x_r coordinates are constant. Furthermore,

$$\pi_{i,j} \circ \gamma(s,t,\ell)(u,v) = (x_i, x_j)$$

where either

(1) $x_i = z_i$ and $x_j = z_\ell + 2v$ with $i \neq \ell$,
(2) $x_i = z_i$ and $x_j = z_\ell + u$ with $i \neq \ell$, or
(3) $x_i = z_\ell + 2v$ or $z_\ell + u$ with $j \neq \ell$.

In either of these three cases $\pi_{i,j} \circ \gamma(s,t,\ell)$ is null-homotopic. Thus, part (1) follows.

    Part (2) is obtained by considering three cases. One case is listed as the others are similar. Thus assume that $t < s < \ell$.

Case 1. $j = t$ and $i = \ell$.

In this case $\pi_{i,j} \circ \gamma(s,t,\ell)(z,w) = (y_\ell + z, y_\ell)$. This last map is evidently homotopic to the map which sends (z,w) to $(y_\ell, y_\ell - z)$ and thus $A_{i,j}$ pulls back to $(-1)^m \iota \otimes 1$.

Case 2. $j = t$ and $i = s$.

In this case $\pi_{i,j} \circ \gamma(s,t,\ell)(z,w) = (y_\ell + z, y_\ell + 2w)$. By shrinking z to 0, this last map is homotopic to the map which sends (z,w) to $(y_\ell, y_\ell + 2w)$. Hence $A_{i,j}$ pulls back to $1 \otimes \iota$.

Case 3. $j = s$ and $i = \ell$.

In this case $\pi_{i,j} \circ \gamma(s,t,\ell)(z,\omega) = (y_\ell + 2w, y_\ell)$ which is homotopic to the map that sends (z,w) to $(y_\ell, y_\ell - w)$. Hence $A_{i,j}$ pulls back to $(-1)^m 1 \otimes \iota$. $\qquad\square$

Let γ denote a fixed choice of fundamental cycle for $H_{m-1}(S^{m-1})$. Consider the natural basis for $H_{m-1}(\mathrm{Conf}(\mathbb{R}^m, k))$ obtained by taking the linear duals $A_{i,j*}$ to the elements $A_{i,j}$ in Lemma 15.2.

Lemma 15.3. (1) If $\ell < t < s$, then

$$\gamma(s,t,\ell)_*(i \otimes 1) = A_{t,\ell*}, \quad and$$
$$\gamma(s,t,\ell)_*(1 \otimes i) = A_{s,\ell*} + A_{s,t*}.$$

(2) If $t < \ell < s$, then

$$\gamma(s,t,\ell)_*(i \otimes 1) = (-1)^m A_{\ell,t*}, \quad and$$
$$\gamma(s,t,\ell)_*(1 \otimes i) = A_{s,\ell*} + A_{s,\ell*}.$$

(3) If $t < \ell < s$, then

$$\gamma(s,t,\ell)_*(i \otimes 1) = (-1)^m A_{\ell,t*}, \quad and$$
$$\gamma(s,t,\ell)_*(1 \otimes i) = A_{s,t*} + (-1)^m A_{\ell,s*}.$$

Next consider the two "axial" inclusions $S^{m-1} \to S^{m-1} \times S^{m-1}$. Passage to adjoints gives two classes x_i in $H_{m-2}(\Omega(S^{m-1})^2; \mathbb{Z})$ such that $H_*(\Omega(S^{m-1})^2; \mathbb{Z})$ is isomorphic to the tensor product of tensor algebras $T[x_1] \otimes T[x_2]$, as an algebra, where

$$[x_1, x_2] = x_1 x_2 - (-1)^m x_2 x_1 = 0.$$

Thus $\Omega\gamma(s,t,\ell)_* = [x_1, x_2] = 0$ by naturality. The infinitesimal braid relations arise by applying this formula to Lemma 15.3.

Corollary 15.4. *If $m \geq 3$, then the following relations hold in*

$$H_*(\Omega\mathrm{Conf}(\mathbb{R}^m, k); \mathbb{Z}) :$$

(1) $[B_{i,j}, B_{i,t} + (-1)^m B_{t,j}] = 0$ *if* $1 \leq j < t < i \leq k$.
(2) $[B_{t,j}, B_{i,j} + B_{i,t}] = 0$ *if* $1 \leq j < t < i \leq k$.

Proof. By Lemma 15.3, and the definition that $B_{i,j}$ is the unique element such that $\sigma_* B_{i,j} = A_{i,j*}$, the following holds:

(i) If $\ell < t < s$, then

$$\Omega\gamma(s,t,\ell)_*(x_1 \otimes 1) = B_{t,\ell}, \text{ and}$$
$$\Omega\gamma(s,t,\ell)_*(1 \otimes x_2) = B_{s,\ell} + B_{s,t}.$$

(ii) If $t < \ell < s$, then

$$\Omega\gamma(s,t,\ell)_*(x_1 \otimes 1) = (-1)^m B_{\ell,t}, \text{ and}$$
$$\Omega\gamma(s,t,\ell)_*(1 \otimes x_2) = B_{s,t} + B_{s,\ell}.$$

(iii) If $t < s < \ell$, then

$$\Omega\gamma(s,t,\ell)_*(x_1 \otimes 1) = (-1)^m B_{\ell,t}, \text{ and}$$
$$\Omega\gamma(s,t,\ell)_*(1 \otimes x_2) = B_{s,t} + (-1)^m B_{\ell,s}.$$

Thus part (i) gives $[B_{t,\ell}, B_{s,\ell} + B_{s,t}] = 0$ for $\ell < t < s$. This is a restatement of equation (2). In addition, part (iii) gives

$$[B_{\ell,t}, (-1)^m B_{s,t} + B_{\ell,s}] = 0 \text{ if } t < s < \ell.$$

This is a restatement of equation (1). The corollary follows.   □

Proposition 15.5. *If $m \geq 3$ and $\{i,j\} \cap \{s,t\} = \emptyset$, then*

$$[B_{i,j}, B_{s,t}] = 0$$

in $H_*(\Omega\mathrm{Conf}(\mathbb{R}^m, k); \mathbb{Z})$.

Proof. If $\{i,j\} \cap \{s,t\} = \emptyset$, then define

$$\theta : S^{m-1} \times S^{m-1} \to \mathrm{Conf}(\mathbb{R}^m, k)$$

by the formula

$$\theta(u,v) = (x_1, \ldots, x_k)$$

where the x_i are defined by the formula (1) $x_v = z_v$ for $v \notin \{i, j, s, t\}$, (2) $x_j = z_j$, $x_i = z_j + u$, (3) $x_s = z_s$, and (4) $x_t = z_s + v$. Then

$$\theta_*(\iota \otimes 1) = A_{i,j_*} \text{ and}$$

$$\theta_*(1 \otimes \iota) = A_{s,t*}.$$

Furthermore, $(\Omega\theta)_*(x_1 \otimes 1) = B_{i,j}$ and $(\Omega\theta)_*(1 \otimes x_2) = B_{s,t}$. Since $[x_1, x_2] = 0$, it follows that $[B_{i,j}, B_{s,t}] = 0$ by naturality. $\qquad\square$

One way in which these relations arise is through the structure of the projection maps $p_i : \mathrm{Conf}(M, k) \to \mathrm{Conf}(M, k-1)$, and "doubling maps" $\sigma_i : \mathrm{Conf}(\mathbb{R}^m, k) \to \mathrm{Conf}(\mathbb{R}^m, k+1)$. These maps correspond to the structure of a simplicial object as developed next. The role of the Lie algebras which satisfy the "horizontal 4T relations" will be described below in this framework.

16. Simplicial Objects, and Δ-Objects

This purpose of this section, as well as Secs. 17 through 21, is to give descriptions of naive properties of configurations spaces and their relationship to simplicial groups. We will consider specific concrete cases which on the one hand give classical structures for describing the homotopy groups of the 2-sphere. The natural connection to the homology of the pointed loop space of the configuration space, given in Theorem 12.7, is also described. One of the main features here is the interplay between the structure of the braid groups, the homology of the loop space of certain configuration spaces, and the Bousfield-Kan spectral sequence associated to the homotopy groups of a simplicial group.

The goal of subsequent sections is to describe the connections between the fundamental groups of configuration spaces, homotopy groups of spheres, Vassiliev invariants and T. Kohno's Lie algebra arising from the infinitesimal braid relations.

Before embarking in this direction, an overview will be given to clarify the connections here. First, consider the projection maps out of configuration spaces

$$p_i : \mathrm{Conf}(M, k) \to \mathrm{Conf}(M, k-1)$$

which are defined in this section by deleting the i-th coordinate. These projection maps satisfy the compatibility condition

$$p_i \circ p_j = p_{j-1} \circ p_i$$

in case $i < j$. With mild conditions concerning base-points for the space M, the analogous formulas are satisfied on the level of fundamental groups with

$$p_{i_*} \circ p_{j_*} = p_{j-1_*} \circ p_{i_*}$$

in case $i < j$.

This compatibility property is precisely the condition for a collection of groups to form a Δ-group, as first developed in [65] and defined below. In addition, in the case $M = \mathbb{R}^m$, the projection maps, together with certain additional maps, give the collection of fundamental groups of the configuration spaces the structure of a simplicial group, also defined below.

The basic combinatorial invariant framework is that of a Δ-set and simplicial set, which model the combinatorics of a simplicial complex. Basic properties of simplicial sets appear in the excellent references [61, 22, 4, 57, 83].

Definition 16.1. A Δ-*set* is a collection of sets

$$K_\bullet = \{K_0, K_1, \ldots\}$$

with functions, *face operations*,

$$d_i : K_t \to K_{t-1} \quad \text{for } 0 \le i \le t$$

which satisfy the identities

$$d_i d_j = d_{j-1} d_i \text{ if } i < j.$$

A Δ-*group* is a Δ-set for which all $d_i : K_t \to K_{t-1}$ are group homomorphisms.

A natural example of a Δ-group arises from the pure braid groups $P_n(S) = \pi_1(\mathrm{Conf}(S, n))$ for a path-connected surface S, as follows, see [5].

Example 16.2. There are $(n + 1)$ homomorphisms

$$d_i : P_{n+1}(S) \to P_n(S),$$

with $0 \le i \le n$, where d_i is obtained by deleting the $(i + 1)$-st strand of a braid in $P_{n+1}(S)$. The homomorphisms d_i are induced on the level of fundamental groups of configuration spaces by the projection maps

$$p_{i+1} : \mathrm{Conf}(S, n + 1) \to \mathrm{Conf}(S, n)$$

given be deleting the $(i+1)$-st coordinate. These satisfy the identities $p_i \circ p_j = p_{j-1} \circ p_i$ for $i < j$ and induce the structure of Δ-group on the collection $\pi_1(\mathrm{Conf}(S, n))$, $n \geq 1$, for a path-connected surface S, as recorded in the next Definition, see [5].

Definition 16.3. Let S be a connected surface. Define

$$\Delta_\bullet(S)$$

by

$$\Delta_n(S) = P_{n+1}(S),$$

the $(n+1)$-st pure braid group for the surface S. By the previous example (together with a check of base-points), $\Delta_\bullet(S)$ is a Δ-group.

In case $S = \mathbb{CP}^1 = S^2$, the associated Δ-group gives basic information about the homotopy groups of the 2-sphere [5]. In case S_g, $g > 1$, is a closed oriented surface of genus g, the Δ-group $\Delta_\bullet(S_g)$ does not admit the structure of a simplicial group as given in the next definition.

Definition 16.4. A *simplicial set* is

(1) a Δ-set

$$K_\bullet = \{K_0, K_1, \ldots\}$$

together with

(2) functions, *degeneracy operations*,

$$s_j : K_t \to K_{t+1} \quad \text{for} \quad 0 \leq j \leq t$$

which satisfy the *simplicial identities*

$$d_i d_j = d_{j-1} d_i \text{ if } i < j, \quad s_i s_j = s_{j+1} s_i \text{ if } i \leq j, \text{ and}$$

$$d_i s_j = \begin{cases} s_{j-1} d_i & \text{if } i < j, \\ \text{identity} & \text{if } i = j \text{ or } i = j+1, \\ s_j d_{i-1} & \text{if } i > j+1. \end{cases}$$

A *simplicial-group*

$$G_\bullet = \{G_0, G_1, \ldots\}$$

is a simplicial-set for which all of the G_i are groups with faces and degeneracies given by group homomorphisms.

Example 16.5. Two examples of simplicial sets are given next.

(1) The simplicial 1-simplex $\Delta[1]$ has two 0-simplices $\langle 0 \rangle$ and $\langle 1 \rangle$. The n-simplices of $\Delta[1]$ are sequences $\langle 0^i, 1^{n+1-i} \rangle$ for $0 \leq i \leq n+1$. The non-degenerate simplices are $\langle 0 \rangle$, $\langle 1 \rangle$, and $\langle 0, 1 \rangle$.

(2) The simplicial circle S^1 is a quotient of the simplicial 1-simplex $\Delta[1]$ obtained by identifying $\langle 0 \rangle$ and $\langle 1 \rangle$. There are exactly two equivalence classes of non-degenerate simplices given by $\langle 0 \rangle$, and $\langle 0, 1 \rangle$. Furthermore, the simplicial circle S^1 is given in degree k by

(a) a single point $\langle 0 \rangle$ in case $k = 0$, and

(b) $n + 1$ points $\langle 0^i, 1^{n+1-i} \rangle$ for $0 \leq i < n+1$ in case $k = n$ for which $\langle 0^{n+1} \rangle$ and $\langle 1^{n+1} \rangle$ are identified.

In what follows below, it is useful to label these simplices by

$$y_{n+1-i} = \langle 0^i, 1^{n+1-i} \rangle$$

for $0 < i \leq n+1$ with $y_0 = s_0^n(\langle 0 \rangle)$.

Classical, elegant constructions for the standard simplicial n-simplex $\Delta[n]$ as well as the n-sphere are given in [4, 22, 57, 83].

Homotopy groups are defined for simplicial sets which satisfy an additional condition known as the (Kan) extension condition.

Example 16.6. A simplicial group $G_\bullet = \{G_0, G_1, \ldots\}$ always satisfies the extension condition, as shown in [61].

An example of a simplicial group obtained naturally from Artin's pure braid groups is described next.

Example 16.7. Consider Δ-groups with $\Delta_n(S) = P_{n+1}(S)$ as given in Example 16.2 for surfaces S. Specialize to the surface

$$S = \mathbb{R}^2.$$

In this case, there are also $n + 1$ homomorphisms

$$s_i \colon P_{n+1} \to P_{n+2},$$

with $0 \leq i \leq n$, where s_i is obtained by "doubling" the $(i + 1)$-st strand. The homomorphisms s_i are induced on the level of fundamental groups by the maps for configuration spaces

$$\mathbb{S}_i : \operatorname{Conf}(\mathbb{R}^2, n + 1) \to \operatorname{Conf}(\mathbb{R}^2, n + 2)$$

defined by the formula

$$\mathbb{S}_i(x_1, \ldots, x_{n+1}) = (x_1, \ldots, x_{i+1}, \lambda(x_{i+1}), x_{i+2}, \ldots, x_{n+1})$$

where $\lambda(x_{i+1}) = x_{i+1} + (\epsilon, 0)$ for $(\epsilon, 0)$ a point in $\mathbb{R}^2$ with

$$\epsilon = (1/2) \cdot \min_{t \neq i+1} ||x_{i+1} - x_t||.$$

The homomorphisms d_i and s_j satisfy the simplicial identities [21, 5].

Thus the pure braid groups, in the case $S = \mathbb{R}^2$, provide an example of a simplicial group, denoted

$$\mathrm{AP}_\bullet,$$

with

$$\mathrm{AP}_n = P_{n+1}$$

for $n = 0, 1, 2, 3, \ldots$.

Consider a pointed topological space $(X, *)$. The pointed loop space of X, $\Omega(X)$, has a natural multiplication coming from "loop sum" which is not associative, but homotopy associative. Milnor proved that the loop space of a connected simplicial complex is homotopy equivalent to a topological group [60]. James [47] proved that the loop space of the suspension of a connected CW-complex is naturally homotopy equivalent to a free monoid as explained in [43], page 282. Milnor [58] realized that the James construction could be translated directly into the language of simplicial groups as described next.

Definition 16.8. Let $K_\bullet$ denote a pointed simplicial set (with base-point $* \in K_0$ and $s_0^n(*) \in K_n$). Define *Milnor's free simplicial group $F[K]_\bullet$* by

$$F[K]_n = F[K_n]/s_0^n(*) = 1$$

for which

$$F[K]$$

denotes the free group generated by the set K.

Then $F[K]_\bullet$ is a simplicial group with face and degeneracy operations given by the natural multiplicative extension of those for $K_\bullet$. In addition, the face and degeneracy operations applied to a generator give either another generator or the identity element.

Example 16.9. An example of $F[K]_\bullet$ is given by $K_\bullet = S^1_\bullet$, the simplicial circle. Notice that $F[S^1]_n = F[y_1, \ldots, y_n]$, the free group on n generators, by Example 16.5.

Milnor defined the geometric realization of a simplicial set $K_\bullet = \{K_0, K_1, \ldots\}$ for which the underlying topology of $K_\bullet$ is discrete [59]. Recall the inclusion of the i-th face $\delta_i \colon \Delta[n-1] \to \Delta[n]$ together with the projection maps to the j-th face $\sigma_j \colon \Delta[n+1] \to \Delta[n]$ [4, 22, 57].

Definition 16.10. The *geometric realization of* $K_\bullet$ is

$$|K_\bullet| = \left(\coprod K_n \times \Delta[n] \right) \Big/ \sim$$

where $\sim$ denotes the equivalence relation generated by requiring:

(1) if $x \in K_{n+1}$ and $\alpha \in \Delta[n]$, then $(d_i(x), \alpha) \sim (x, \delta_i(\alpha))$, and
(2) if $y \in K_n$ and $\beta \in \Delta[n+1]$, then $(x, \sigma_j(\beta)) \sim (s_j(x), \beta)$.

Theorem 16.11. *If $K_\bullet$ is a reduced simplicial set (that is K_0 is equal to a single point $\{*\}$), then the geometric realization $|F[K]_\bullet|$ is homotopy equivalent to $\Omega\Sigma|K_\bullet|$. Thus the homotopy groups of $F[K]_\bullet$ (as given in [61] and mentioned above in Example 16.6) are isomorphic to the homotopy groups of the space $\Omega\Sigma|K_\bullet|$.*

Example 16.12. Consider the special case of $K_\bullet = S^1_\bullet$. Then the geometric realization $|F[S^1]_\bullet|$ is homotopy equivalent to ΩS^2, and there are isomorphisms

$$\pi_n(F[S^1]_\bullet) \to \pi_n \Omega S^2 \cong \pi_{n+1} S^2.$$

A partial synthesis of this information is given in Secs. 19 through 21.

17. Pure Braid Groups, and Vassiliev Invariants

The section addresses a naive construction with the braid groups arising as a "cabling" construction. This construction is interpreted in later sections in terms of the structure of braid groups, Vassiliev invariants of pure braids as developed by Toshitake Kohno [49, 51], associated Lie algebras and the homotopy groups of the 2-sphere [21, 5, 81].

Recall from Definition 8.1, B_k denotes Artin's k-stranded braid group while P_k denotes the pure k-stranded braid group. Furthermore, the group B_k is the fundamental group of the orbit space

$$\mathrm{Conf}(\mathbb{R}^2, k)/\Sigma_k,$$

and the pure braid group $P_k(S)$ is the fundamental group

$$\pi_1(\mathrm{Conf}(S, k)).$$

The pure braid groups P_k and $P_k(S^2)$ are closely related to the loop space of the 2-sphere as elucidated below in Sec. 16. Similar properties are satisfied for any sphere, as described in Sec. 21. We will now start to address this point.

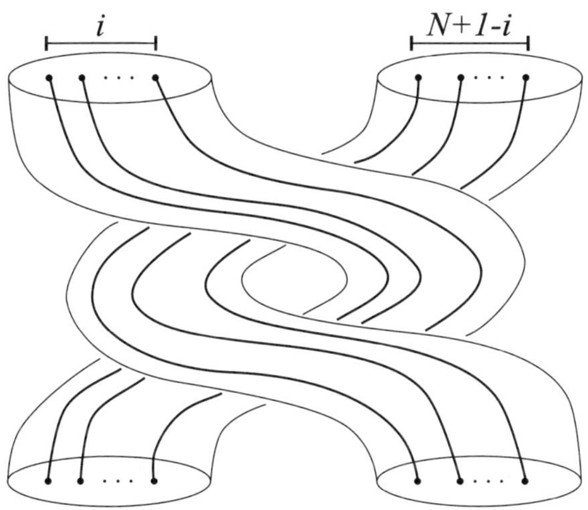

Fig. 2.   The braid x_i in P_{N+1}.

Consider the free group on N letters $F_N = F[y_1, \ldots, y_N]$ and elements x_i in P_{N+1} for $1 \le i \le N$, with x_i given by the naive "cabling" pictured in Fig. 2 above. The braid x_1 with $N = 1 = i$ in Fig. 2 is Artin's generator $A_{1,2}$ of P_2. The braids x_i yield homomorphisms from F_N to P_{N+1},

$$\Theta_N : F[y_1, \ldots, y_N] \to P_{N+1}$$

defined on generators y_i in F_N by the formula

$$\Theta_N(y_i) = x_i.$$

The maps Θ_N are the subject of [21] where it is shown that $\Theta_N : F_N \to P_{N+1}$ is faithful for every N. Three natural questions arise: (1) Why would one want to know whether Θ_n is faithful, (2) are there sensible applications and (3) why is Θ_n faithful? The answers to these three questions provide the main content of this expository article.

18. On Θ_n

This section addresses one reason why the map Θ_n is faithful [21]. The method of proof is to appeal to the structure of the Lie algebras obtained from the descending central series for both the source and the target of Θ_n. The structure of these Lie algebras is reviewed below.

Recall the descending central series of a discrete group π, given by

$$\pi = \Gamma^1(\pi) \geq \Gamma^2(\pi) \geq \cdots$$

where $\Gamma^i(\pi)$ is the subgroup of π generated by all commutators

$$[[\cdots [[x_1, x_2]x_3] \cdots]x_t]$$

for $t \geq i$ with $x_i \in \pi$. The group $\Gamma^i(\pi)$ is a normal subgroup of π with the successive sub-quotients

$$\mathrm{gr}_i(\pi) = \Gamma^i(\pi)/\Gamma^{i+1}(\pi),$$

which are abelian groups, having additional structure as follows [55].

Consider the direct sum of all of the $\mathrm{gr}_i(\pi) = \Gamma^i(\pi)/\Gamma^{i+1}(\pi)$ denoted

$$\mathrm{gr}_*(\pi) = \oplus_{i \geq 1}\Gamma^i(\pi)/\Gamma^{i+1}(\pi).$$

The commutator function

$$[-,-] : \pi \times \pi \to \pi,$$

given by

$$[x, y] = xyx^{-1}y^{-1},$$

passes to quotients to give a bilinear map

$$[-,-] : \mathrm{gr}_s(\pi) \otimes_{\mathbb{Z}} \mathrm{gr}_t(\pi) \to \mathrm{gr}_{s+t}(\pi)$$

which satisfies both the antisymmetry law and Jacobi identity for a Lie algebra.

Remark 18.1. The abelian group $\mathrm{gr}_*(\pi)$ is both a graded abelian group and a Lie algebra, but not a graded Lie algebra as the sign conventions do not work properly in this context. This situation can be remedied by doubling all degrees of elements in $\mathrm{gr}_*(\pi)$.

The associated graded Lie algebra obtained from the descending central series for the target yields Vassiliev invariants of pure braids, by work of Kohno [49, 51]. This Lie algebra has been used by both Kohno and Drinfel'd [25] in their work on the KZ equations. The Lie algebra obtained from the descending central series of the free group F_N is a free Lie algebra, by a classical result due to P. Hall [41, 67].

The proof described next yields more information than just the fact that Θ_N is faithful. The method of proof gives a natural connection of Vassiliev invariants of braids to a classical spectral sequence abutting to the homotopy groups of the 2-sphere. Sections 19 through 21 below provide an elucidation of this interconnection.

A discrete group π is said to be residually nilpotent provided

$$\bigcap_{i \geq 1} \Gamma^i(\pi) = \{\text{identity}\}$$

where $\Gamma^i(\pi)$ denotes the i-th stage of the descending central series for π. Examples of residually nilpotent groups are free groups, and P_n.

Lemma 18.2. (1) *Assume that π is a residually nilpotent group. Let*

$$\alpha \colon \pi \to G$$

be a homomorphism of discrete groups such that the morphism of associated graded Lie algebras

$$\mathrm{gr}_*(\alpha) : \mathrm{gr}_*(\pi) \to \mathrm{gr}_*(G)$$

is a monomorphism. Then α is a monomorphism.
(2) *If π is a free group, and $\mathrm{gr}_*(\alpha)$ is a monomorphism, then α is a monomorphism.*

Thus one step in the proof of Theorem 18.3 below is to describe the map

$$\Theta_n \colon F[y_1, y_2, \ldots, y_n] \to P_{n+1}$$

on the level of associated graded Lie algebras

$$\mathrm{gr}_*(\Theta_n) \colon \mathrm{gr}_*(F[y_1, y_2, \ldots, y_n]) \to \mathrm{gr}_*(P_{n+1}).$$

Recall Artin's generators $A_{i,j}$ for P_{n+1} together with the projections of the $A_{i,j}$ to $\mathrm{gr}_*(P_{n+1})$, labeled $B_{i,j}$ [21]. The next theorem was proven in [21] by a direct computation.

Theorem 18.3. *The induced morphism of Lie algebras*

$$\mathrm{gr}_*(\Theta_n) \colon \mathrm{gr}_*(F[y_1, y_2, \ldots, y_n]) \to \mathrm{gr}_*(P_{n+1})$$

satisfies the formula

$$\mathrm{gr}_*(\Theta_n)(y_q) = \Sigma_{1 \leq i \leq n-q+1 < j \leq n+1} B_{i,j}.$$

Examples of this theorem are listed next.

Example 18.4.   (1) If $q = 1$, then

$$\mathrm{gr}_*(\Theta_n)(y_1) = B_{1,n+1} + B_{2,n+1} + \cdots + B_{n,n+1}.$$

Thus if $q = 1$, and $n = 3$,

$$\mathrm{gr}_*(\Theta_3)(y_1) = B_{1,4} + B_{2,4} + B_{3,4}.$$

(2) If $q = 2$, then

$$\mathrm{gr}_*(\Theta_n)(y_2) = (B_{1,n+1} + B_{2,n+1} + \cdots + B_{n-1,n+1})$$
$$+ (B_{1,n} + B_{2,n} + \cdots + B_{n-1,n}).$$

Thus if $q = 2$, and $n = 3$,

$$\mathrm{gr}_*(\Theta_3)(y_2) = (B_{1,4} + B_{2,4}) + (B_{1,3} + B_{2,3}).$$

(3) In general,

$$\mathrm{gr}_*(\Theta_n)(y_q) = V_{n-q+2} + V_{n-q+3} + \cdots + V_{n+1}$$

where

$$V_r = B_{1,r} + B_{2,r} + \cdots + B_{r-1,r}.$$

Thus if $q = 3$, and $n = 3$,

$$\mathrm{gr}_*(\Theta_3)(y_3) = B_{1,2} + B_{1,3} + B_{1,4}.$$

To determine the map of Lie algebras with a more global view, the structure of the Lie algebra $\mathrm{gr}_*(P_n)$ is useful, and is given as follows. Let $L[S]$ denote the free Lie algebra over $\mathbb{Z}$ generated by a set S. The next theorem was given in work of Kohno [49, 51] and Falk-Randell [34].

Theorem 18.5. *The Lie algebra* $\mathrm{gr}_*(P_n)$ *is the quotient of the free Lie algebra generated by* $B_{i,j}$ *for* $1 \leq i < j \leq n$ *given by*

$$\mathrm{gr}_*(P_n) = L[B_{i,j}\mid 1 \leq i < j \leq n]/I$$

where I *denotes the 2-sided (Lie) ideal generated by the infinitesimal braid relations listed next:*

(1) $[B_{i,j}, B_{s,t}] = 0$, *if* $\{i, j\} \cap \{s, t\} = \emptyset$.
(2) $[B_{i,j}, B_{i,s} + B_{s,j}] = 0$.
(3) $[B_{i,j}, B_{i,t} + B_{j,t}] = 0$.

Remark 18.6. The Lie algebra in Theorem 18.5, apart from a natural degree shift, is precisely the one arising in Theorem 12.7 and gives the homology of the loop space of the configuration space. We will see below, in Sec. 20, that this same Lie algebra gives the E^1-term of the Bousfield-Kan spectral sequence abutting to certain homotopy groups.

A computation with these maps gives the following result of [21]. Further connections regarding this result are elucidated in Secs. 19 and 21.

Theorem 18.7. *The maps* $\Theta_n \colon F[y_1, y_2, \ldots, y_n] \to P_{n+1}$ *on the level of associated graded Lie algebras*

$$\mathrm{gr}_*(\Theta_n) \colon \mathrm{gr}_*(F[y_1, y_2, \ldots, y_n]) \to \mathrm{gr}_*(P_{n+1})$$

are monomorphisms. Thus the maps Θ_n *are monomorphisms.*

Remark 18.8. Two remarks concerning Θ_n are given next.

(1) That Θ_n is a monomorphism identifies $F[y_1, y_2, \ldots, y_n]$ as a free subgroup of rank n in P_{n+1}. However, there are other, natural free groups of rank n in P_{n+1}. These arise from the fibrations of Fadell and Neuwirth given by the projection maps

$$p_{i+1} : \mathrm{Conf}(\mathbb{R}^2, n+1) \to \mathrm{Conf}(\mathbb{R}^2, n)$$

which delete the $(i+1)$-st coordinate and have fiber of the homotopy type of an n-fold wedge of circles $\vee_n S^1$ [32].

Recall that $d_i : P_{n+1} \to P_n$ denotes the map induced by p_{i+1} on the level of fundamental groups. The kernel of d_i is a free group of rank n.

The image of Θ_n has a contrasting feature: Any composite of the natural projection maps, $d_I : P_{n+1} \to P_2$, precomposed with Θ_n,

$$F[y_1, y_2, \ldots, y_n] \xrightarrow{\Theta_n} P_{n+1} \xrightarrow{d_I} P_2,$$

is a surjection.

(2) The combinatorial behavior of the map $\mathrm{gr}_*(\Theta_n)$ is intricate even though the definition is elementary as well as natural. For example, various powers of 2 arise in the computation of the map

$$\mathrm{gr}_*(\Theta_n) \colon \mathrm{gr}_*(F[y_1, y_2, \ldots, y_n]) \to \mathrm{gr}_*(P_{n+1})$$

for $n > 2$. One example is listed next.

Example 18.9. $\Theta_3([[[y_1, y_2]y_3]y_2]) = -[[[\gamma_1, \gamma_2]\gamma_3]\gamma_2] + 2[[[\gamma_1, \gamma_3]\gamma_2]\gamma_2] + \delta$ where δ is independent of the other terms with $\gamma_1 = B_{1,4} + B_{2,4} + B_{3,4}$,

$\gamma_2 = B_{3,4}$ and $\gamma_3 = B_{2,4} + B_{3,4}$. At first glance, these elements may appear to be "random". However, this formula represents a systematic behavior which arises naturally from kernels of certain morphisms of Lie algebras.

The crucial feature which makes the computations effective is the "infinitesimal braid relations". In addition, the behavior of the map $\mathrm{gr}_*(\Theta_n)$ is more regular after restricting to certain sub-Lie algebras arising in the third stage of the descending central series [21]. Finally, the maps Θ_n also induce monomorphisms of restricted Lie algebras on passage to the Lie algebras obtained from the mod-p descending central series [21].

19. Pure Braid Groups of Surfaces as Simplicial Groups and Δ-Groups

The homomorphism $\Theta_n : F[y_1, y_2, \ldots, y_n] \to P_{n+1}$ which arises from the cabling operation described in Fig. 2 satisfies the following properties:

(1) The homomorphisms $\Theta_n : F[y_1, y_2, \ldots, y_n] \to P_{n+1}$ give a morphism of simplicial groups

$$\Theta : F[S^1]_\bullet \to \mathrm{AP}_\bullet$$

for which the homomorphism Θ_n is the restriction of Θ to $F[S^1]_n$.

(2) By Theorem 18.7, the homomorphisms $\Theta_n : F[y_1, y_2, \ldots, y_n] \to P_{n+1}$ are monomorphisms and so the morphism $\Theta : F[S^1] \to \mathrm{AP}_\bullet$ is a monomorphism of simplicial groups.

(3) There is exactly one morphism of simplicial groups Θ with the property that $\Theta_1(y_1) = A_{1,2}$.

Thus, the picture given in Fig. 2 is a description for generators of $F[S^1]_n$ in the simplicial group $F[S^1]_\bullet$. These features are summarized next.

Theorem 19.1. *The homomorphisms*

$$\Theta_n : F[y_1, y_2, \ldots, y_n] \to P_{n+1}$$

("pictured" in Fig. 2) give the unique morphism of simplicial groups

$$\Theta : F[S^1]_\bullet \to \mathrm{AP}_\bullet$$

with $\Theta_1(y_1) = A_{1,2}$. The map Θ is an embedding. Hence the n-th homotopy group of $F[S^1]$, isomorphic to $\pi_{n+1}(S^2)$, is a natural sub-quotient of AP_n. Furthermore, the smallest sub-simplicial group of $\mathrm{AP}_\bullet$ which contains the element $\Theta_1(y_1) = A_{1,2}$ is isomorphic to $F[S^1]_\bullet$.

On the other hand, the homotopy sets for the Δ-group $\Delta_\bullet(S^2)$ also give the homotopy groups of the 2-sphere, via a different occurrence of $F[S^1]_\bullet$. The homeomorphism of spaces

$$\mathrm{Conf}(S^2, k) \to PGL(2, \mathbb{C}) \times \mathrm{Conf}(S^2 - Q_3, k - 3),$$

for $k \geq 3$ and where Q_3 denotes a set of three distinct points in S^2, is basic for the next theorem [5].

Theorem 19.2. *If $S = S^2$ and $n \geq 4$, then there are isomorphisms*

$$\pi_n(\Delta_\bullet(S^2)) \to \pi_n(S^2).$$

The descriptions of homotopy groups implied by these theorems admit interpretations in terms of classical, well-studied features of the braid groups as given in the next section. An extension to all spheres is given in [21], as pointed out in Sec. 21.

20. Brunnian Braids, "Almost Brunnian" Braids, and Homotopy Groups

The homotopy groups of a simplicial group, or the homotopy sets of a Δ-group, admit a combinatorial description, as discussed in Lemma 20.3 below. These homotopy sets are the set of left cosets Z_n/B_n where Z_n is the group of n-cycles and B_n is the group of n-boundaries for the Δ-group.

Recall Example 16.2 in which the Δ-group $\Delta_\bullet(S)$ is specified by $\Delta_n(S) = P_{n+1}(S)$, the $(n+1)$-stranded pure braid group for a connected surface S. The main point of this section is that the n-cycles Z_n are given by the "Brunnian braids" while the n-boundaries B_n are given by the "almost Brunnian braids", subgroups considered next which are also important in other applications [56].

Definition 20.1. Consider the n-stranded pure braid group for any (connected) surface S, the fundamental group of $\mathrm{Conf}(S, n)$. The group of Brunnian braids $\mathrm{Brun}_n(S)$ is the subgroup of $P_n(S)$ given by those braids which become trivial after deleting any single strand. That is,

$$\mathrm{Brun}_n(S) = \bigcap_{0 \leq i \leq n-1} \ker(d_i \colon P_n(S) \to P_{n-1}(S))$$

for which

$$d_i = (p_{i+1})_* : P_n(S) \to P_{n-1}(S).$$

The "almost Brunnian" $(n+1)$-stranded braid group is

$$\text{QBrun}_{n+1}(S) = \bigcap_{1 \leq i \leq n} \ker(d_i \colon P_{n+1}(S) \to P_n(S)).$$

The subgroup $\text{QBrun}_{n+1}(S)$ of $P_{n+1}(S)$ consists of those braids which are trivial after deleting any one of the strands $2, 3, \ldots, n+1$, but not necessarily the first.

Example 20.2. Consider the simplicial group $\text{AP}_\bullet$ with

$$\text{AP}_n = P_{n+1}$$

for $n = 0, 1, 2, 3, \ldots$ as given in Example 16.7.

In this case, notice that the map $d_0 \colon \text{QBrun}_{k+2} \to \text{Brun}_{k+1}$ is a split epimorphism. Thus the homotopy groups of the simplicial group $\text{AP}_\bullet$ are all trivial.

An inspection of definitions gives the next lemma.

Lemma 20.3. *Let S denote a connected surface with associated Δ-group $\Delta_\bullet(S)$ (as given in Example 16.2). Then the following hold.*

(1) *The group $\text{Brun}_{n+1}(S)$ is equal to the group of n-cycles $Z_n(S)$.*
(2) *The group of boundaries $B_n(S)$ is $d_0(\text{QBrun}_{n+2}(S))$.*
(3) *There is an isomorphism*

$$\pi_k(\text{AP}_\bullet) \to \text{Brun}_{k+1}/d_0(\text{QBrun}_{k+2}).$$

Furthermore, $\pi_k(\text{AP}_\bullet)$ is the trivial group.
(4) *There is an isomorphism of left cosets which is natural for pointed embeddings of connected surfaces S*

$$\pi_k(\Delta_*(S)) \to \text{Brun}_{k+1}(S)/d_0(\text{QBrun}_{k+2}(S)).$$

Properties of the Δ-group for the 2-sphere $S = \mathbb{CP}^1 = S^2$ is the main subject of [5] where the next result is proven.

Theorem 20.4. *If $S = S^2$ and $k \geq 4$, then*

$$\pi_k(\Delta_\bullet(S^2)) = \text{Brun}_{k+1}(S^2)/d_0(\text{QBrun}_{k+2}(S^2))$$

is a group which is isomorphic to the classical homotopy group $\pi_k(S^2)$.
Furthermore, if $k \geq 4$, there is an exact sequence of groups

$$1 \to \text{Brun}_{k+2}(S^2) \to \text{Brun}_{k+1}(\mathbb{R}^2) \to \text{Brun}_{k+1}(S^2) \to \pi_k(S^2) \to 1.$$

Remark 20.5. Recently, the authors have proven (unpublished) that the Brunnian braid group $\mathrm{Brun}_4(S^2)$ is isomorphic to the principal congruence subgroup of level 4 in $PSL(2,\mathbb{Z})$ [5]. (This fact is checked in the appendix here.)

This identification may admit an extension by considering the Brunnian braid groups $\mathrm{Brun}_{2g}(S^2)$ as natural subgroups of mapping class groups for genus g surfaces. The subgroups $\mathrm{Brun}_{2g}(S^2)$ may embed naturally in $Sp(2g,\mathbb{Z})$ via classical surface topology using branched covers of the 2-sphere (work in progress). It seems reasonable to conjecture that this is correct.

The next lemma follows by a direct check of the long exact homotopy sequence obtained from the Fadell-Neuwirth fibrations for configuration spaces [32, 31].

Lemma 20.6. *If S is a surface not homeomorphic to either S^2 or $\mathbb{RP}^2$, and $k \geq 3$, then $\mathrm{Brun}_k(S)$ and $\mathrm{QBrun}_k(S)$ are free groups. If S is any surface, and $k \geq 4$, then $\mathrm{Brun}_k(S)$ and $\mathrm{QBrun}_k(S)$ are free groups.*

Lemma 20.7. *If $k \geq 3$, then $\Theta_k(F_k) \cap \mathrm{Brun}_{k+1}$ as well as $\Theta_k(F_k) \cap d_0(\mathrm{QBrun}_{k+2})$ are countably infinitely generated free groups.*

The standard Hall collection process or natural variations can be used to give inductive recipes rather than closed forms for generators. T. Stanford has given a related elegant exposition of the Hall collection process [74]. The analogous process was applied in joint work of Ran Levi and the author to give group theoretic models for iterated loop spaces (available on Levi's website).

The connection to the homotopy groups of S^2, as well as to the Lie algebra attached to the descending central series of the pure braid groups, is discussed next.

Theorem 20.8. *The group $\Theta_k(F_k) \cap d_0(\mathrm{QBrun}_{k+2})$ is a normal subgroup of $\Theta_k(F_k) \cap \mathrm{Brun}_{k+1}$. There are isomorphisms*

$$\Theta_k(F_k) \cap \mathrm{Brun}_{k+1}/\Theta_k(F_k) \cap d_0(\mathrm{QBrun}_{k+2}) \to \pi_{k+1}S^2.$$

The method of proving that the maps $\Theta_n \colon F[y_1, y_2, \ldots, y_n] \to P_{n+1}$ are monomorphisms via Lie algebras admits an interpretation in terms of classical homotopy theory. The method is to filter both simplicial groups $F[S^1]_\bullet$ and $\mathrm{AP}_\bullet$ via the descending central series, and then to analyze the natural map on the level of associated graded Lie algebras.

On the other hand, the Lie algebra arising from filtering any simplicial group by its descending central series gives the E^0-term of the Bousfield-Kan spectral sequence for the simplicial group in question [4]. Similarly, filtering via the mod-p descending central series gives the classical unstable Adams spectral sequence [4, 22, 81].

Thus the method of proof of Theorem 18.7 is precisely an analysis of the behavior of the natural map $\Theta \colon F[S^1]_\bullet \to \mathrm{AP}_\bullet$ on the level of the E^0-term of the Bousfield-Kan spectral sequence. This method exhibits a close connection between Vassiliev invariants of pure braids and these natural spectral sequences. The next result is restatement of Theorem 18.7 proven in [21].

Corollary 20.9. *The maps* $\Theta_n \colon F[y_1, y_2, \ldots, y_n] \to P_{n+1}$ *on the level of associated graded Lie algebras*

$$\mathrm{gr}_*(\Theta_n) \colon \mathrm{gr}_*(F[y_1, y_2, \ldots, y_n]) \to \mathrm{gr}_*(P_{n+1})$$

are monomorphisms. Thus the maps Θ_n *induce embeddings on the level of the* E^0*-term of the Bousfield-Kan spectral sequences for* $E^0(\Theta) \colon E^0(F[S^1]_\bullet) \to E^0(\mathrm{AP}_\bullet)$.

21. Other Connections

Connections to other spheres: The work above has been extended to all spheres, as well as other connected CW-complexes [21]. One way in which other spheres arise is via the induced embedding of free products of simplicial groups

$$\Theta \amalg \Theta \colon F[S^1]_\bullet \amalg F[S^1]_\bullet \to \mathrm{AP}_\bullet \amalg \mathrm{AP}_\bullet.$$

The geometric realization of $F[S^1]_\bullet \amalg F[S^1]_\bullet$ is homotopy equivalent to $\Omega(S^2 \vee S^2)$ by Milnor's theorem stated above as Theorem 16.11. Furthermore, $\Omega(S^2 \vee S^2)$ is homotopy equivalent to a weak infinite product of spaces $\Omega(S^n)$ for all $n > 1$.

Connection to certain Galois groups: Consider automorphism groups $\mathrm{Aut}(H)$ where H is one of F_n, the pro-finite completion $\widehat{F_n}$ or the pro-ℓ completion $\widehat{(F_n)}_\ell$. Certain Galois groups G are identified as natural subgroups of these automorphism groups in [2, 23, 25, 26, 45, 46, 68]. One example is Drinfel'd's Grothendieck-Teichmüller Galois group $G = \widehat{GT}$, a subgroup of $\mathrm{Aut}(\widehat{F_2})$.

Let $\mathrm{Der}(L^R[V_n])$ denote the Lie algebra of derivations of the free Lie algebra $L^R[V_n]$ where V_n denotes a free module of rank n over R a commutative ring with identity. Two natural morphisms of Lie algebras, which take values in $\mathrm{Der}(L^R[V_n])$, occur in this context, as follows.

Recall the adjoint representation

$$Ad : \mathrm{gr}_*(P_{n+1}) \to \mathrm{Der}(\mathrm{gr}_*(P_{n+1}))$$

defined by the equation $Ad(x)(y) = [y, x]$. Observe that the infinitesimal braid relations as stated in Theorem 18.5 give that the map $Ad(x)(-)$ restricted to $L^{\mathbb{Z}}[V_n]$ also preserves $L^{\mathbb{Z}}[V_n]$. Thus there is an induced second natural map

$$Ad : \mathrm{gr}_*(P_{n+1}) \to \mathrm{Der}(L^{\mathbb{Z}}[V_n])$$

whose kernel is precisely the center of $\mathrm{gr}_*(P_{n+1})$ [17]. Combining this last fact with Theorem 18.3 gives properties of the composite morphism of Lie algebras

$$\mathrm{gr}_*(F_n) \xrightarrow{\mathrm{gr}_*(\Theta_n)} \mathrm{gr}_*(P_{n+1}) \xrightarrow{Ad} \mathrm{Der}(L^{\mathbb{Z}}[V_n]).$$

Proposition 21.1. *If $n \geq 2$, the induced morphism of Lie algebras*

$$Ad \circ \mathrm{gr}_*(\Theta_n) \colon \mathrm{gr}_*(F[y_1, y_2, \dots, y_n]) \to \mathrm{Der}(L^{\mathbb{Z}}[V_n])$$

is a monomorphism.

In addition, certain Galois groups G above are filtered with induced morphisms of Lie algebras

$$\mathrm{gr}_*(G) \to \mathrm{Der}(L^{\widehat{\mathbb{Z}}}[V_n])$$

where $\widehat{\mathbb{Z}}$ denotes the pro-finite completion of the integers. One example is $G = \widehat{GT}$ with

$$\mathrm{gr}_*(\widehat{GT}) \to \mathrm{Der}(L^{\widehat{\mathbb{Z}}}[V_2]),$$

as given in [23, 45, 46, 68].

This raises the question of (i) whether the images of

$$Ad \circ \mathrm{gr}_*(\Theta_2),$$

and

$$\mathrm{gr}_*(\widehat{GT})$$

in $\mathrm{Der}(L^{\widehat{\mathbb{Z}}}[V_2])$ have a non-trivial intersection, or (ii) whether this intersection is meaningful.

22. Questions

The point of this section is to consider whether the connections between the braid groups and homotopy groups above are useful. Some natural, as well as speculative, problems are listed next.

The combinatorial problem of distinguishing elements in the pure braid groups has been well-studied. For example, the Lie algebra associated to the descending central series of the pure braid group P_n has been connected with Vassiliev theory and has been shown to be a complete set of invariants which distinguish all elements in P_n [51]. Furthermore, these Lie algebras have been applied to other questions arising from the classical KZ-equations [49, 25] as well as the structure of certain Galois groups [45, 23, 25, 26].

Question 1. One description of homotopy groups is given in Theorem 20.8. This relation is coarser than that given by Vassiliev invariants. Give methods of understanding this coarser relation.

Question 2. Consider Brunnian braids Brun_k. Fix a braid γ with image in the k-th symmetric group Σ_k given by a k-cycle. For any braid α in Brun_k, the braid closure of $\alpha \circ \gamma$ is a knot. Describe features of these knots or those obtained from the analogous constructions for $\Theta_k(F_{k-1}) \cap \mathrm{Brun}_k$. Where do these fit in Budney's description of the space of long knots [8]?

Question 3. Give combinatorial properties of the natural map $\mathrm{Brun}_{k+1}(\mathbb{R}^2) \to \mathrm{Brun}_{k+1}(S^2)$ which provide information about the cokernel. Two concrete problems are stated next.

(1) Give group theoretic reasons why the order of the 2-torsion in $\pi_*(S^2)$ is bounded above by 4 and why the p-torsion for an odd prime p is bounded above by p.
(2) If $k + 1 \geq 5$, the image of $\mathrm{Brun}_{k+1}(\mathbb{R}^2) \to \mathrm{Brun}_{k+1}(S^2)$ is a normal subgroup of finite index.

    This fact follows from Serre's classical theorem that $\pi_k(S^2)$ is finite for $k > 3$ and Theorems 19.2 and 20.4, proven in [5].

    Do natural features of the braid groups imply this result?

Question 4. Let F_n denote the image of $\Theta_n(F_n)$. Observe that the groups $\mathrm{QBrun}_{n+2} \cap F_{n+1}$, and $\mathrm{Brun}_{n+1} \cap F_n$ are free. Furthermore, there is a short exact sequence of groups

$$(\text{Extension 1}): \ 1 \to F_n \cap d_0(\mathrm{QBrun}_{n+2}) \to F_n \cap \mathrm{Brun}_{n+1} \to \pi_{n+1}S^2 \to 1$$

as well as isomorphisms

$$F_n \cap \mathrm{Brun}_{n+1}/(F_n \cap d_0(\mathrm{QBrun}_{n+2})) \to \pi_{n+1}S^2,$$

by Theorem 20.8.

Consider the Serre 5-term exact sequence for the group extension given by Extension 1 directly above to obtain information about the induced surjection

$$H_1(F_n \cap \mathrm{Brun}_{n+1}) \to \pi_{n+1}(S^2).$$

This 5-term exact sequence specializes to

$$H_2(\pi_{n+1}(S^2)) \to H_1(F_n \cap d_0(\mathrm{QBrun}_{n+2}))_{\pi_{n+1}(S^2)}$$
$$\to H_1(F_n \cap \mathrm{Brun}_{n+1}) \to \pi_{n+1}(S^2)$$

where A_π denotes the group of co-invariants of a π-module A. Thus $\pi_{n+1}(S^2)$ is a quotient of the free abelian group $H_1(F_n \cap \mathrm{Brun}_{n+1})$ with relations given by the image of the co-invariants $H_1(F_n) \cap d_0(\mathrm{QBrun}_{n+2})_{\pi_{n+1}(S^2)}$.

Give combinatorial descriptions of the induced map on the level of the first homology groups

$$H_1(F_n \cap d_0(\mathrm{QBrun}_{n+2})) \to H_1(F_n \cap \mathrm{Brun}_{n+1}).$$

A similar problem arises with the epimorphism $\mathrm{Brun}_{n+1}(S^2) \to \pi_n S^2$ with kernel in the image of $\mathrm{Brun}_{k+1}(\mathbb{R}^2)$ for $n + 1 \geq 5$.

Question 5. Let $L[V]$ denote the free Lie algebra over the integers generated by the free abelian group V. Let $\mathrm{Der}(L[V])$ denote the Lie algebra of derivations of $L[V]$ and consider the classical adjoint representation

$$\mathrm{Ad} \colon L[V] \to \mathrm{Der}(L[V]).$$

Recall that the map $\Theta_k \colon F_k \to P_{k+1}$ induces a monomorphism of Lie algebras

$$\mathrm{gr}_*(\Theta_k) \colon \mathrm{gr}_*(F_k) \to \mathrm{gr}_*(P_{k+1})$$

where $\mathrm{gr}_*(F_k)$ is isomorphic to the free Lie algebra $L[V_k]$ with V_k a free abelian group of rank k. In addition, properties of the "infinitesimal braid relations" give a representation

$$Ad \colon \mathrm{gr}_*(P_{k+1}) \to \mathrm{Der}(L[V_k])$$

as given in Sec. 21 and appearing in work on certain Galois groups [45] (with the integers $\mathbb{Z}$ replaced by the pro-finite completion of $\mathbb{Z}$).

Identify F_k with $\Theta_k(F_k)$ in what follows below. Give methods to describe combinatorial properties of the composite

$$\mathrm{gr}_*(F_k \cap \mathrm{Brun}_{k+1}) \xrightarrow{\mathrm{gr}_*(i_k)} \mathrm{gr}_*(F_k) \xrightarrow{\mathrm{gr}_*(\Theta_k)} \mathrm{gr}_*(P_{k+1}) \xrightarrow{Ad} \mathrm{Der}(L[V_k])$$

where $i_k : F_k \cap \mathrm{Brun}_{k+1} \to F_k$ is the natural inclusion. Let Φ_{k+1} denote this composite. When restricted to $\mathrm{gr}_*(F_k) = L[V_k]$, this map is a monomorphism. Give methods to describe the sub-quotient

$$\Phi_{k+1}(\mathrm{gr}_*(F_k \cap \mathrm{Brun}_{k+1}))/\Phi_{k+1}(\mathrm{gr}_*(F_k \cap d_0(\mathrm{QBrun}_{k+2}))).$$

Question 6. Assume that the pure braid groups $P_n(S)$ are replaced by either their pro-finite completions $\widehat{P_n(S)}$ or their pro-ℓ completions. Describe the associated changes for the homotopy groups arising in Theorems 19.2, 20.4, or 19.1. For example, is the torsion in these homotopy groups left unchanged by replacing $P_n(S)$ by $\widehat{P_n(S)}$?

23. Appendix. Brunnian Braids and Principal Congruence Subgroups

The Brunnian braid group has features in common with the Borromean rings dating back at least to Carlo Borromeo in 1560. Related structures for Brunnian links are given in [56].

The purpose of this section is to record an observation concerning Brunnian braid groups for the 2-sphere and to pose a related question as well as to describe a starting point to appear in joint work with Berrick, Wong and Wu.

Recall that there are classical maps

$$r : B_4 \to B_3$$

and

$$\Theta : B_3 \to SL(2, \mathbb{Z}).$$

The map $r : B_4 \to B_3$ is defined by the formula

$$r(\sigma_i) = \begin{cases} \sigma_i & \text{if } i = 1 \text{ or } i = 2, \\ \sigma_1 & \text{if } i = 3. \end{cases}$$

The map

$$\Theta : B_3 \to SL(2, \mathbb{Z})$$

is defined by the formula

$$\Theta(\sigma_1) = \begin{pmatrix} 1 & 0 \\ -1 & 1 \end{pmatrix}$$

and

$$\Theta(\sigma_2) = \begin{pmatrix} 1 & 1 \\ 0 & 1 \end{pmatrix}.$$

The map Θ arises from a map of B_3 to the mapping class group for genus 1 surfaces, $SL(2,\mathbb{Z})$, obtained via two Dehn twists along two circles which intersect in a single point [3].

Recall that $\Gamma(2,r)$ denotes the kernel of the mod-r reduction map

$$\rho_r : SL(2,\mathbb{Z}) \to SL(2,\mathbb{Z}/r\mathbb{Z})$$

(in Sec. 10). Similarly, let $P\Gamma(2,r)$ denote the kernel of the mod-r reduction map

$$\rho_r : PSL(2,\mathbb{Z}) \to PSL(2,\mathbb{Z}/r\mathbb{Z}).$$

The following is the main result of this section in which D_8 denotes the dihedral group of order 8.

Theorem 23.1. *The classical maps*

$$r : B_4 \to B_3$$

and

$$\Theta : B_3 \to SL(2,\mathbb{Z})$$

induce maps which give

(1) *a homomorphism $B_4(S^2) \to PSL(2,\mathbb{Z})$ together with a short exact sequence of groups*

$$1 \to D_8 \to B_4(S^2) \to PSL(2,\mathbb{Z}) \to 1,$$

(2) *a homomorphism $P_4(S^2) \to P\Gamma(2,2)$ together with a split short exact sequence of groups*

$$1 \to \mathbb{Z}/2\mathbb{Z} \to P_4(S^2) \to P\Gamma(2,2) \to 1$$

with an isomorphism

$$P_4(S^2) \to \Gamma(2,2),$$

and

(3) *a homomorphism* $\mathrm{Brun}_4(S^2) \to P\Gamma(2,4) = \Gamma(2,4)$ *which is an isomorphism.*

The previous theorem may admit an extension to other Brunnian braid groups

$$\mathrm{Brun}_{2g+2}(S^2).$$

Preparation for this possible extension is given by two digressions, one concerning the "hyperelliptic mapping class group" given by the central extension Δ_{2g+2}, described in Example 9.10. The second digression concerns principle congruence subgroups.

Recall the "hyperelliptic mapping class group" Δ_{2g+2}, a subgroup of Γ_g the mapping class group for genus g surfaces. By Example 9.10 or [11], there is a central extension

$$1 \to \mathbb{Z}/2\mathbb{Z} \to \Delta_{2g+2} \to \Gamma_0^{2g+2}(S^2) \to 1.$$

This extension, as well as information about its characteristic class, is given in [11]. A construction for the classifying space $B\Delta_{2g+2}$ arises from complex 2-plane bundles as described in Example 9.10 or [11], given by

$$EU(2) \times_{U(2)} \mathrm{Conf}(S^2, 2g+2) \times_{\Sigma_{2g+2}} S^1$$

for g even.

Definition 23.2. Given $n \geq 3$, define

(1) $\mathbb{X}_n = EU(2) \times_{U(2)} \mathrm{Conf}(S^2, n) \times S^1$, and
(2) $\mathbb{Y}_n = EU(2) \times_{U(2)} \mathrm{Conf}(S^2, n) \times_{\Sigma_n} S^1$ with
(3) $q_n : \mathbb{X}_n \to \mathbb{Y}_n$ the standard Σ_n-cover.

Then there are n natural projection maps

$$p_i : \mathbb{X}_n \to \mathbb{X}_{n-1}$$

which are also bundle projections with fiber $S^2 - Q_{n-1}$ for which Q_{n-1} denotes a set of $(n-1)$ distinct points in S^2 by Theorem 3.3.

Define

$$\mathrm{Brun}_n(\mathbb{X})$$

as the intersection of the kernels of the induced maps

$$p_{i*} : \pi_1(\mathbb{X}_n) \to \pi_1(\mathbb{X}_{n-1})$$

for $1 \leq i \leq n$.

Further, define $P\Delta_{2g+2}$ as the kernel of the natural map $\Delta_{2g+2} \to \Sigma_{2g+2}$.

Then consider the classical maps

$$\Delta_{2g+2} \to \Gamma_g \to Sp(2g, \mathbb{Z}).$$

Notice that mod-2 reduction gives a map

$$\rho_2 \colon Sp(2g, \mathbb{Z}) \to Sp(2g, \mathbb{Z}/2\mathbb{Z})$$

and that the composite map

$$\Delta_{2g+2} \to Sp(2g, \mathbb{Z}) \to Sp(2g, \mathbb{Z}/2\mathbb{Z})$$

factors through

$$\Sigma_{2g+2} \subset Sp(2g, \mathbb{Z}/2\mathbb{Z}).$$

The second digression concerns classical principle congruence subgroups of $Sp(2g, \mathbb{Z})$ as follows where one example is the kernel of the map $\rho_2 \colon Sp(2g, \mathbb{Z}) \to Sp(2g, \mathbb{Z}/2\mathbb{Z})$.

Definition 23.3. Let

$$\Gamma^{Sp}(2g, r)$$

denote the principle congruence subgroup of level r, the kernel of the mod-r reduction map

$$\rho_r \colon Sp(2g, \mathbb{Z}) \to Sp(2g, \mathbb{Z}/r\mathbb{Z}).$$

Regarding $SL(2, \mathbb{Z}) = Sp(2, \mathbb{Z})$, the kernel of $\rho_r \colon SL(2, \mathbb{Z}) \to SL(2, \mathbb{Z}/r\mathbb{Z})$, denoted $\Gamma(2, r)$ above, is equal to $\Gamma^{Sp}(2, r)$.

A sketch of a possible natural extension of Theorem 23.1 is given next. This sketch consists of 5 steps.

(1) Consider the classical map

$$\Delta_{2g+2} \to \Gamma_g \to Sp(2g, \mathbb{Z})$$

and the "mod-2 reduction map" $Sp(2g, \mathbb{Z}) \to Sp(2g, \mathbb{Z}/2\mathbb{Z})$ together with the composite map $\Delta_{2g+2} \to Sp(2g, \mathbb{Z}) \to Sp(2g, \mathbb{Z}/2\mathbb{Z})$. Restricting this composite to $P\Delta_{2g+2}$ gives a factorization of the composite

$$P\Delta_{2g+2} \to Sp(2g, \mathbb{Z})$$

through $\Gamma^{Sp}(2g, 2)$ as described above.

(2) Inspection of n elements $\{x_1, \ldots, x_n\}$ in $\Gamma^{Sp}(2g, 2)$ gives that their commutator $[[\cdots [[x_1, x_2], x_3] \cdots,]x_n]$ is in the principal congruence subgroup of level 2^n in $Sp(2g, \mathbb{Z})$, $\Gamma^{Sp}(2g, 2^n)$. This fact is a special case of the statement that the commutator

$$[x, y]$$

of an element x in the principal congruence subgroup of level p, and an element y in the principal congruence subgroup of level q, is in the principal congruence subgroup level pq [24].

(3) A direct computation gives an isomorphism

$$\mathrm{Brun}_{2g+2}(\mathbb{X}) \to \mathrm{Brun}_{2g+2}(S^2) \times \mathbb{Z}/2\mathbb{Z}.$$

(4) Observe that $\mathrm{Brun}_{2g+2}(\mathbb{X})$ is generated by commutators of length at least $(g+1)$ as an iterated application of the fibrations $p_i : \mathbb{X}_n \to \mathbb{X}_{n-1}$ together with the analogous argument in [21] in the case of pure braid groups. Thus the image of $\mathrm{Brun}_{2g+2}(\mathbb{X})$ in $Sp(2g, \mathbb{Z})$ lies in the principle congruence subgroup of level 2^{g+2}.

(5) It is natural to conjecture that $\mathrm{Brun}(S^2, 2g + 2)$ is isomorphic to a subgroup of $\Gamma^{Sp}(2g, 2^{g+1})$ in $Sp(2g, \mathbb{Z})$, thus extending Theorem 23.1.

If this conjecture is, in fact, correct, is there some natural additional geometry associated to the homotopy groups of ΩS^2, arising from this connection to $\Gamma^{Sp}(2g, 2^{g+1})$, which informs on the associated homotopy groups?

A proof of Theorem 23.1 is given via a sequence of lemmas.

Lemma 23.4. *The image of the natural map*

$$\pi_1(\mathrm{Conf}(S^2, 4)) \to \pi_1(\mathrm{Conf}(S^2, 3)^4)$$

is exactly

$$\oplus_3 \mathbb{Z}/2\mathbb{Z}.$$

Thus the 4-stranded Brunnian braid group

$$\mathrm{Brun}_4(S^2),$$

which is the kernel of $\pi_1(\mathrm{Conf}(S^2, 4)) \to \pi_1(\mathrm{Conf}(S^2, 3)^4)$, is equal to the kernel of

$$P_4(S^2) \to \oplus_3 \mathbb{Z}/2\mathbb{Z}.$$

Proof. Recall that there are homeomorphisms

$$\theta_q : PSL(2,\mathbb{C}) \times \mathrm{Conf}(S^2 - \{0,1,\infty\}, q) \to \mathrm{Conf}(S^2, q+3)$$

given by

$$\theta_q(\rho, (z_1, z_2, \ldots, z_q)) = (\rho(0), \rho(1), \rho(\infty), \rho(z_1), \rho(z_2), \ldots, \rho(z_q)).$$

In addition, the group $\mathrm{Brun}_n(S^2)$ is the kernel of the map

$$\pi_1(\mathrm{Conf}(S^2, n)) \to \pi_1(\mathrm{Conf}(S^2, n-1)^n)$$

induced by the n different choices of projection maps

$$p_i : \mathrm{Conf}(S^2, n) \to \mathrm{Conf}(S^2, n-1)$$

where the projection p_i deletes the i-th coordinate. Thus the map

$$\pi_1(\mathrm{Conf}(S^2, 4)) \to \pi_1(\mathrm{Conf}(S^2, 3)^4)$$

is given by

$$\mathbb{Z}/2\mathbb{Z} \times F_2 \to \oplus_4 \mathbb{Z}/2\mathbb{Z}.$$

The next step is to identify the image of this map a subgroup generated by three elements, and thus is a vector space of 2-rank at most 3.

Notice that each projection map $p_i : \mathrm{Conf}(S^2, 4) \to \mathrm{Conf}(S^2, 3)$ induces an epimorphism $\pi_1(\mathrm{Conf}(S^2, 4)) \to \pi_1(\mathrm{Conf}(S^2, 3))$ by comparing the homeomorphisms

$$\theta_q : PSL(2,\mathbb{C}) \times \mathrm{Conf}(S^2 - \{0,1,\infty\}, q) \to \mathrm{Conf}(S^2, q+3)$$

and various choices of projection maps. Thus to analyze the requisite image on the level of fundamental groups, it suffices to check the behavior of the map

$$\mathbb{R}^2 - \{0,1\} \to \mathrm{Conf}(S^2, 3)^4$$

on the level of fundamental groups where $\mathbb{R}^2 - \{0,1\}$ is identified as the subspace of $\mathrm{Conf}(S^2, 4))$ given by

$$\{(\infty, 0, 1, z) \mid z \in \mathbb{R}^2 - \{0,1\}\}.$$

The image of the free group of rank 2, F_2, is given by the image of this map on the level of fundamental groups.

Notice that the image $\pi_1(\mathrm{Conf}(S^2, 4)) \to \pi_1(\mathrm{Conf}(S^2, 3)^4)$ is exactly $\oplus_3 \mathbb{Z}/2\mathbb{Z}$ as follows:

(1) The induced map $p_{1*} : \pi_1(\mathrm{Conf}(S^2, 4)) \to \pi_1(\mathrm{Conf}(S^2, 3))$ is an epimorphism by comparing to θ_1 via the homeomorphism

$$\theta_q : PSL(2, \mathbb{C}) \times \mathrm{Conf}(S^2 - \{0, 1, \infty\}, q) \to \mathrm{Conf}(S^2, q + 3)$$

and the projection maps. That is, $p_1((\infty, 0, 1, z)) = (0, 1, z)$ represents a generator γ_1 of $\pi_1(\mathrm{Conf}(S^2, 3))$ in $\pi_1(\mathrm{Conf}(S^2, 3)^4)$.

(2) The induced map $p_{2*} : \pi_1(\mathrm{Conf}(S^2, 4)) \to \pi_1(\mathrm{Conf}(S^2, 3))$ is an epimorphism by comparing θ_2 and the projection maps. That is, $p_1((\infty, 0, 1, z)) = (\infty, 1, z)$ which carries a generator x of $F[x, y]$ in $\pi_1(\mathrm{Conf}(S^2, 4))$ to an independent generator γ_2 in $\pi_1(\mathrm{Conf}(S^2, 3)^4)$.

(3) The induced map $p_{3*} : \pi_1(\mathrm{Conf}(S^2, 4)) \to \pi_1(\mathrm{Conf}(S^2, 3))$ is an epimorphism by comparing θ_3 and the projection maps. That is, $p_1((\infty, 0, 1, z)) = (\infty, 0, z)$ which carries a generator y of $F[x, y]$ to an independent generator γ_3 in $\pi_1(\mathrm{Conf}(S^2, 3)^4)$.

(4) The induced map

$$p_{4*} : \pi_1(\mathrm{Conf}(S^2, 4)) \to \pi_1(\mathrm{Conf}(S^2, 3))$$

is trivial as $p_4((\infty, 0, 1, z)) = (\infty, 0, 1)$ which is constant.

(5) Thus the image of $\pi_1(\mathrm{Conf}(S^2, 4)) \to \pi_1(\mathrm{Conf}(S^2, 3)^4)$ has 2-rank three.

The lemma follows. $\qquad\qquad\qquad\qquad\qquad\qquad\qquad\qquad\qquad\qquad$ $\square$

Lemma 23.5. *The kernels of the natural maps*

$$SL(2, \mathbb{Z}) \to SL(2, \mathbb{Z}/4\mathbb{Z})$$

and

$$PSL(2, \mathbb{Z}) \to PSL(2, \mathbb{Z}/4\mathbb{Z})$$

are equal.

Proof. Observe that there is a commutative diagram

$$
\begin{array}{ccccc}
\{1\} & \longrightarrow & \Gamma(2, 2^2) & \longrightarrow & P\Gamma(2, 2^2) \\
\downarrow & & \downarrow & & \downarrow \\
\mathbb{Z}/2\mathbb{Z} & \longrightarrow & SL(2, \mathbb{Z}) & \longrightarrow & PSL(2, \mathbb{Z}) \\
\downarrow{\scriptstyle 1} & & \downarrow & & \downarrow \\
\mathbb{Z}/2\mathbb{Z} & \longrightarrow & SL(2, \mathbb{Z}/4\mathbb{Z}) & \longrightarrow & PSL(2, \mathbb{Z}/4\mathbb{Z}).
\end{array}
$$
$\qquad$ $\square$

Lemma 23.6. *The map*

$$\Theta : B_4 \to SL(2, \mathbb{Z})$$

induces a map

$$\Phi : B_4(S^2) \to PSL(2, \mathbb{Z})$$

together with a commutative diagram

$$
\begin{array}{ccccc}
\mathbb{Z}/2\mathbb{Z} & \longrightarrow & P_4(S^2) & \longrightarrow & P\Gamma(2,2) \\
\downarrow & & \downarrow & & \downarrow \\
K & \longrightarrow & B_4(S^2) & \longrightarrow & PSL(2,\mathbb{Z}) \\
\downarrow{\scriptstyle 1} & & \downarrow & & \downarrow \\
\oplus_2 \mathbb{Z}/2\mathbb{Z} & \longrightarrow & \Sigma_4 & \longrightarrow & PSL(2, \mathbb{Z}/2\mathbb{Z})
\end{array}
$$

where K denotes the kernel of the map

$$\Phi : B_4(S^2) \to PSL(2, \mathbb{Z}).$$

Proof. Observe that the kernel of

$$B_4 \to B_4(S^2)$$

is generated by the two elements $(\sigma_1 \sigma_2 \sigma_3)^4$ and $(\sigma_1 \sigma_2)^3$. Furthermore

$$\Theta((\sigma_1 \sigma_2)^3) = -Id$$

where

$$-Id = \begin{pmatrix} -1 & 0 \\ 0 & -1 \end{pmatrix}.$$

In addition,

$$\Theta((\sigma_1 \sigma_2 \sigma_3)^4) = \Theta((\sigma_1 \sigma_2 \sigma_1)^4) = \Theta((\sigma_1 \sigma_2)^6) = Id.$$

Thus there is an induced epimorphism

$$\Phi : B_4(S^2) \to PSL(2, \mathbb{Z})$$

together with the commutative diagram stated in the lemma.

Notice that

(1) The kernel of $\Sigma_4 \to PSL(2, \mathbb{Z}/2\mathbb{Z})$ is $\oplus_2 \mathbb{Z}/2\mathbb{Z}$, generated by the images of the two elements $A = \sigma_2 \sigma_1 \sigma_3^{-1} \sigma_2$ and $B = \sigma_1 \sigma_3^{-1}$.

(2) The kernel of $P_4(S^2) \to P\Gamma(2,2)$ is generated by the image of the element $C = (\sigma_1\sigma_2)^3$.

(3) The elements A, B, and C generate a subgroup of $B_4(S^2)$ isomorphic to D_8 by a direct check.

The lemma follows. $\qquad\qquad\qquad\qquad\qquad\qquad\qquad\qquad\qquad\qquad\square$

One proof of Theorem 23.1 is as follows. To check the first assertion, the kernel of the natural epimorphism

$$B_4(S^2) \to PSL(2, \mathbb{Z}),$$

denoted K in Lemma 23.6, is isomorphic to D_8. Thus, the first assertion follows.

By the proof of Lemma 23.6, there is a central extension

$$1 \to \mathbb{Z}/2\mathbb{Z} \to P_4(S^2) \to P\Gamma(2,2) \to 1$$

(as the $\mathbb{Z}/2\mathbb{Z}$ is generated by $(\sigma_1\sigma_2)^3$). Since $P\Gamma(2,2)$ is free on two generators, the extension is split. Notice that this is overkill as Lemma 23.4 has been re-proven. By inspection, there is a commutative diagram

$$
\begin{array}{ccccc}
\{1\} & \longrightarrow & \mathrm{Brun}_4(S^2) & \longrightarrow & P\Gamma(2,2^2) \\
\downarrow & & \downarrow & & \downarrow \\
\mathbb{Z}/2\mathbb{Z} & \longrightarrow & P_4(S^2) & \longrightarrow & P\Gamma(2,2) \\
\downarrow{\scriptstyle 1} & & \downarrow & & \downarrow \\
\mathbb{Z}/2\mathbb{Z} & \longrightarrow & \oplus_3 \mathbb{Z}/2\mathbb{Z} & \longrightarrow & \oplus_2 \mathbb{Z}/2\mathbb{Z}
\end{array}
$$

where the rows and columns are all group extensions (using the fact that $P\Gamma(2,2)$ is generated by the two matrices

$$x = \begin{pmatrix} 1 & 2 \\ 0 & 1 \end{pmatrix}$$

and

$$y = \begin{pmatrix} 1 & 0 \\ 2 & 1 \end{pmatrix}.$$

This suffices.

Acknowledgments

The author would like to express appreciation for support from the National University of Singapore for this interesting as well as enjoyable conference. The work here was partially done while the author was visiting the Institute for Mathematical Sciences, National University of Singapore in 2007. The visit was supported by the Institute.

The author would also like to thank the National Science Foundation under Grant No. 9704410, and Darpa Grant number 2006-06918-01 for partial support.

The author would like to thank several friends for help in the preparation of these notes, notably Liz Hanbury who did a wonderful job of proofreading, Benson Farb, Jon Berrick, Sam Gitler, Larry Taylor, Yan Loi Wong and Jie Wu.

References

1. E. Artin, *Theorie die Zöpfe*, Hamburg Abhandlung (1924), Abhandlung Math. Semin. Univ. Hamburg **4**(1925), 47–72.
2. V. Belyi, *On Galois extensions of a maximal cyclotomic field*, Math. USSR Izv. **14**(1980), 247–256.
3. J. Birman, *Braids, Links and Mapping Class Groups*, Ann. of Math. Studies **82**, Princeton Univ. Press, Princeton, N.J. (1975).
4. A. K. Bousfield and D. M. Kan, *Homotopy Limits, Completions and Localizations*, Lect. Notes in Math. **304**, Springer-Verlag, Berlin-New York (1972).
5. J. Berrick, F. R. Cohen, Y. L. Wong and J. Wu, *Configurations, braids, and homotopy groups*, J. Amer. Math. Soc. **19**(2006), 265–326.
6. D. Benson and F. R. Cohen, *Mapping class groups of low genus*, Memoirs of the AMS **441** (1991).
7. R. Budney, *The topology of knotspaces in dimension* 3, math.GT/0506524.
8. ______, *Little cubes and long knots*, Topology. An International Journal of Mathematics **46**(2007), 1–27.
9. R. Budney and F. R. Cohen, *On the homology of the space of knots*, math.GT/0504206, to appear in Geometry & Topology.
10. R. Chari and A. Pressley, *A Guide to Quantum Groups*, Cambridge University Press (1994).
11. F. R. Cohen, *On the hyperelliptic mapping class groups, $SO(3)$, and $Spin^c(3)$*, American J. Math. **115**(1993), 389–434.
12. ______, *On genus one mapping class groups, function spaces, and modular forms*, in Topology, Geometry, and Algebra: Interactions and New Directions, Cont. Math., R. J. Milgram, A. Adem, G. Carlsson and R. L. Cohen (eds) **279**, Amer. Math. Soc. (2001) 103–128.
13. F. R. Cohen and S. Gitler, *On loop spaces of configuration spaces*, Trans. Amer. Math. Soc. **354**(2002), 1705–1748.

14. F. R. Cohen, T. Kohno and M. Xicoténcatl, *Orbit configuration spaces associated to discrete subgroups of* $PSL(2, \mathbb{R})$, to appear in J. Pure Appl. Math.

15. F. R. Cohen, T. J. Lada and J. P. May, *The Homology of Iterated Loop Spaces*, Lecture Notes in Math. **533**, Springer-Verlag (1976).

16. F. R. Cohen, J. C. Moore and J. A. Neisendorfer, *The double suspension and exponents in the homotopy groups of spheres*, Ann. of Math. **110**(1979), 121–168.

17. F. R. Cohen and S. Prassidis, *On injective homomorphisms for pure braid groups, and associated Lie algebras*, Jour. of Algebra **298**(2006) 363–370.

18. F. R. Cohen and M. Xicoténcatl, *On orbit configuration spaces for a torus.*

19. F. R. Cohen and T. Sato, *On groups of coalgebra maps*, preprint, electronic copies available.

20. F. R. Cohen and L. R. Taylor, *Notes on configuration spaces*, in preparation.

21. F. R. Cohen and J. Wu, *On Braid Groups, Free Groups, and the Loop Space of the 2-Sphere*, Progress in Mathematics **215**, Birkhaüser (2003) 93–105, and *Braid groups, free groups, and the loop space of the 2-sphere*, math.AT/0409307, *On braid groups and homotopy groups*, Geometry & Topology Monographs **13**(2008) 169–193.

22. E. B. Curtis, *Simplicial homotopy theory*, Adv. in Math. **6**(1971), 107–209.

23. P. Deligne, *Le groupe fondamental de la droite projective moins trois points*, in "Galois groups over $\mathbb{Q}$," Publ. MSRI **16**(1989), 79–298.

24. J. D. Dixon, M. P. F. Du Sautoy, A. Mann and D. Segal, *Analytic Pro-P Groups*, Cambridge Studies in Advanced Mathematics, Cambridge Univ. Press (2003).

25. V. G. Drinfel'd, *Quasi-Hopf algebras and the Knishnik-Zamolodchikov equations*, in Problems of Modern Quantum Field Theory, A. A. Belavin, A. U. Klimyk, A. B. Zamolodchikov (eds) Springer, Berlin (1989) 1–13.

26. ———, *On quasitriangular Hopf algebras and a group closely connected with* $Gal(\bar{\mathbb{Q}}/\mathbb{Q})$, Leningrad Math. Jour., No. 4 **2**(1991), 829–860.

27. C. Earle and J. Eells, *The diffeomorphism group of a compact Riemann surface*, Bull. Amer. Math. Soc., No. 4 **73**(1967), 557–559.

28. ———, *A fiber bundle description of Teichmüller theory*, J. Differential Geometry **3**(1969), 19–43.

29. M. Eichler, *Eine Verallgemeinerung der Abelschen Integrale*, Math. Zeit. **67**(1957), 267–298.

30. M. Eichler and D. Zagier, *The Theory of Jacobi Forms*, Birkhaüser (1985).

31. E. Fadell and S. Husseini, *Geometry and Topology of Configuration Spaces*, Springer Monographs in Mathematics, Springer (2001).

32. E. Fadell and L. Neuwirth, *Configuration Spaces*, Math. Scand. **10**(1962), 111–118.

33. E. Fadell and J. Van Buskirk, *The braid groups of* E^2 *and* S^2, Duke Math. Jour. **29**(1962), 243–258.

34. M. Falk and R. Randell, *The lower central series of a fiber-type arrangement*, Invent. Math. **82**(1985), 77–88.

35. B. Farb, *The Johnson homomorphism for* $\mathrm{Aut}(F_n)$, in preparation (2003).

36. R. H. Fox, *Homotopy groups and torus homotopy groups*, Ann. of Math. **49**(1948), 471–510.

37. R. H. Fox and L. P. Neuwirth, *The braid groups*, Math. Scand. **10**(1962), 119–126

38. H. Frasch, *Die Erzeugenden der Hauptkongruenzgruppen für Primastahlstufen*, Math. Annalen **108**(1933), 229–252.

39. W. Fulton and R. MacPherson, *A compactification of configuration spaces*, Ann. of Math. **139**(1994), 183–225.

40. M. Furusawa, M. Tezuka and N. Yagita, *On the cohomology of classifying spaces of torus bundles, and automorphic forms*, J. London Math. Soc. (2) **37**(1988), 528–543.

41. P. Hall, *A contribution to the theory of groups of prime power order*, Proc. Lond. Math. Soc. Series 2 **36**(1933), 29–95.

42. M. E. Hamstrom, *Homotopy groups of the space of homeomorphisms on a 2-manifold*, Ill. J. Math. **10**(1966), 563–573.

43. A. Hatcher, *Algebraic Topology*, Cambridge University Press (2001).

44. Y. Ihara, *Galois group and some arithmetic functions*, Proceedings of the International Congress of mathematicians, Kyoto, 1990, Springer (1991), 99–120.

45. Y. Ihara, *On the stable derivation algebra associated with some braid groups*, Israel J. Math., Nos. 1–2 **80**(1992), 135–153.

46. ———, *On the embedding of $Gal(\bar{\mathbb{Q}}/\mathbb{Q})$ in $\widehat{GT}$*, in The Grothendieck Theory of Dessins d'Enfants, London Mathematical Society Lecture Notes, Vol. 200, Cambridge Univesity Press (1994).

47. I. James, *Reduced product spaces*, Ann. of Math. **62**(1955), 170–197.

48. F. Klein, *Gesammelte mathematische Abhandlungen*, Springer-Verlag, Berlin (1923).

49. T. Kohno, *Linear representations of braid groups and classical Yang-Baxter equations*, Cont. Math. **78**(1988), 339–363.

50. T. Kohno, *Série de Poincaré-Koszul associée aux groupes de tresses pure*, Invent. Math. **82**(1985), 57–75.

51. ———, *Vassiliev invariants and de Rham complex on the space of knots*, in Symplectic Geometry and Quantization, Contemp. Math. **179**, Amer. Math. Soc., Providence, RI (1994) 123–138.

52. ———, *Vassiliev invariants and de Rham complex on the space of knots*, in Symplectic Geometry and Quantization, Contemp. Math. **179**, Amer. Math. Soc., Providence, RI (1994) 123–138.

53. W. Magnus, *Über n-dimensionale Gittertransformationen*, Acta Math. **64**(1934), 353–367.

54. ———, *Über Automorphismen von Fundamental-Gruppen berandeter Flächen*, Math. Ann. **109**(1934), 617–646.

55. W. Magnus, A. Karass and D. Solitar, *Combinatorial Group Theory*, Wiley (1966).

56. B. Mangum and T. Stanford, *Brunnian links are determined by their complements*, Algebraic and Geometric Topology **1**(2001), 143–152, and arXiv. GT/9912006.

57. J. P. May, *Simplicial Objects in Algebraic Topology*, Van Nostrand Mathematical studies **11**(1967).

58. J. Milnor, *On the construction F[K]*, in A Student's Guide to Algebraic Topology, J. F. Adams (ed) Lecture Notes of the London Mathematical Society **4**(1972), 119–136.

59. ———, *The geometric realization of a semi-simplicial complex*, Ann. of Math. **65**(1957), 357–362.

60. ———, *Construction of universal bundles, I*, Ann. of Math. **63**(1956), 272–284.

61. J. C. Moore, *Homotopie des complexes monoïdaux*, Séminaire H. Cartan (1954/55).

62. R. Palais, *Local triviality of the restriction map for embeddings*, Comment. Math. Helv. **34**(1960), 305–312.

63. L. Paris, *Surface braid groups*, T.A.M.S.

64. D. Quillen, *The spectrum of an equivariant cohomology ring: I*, Ann. of Math., Second Series **94**(1971), 549–572.

65. C. P. Rourke and B. J. Sanderson, Δ-*sets I: Homotopy theory*, Quart. Jour. Math. **22**(1971) 321–338.

66. P. Selick, *Introduction to Homotopy Theory*, Fields Institute Report (1996).

67. J. P. Serre, *Lie Groups and Lie Algebras*, Benjamin, New York (1965).

68. L. Schneps, *Groupe de Grothendieck-Teichmüller et automorphismes de groupes de tresses*, on website.

69. G. P. Scott, *The space of homeomorphisms of a 2-manifold*, Topology **9**(1970), 97–109.

70. ———, *Braid groups and the group of homeomorphisms of a surface*, Proc. Camb. Phil. Soc. **68**(1970), 605–617.

71. G. Shimura, *Introduction to the Arithmetic Theory of Automorphic Forms*, Publications of the Mathematical Society of Japan **11**, Iwanami Shoten, Tokyo; University Press, Princeton (1971).

72. S. Smale, *Diffeomorphisms of the 2-sphere*, P.A.M.S. **10**(1959), 621–626.

73. E. Spanier, *Algebraic Topology*, McGraw Series in Higher Mathematics, McGraw-Hill, New York (1966).

74. T. Stanford, *Brunnian braids and some of their generalizations*, Bull. of the London Math. Soc., to appear, arXiv math.GT/9907072.

75. N. E. Steenrod, *The Topology of Fibre Bundles*, Princeton University Press (1951).

76. ———, *A convenient category of topological spaces*, Michigan Math. Jour., Issue 2 **35**(1967), 133–152.

77. B. Totaro, *Configuration spaces of algebraic varieties*, Topology, No. 4 **35**(1996), 1057–1067.

78. V. V. Vershinin, *Braid groups and loop spaces*, (Russian) Uspekhi Mat. Nauk, **54**(1999), No. 2(326), 3–84; translation in Russian Math. Survey, **54**(1999), 273–350.

79. ———, *On braid groups in handlebodies*, Siberian Math. J., No. 4 **39**(1998), 645–654.

80. J. Wang, *On the braid groups for* $\mathbb{RP}^2$, J.P.A.A. **166**(2002), 203–227.

81. J. Wu, *On combinatorial descriptions of the homotopy groups of certain spaces*, Math. Proc. Camb. Philos. Soc., No. 3 **130**(2001), 489–513.

82. ______, *A braided simplicial group*, Proc. London Math. Soc., (3) **84**(2002), 645–662.

83. ______, *Lecture notes on simplicial sets*, proceedings.

84. ______, *On fibrewise simplicial monoids and Milnor-Carlsson's constructions*, Topology **37**(1998), 1113–1134.

85. M. A. Xicoténcatl, *The Lie algebra of the pure braid group*, Bol. Soc. Mat. Mexicana **6**(2000), 55–62.

86. ______, *Product decomposition of loop spaces of configuration spaces*, Topology and its Applications **15**(2002), 33–38.

87. ______, *Ph.D. Thesis*, University of Rochester (1997).

CONFIGURATION SPACES, BRAIDS, AND ROBOTICS

Robert Ghrist

Departments of Mathematics and Electrical & Systems Engineering
University of Pennsylvania
Philadelphia, PA, USA
E-mail: ghrist@math.upenn.edu

Braids are intimately related to configuration spaces of points. These configuration spaces give a useful model of autonomous agents (or robots) in an environment. Problems of relevance to autonomous engineering systems (e.g., motion planning, coordination, cooperation, assembly) are directly related to topological and geometric properties of configuration spaces, including their braid groups. This chapter details this correspondence, and explore several novel examples of configuration spaces relevant to applications in robotics. No familiarity with robotics is assumed.

1. Configuration Spaces and Braids

These notes outline an elegant relationship between braids, configuration spaces, and applications across several engineering disciplines associated with robotics and coordination. We begin with a few standard definitions.

Definition 1.1. The **configuration space** of n distinct labeled points on a topological space X, denoted[a] $\mathcal{C}^n(X)$, is the space

$$\mathcal{C}^n(X) = \prod_1^n X - \Delta, \tag{1.1}$$

where Δ denotes the **diagonal**

$$\Delta = \{(x_i)_1^n : x_i = x_j \text{ for some } i \neq j\}. \tag{1.2}$$

---

[a]Notation varies greatly with author: $F(X, n)$ is a common alternative to $\mathcal{C}^n(X)$.

The **unlabeled configuration space**, denoted $\mathcal{UC}^n(X)$, is defined to be the quotient of $\mathcal{C}^n(X)$ by the natural action of the symmetric group S_n which permutes the ordered points in X. The n-strand **braid group** of a space X is defined to be $B_n(X) = \pi_1(\mathcal{UC}^n(X))$, whereas the n-strand **pure braid group** of X is $P_n = \pi_1(\mathcal{C}^n(X))$. (We generally deal with connected configuration spaces and thus ignore basepoints.)

Example 1.2. The simplest and best-known (nontrivial) examples are the configuration spaces of the plane $\mathbb{R}^2$, whose fundamental groups yield the classical **Artin braid groups** $B_n = \pi_1(\mathcal{UC}^n(\mathbb{R}^2))$ and their pure cousins $P_n = \pi_1(\mathcal{C}^n(\mathbb{R}^2))$.

Planar configuration spaces are not easy to visualize for arbitrary n, but the elements of the braid groups are eminently intuitive — almost tactile — objects.

Exercise 1.3. Show that $\mathcal{C}^2(\mathbb{R}^2)$ is homeomorphic to $\mathbb{R}^3 \times S^1$, where S^1 denotes the unit circle in $\mathbb{R}^2$. Hint: Think about placing tokens on the table one at a time. Does your method of proof give a simple representation for $\mathcal{C}^3(\mathbb{R}^2)$?

Configuration spaces of the plane have a number of excellent and elegant algebraic properties, including:

(1) They are **Eilenberg-MacLane spaces** of type $K(\pi_1, 1)$, meaning that the fundamental group determines the homotopy type of the space and all higher homotopy groups vanish.
(2) The fundamental groups (the Artin braid groups B_n and P_n) are all torsion-free.

The method of **iterated fibrations** — in which one fixes the location of one distinguished point at a time and builds fiber bundles with these restrictions as projection maps — makes it easy to unlock topological properties of configuration spaces of points on the plane. See, e.g., [7, 26] for configuration spaces of manifolds treated in this manner.

2. Planning Problems in Robotics

It is a truth perhaps not universally acknowledged that an outstanding place to find rich topological objects is within the walls of an automated warehouse or factory.

2.1. *Motion planning*

Consider an automated factory equipped with a cadre of **automated guided vehicles** (AGVs), or mobile robots, which transport items from place to place. A common goal is to place several, say n, of these robots in motion simultaneously, controlled by an algorithm that either guides the robots from initial positions to goal positions (in a warehousing application), or executes a cyclic pattern (in manufacturing applications). These robots are costly and cannot tolerate collisions (with obstacles or with each other) without a loss of performance.

Anyone who shops at a large supermarket with wide aisles is familiar with this problem and a solution. If two carts are headed toward each other, a slight swerve is sufficient to avoid a collision, assuming the other does not move in the same direction. *The resolution of collisions on* $\mathbb{R}^2$ *is a local phenomenon.* This does not imply that planning coordinated motions is a simple task: it requires an extraordinary effort to coordinate air traffic at a busy airport, a scenario in which a collision is to be avoided at all costs.

Modeling the factory floor as $\mathbb{R}^2$ and the robots as points, one often wishes to find paths or cycles in $C^n(\mathbb{R}^2)$ to enact specific behaviors. Obstacles can easily be incorporated into these models — there is a vast literature on this subject (e.g., [54, 55]). Executing cyclic motions is more complex but can at first be approximated by composed point-to-point motions. Various kinematic issues (e.g., steering geometry) and other physical features of real autonomous systems must be addressed in general, but the configuration space model is an appropriate approximation to reality. Of course, since the robots are not truly points, and since no control algorithm implementation is of infinite precision, we require that the control path reside outside of a neighborhood of the diagonal Δ in $(\mathbb{R}^2)^n$.

It is possible to construct safe control schemes using configuration spaces. The work of Koditschek and Rimon [51] provides one example of a concrete solution: they write out explicit vector fields on these configuration spaces that can be used to flow from initial to goal positions in the presence of certain types of obstacles. By arranging these vector fields so that they strongly push away from the vestiges of the diagonal Δ on the boundary of $C^n(\mathbb{R}^2)$, the control scheme is *provably* safe from collisions (as opposed to being *statistically* safe via computer simulations): no path can ever intersect the diagonal. Furthermore, since a neighborhood of the diagonal is repelling, the control scheme is *stable* with respect to perturbations to the system. This is quite important, as mechanical systems have an annoying tendency to malfunction occasionally. Placing an appropriate vector field on a configuration space yields an excellent method of self-correction.

This is a clean, direct application of topological and dynamical ideas to a matter of great practical relevance, and it is currently used in various industrial settings.

The robotics community, largely independently of the topology community, has enjoyed great success at identifying and manipulating configuration spaces to their advantage in control problems: see the books [54, 55] and references therein. There are, however, several classes of simple, physically relevant scenarios where there in an underlying configuration space which has remained untapped. We outline a few examples in the following subsections.

2.2. *Motion planning on tracks*

Suppose the robots must move about on a collection of tracks embedded in the floor, or via a path of electrified guide-wires from the ceiling; see [14] for examples. Such a restricted network is quite common, mainly because it is more cost-effective than a full two degree-of-freedom steering system for robots. In this setting, the state of the system at any instant of time is a point in the configuration space of the graph Γ:

$$\mathcal{C}^n(\Gamma) = (\Gamma \times \cdots \times \Gamma) - \Delta.$$

As before, to navigate safely on a graph, one must construct appropriate paths that remain strictly within $\mathcal{C}^n(\Gamma)$ and are repulsed by any boundaries near Δ. But several problems seem to prevent an analogous solution, including the following:

(1) What do these spaces look like?
(2) How does one resolve an impending collision?

A significant difference between this setting and that of motion planning on the full plane is that collisions within a track are no longer locally resolvable. Imagine that the aisles of a grocery store are only as wide as the shopping carts, so that passing another person is impossible. A store full of shoppers (using carts) would pose a difficult coordinated control problem. How can carts avoid a collision in the interior of an aisle? Clearly, at least one of the participants must make a large-scale change in plans and back up to the end of the aisle. *The resolution of a collision on a graph is a non-local phenomenon.*

2.3. *Shape planning for metamorphic robots*

In recent years, several groups in the robotics community have been modeling and building **reconfigurable** or, more specifically, **metamorphic** robots (e.g., [13, 16, 52, 61, 63, 72, 86, 87]). Such a system consists of multiple identical robotic cells in an underlying lattice structure which can disconnect/reconnect with adjacent neighbors, and slide, pivot, or otherwise locomote to neighboring lattice points following prescribed rules. There are as many models for such robots as there are researchers in the sub-field: 2-d and 3-d lattices; hexagonal, square, and dodecahedral cells; pivoting or sliding motion: see, e.g., [16, 52, 58, 61, 62, 10, 64, 86, 87] and the references therein. The common feature of these robots is an aggregate of lattice-based cells having prescribed local transitions from one shape to another.

The primary challenge for such systems is **shape-planning**: how to move from one shape to another via legal moves. One centralized approach [17, 18, 66] is to build a **transition graph** whose vertices are the various shapes and whose edges are elementary legal moves from one shape to the next. It is easily demonstrated that the size of this graph is exponential in the number of cells. For this and other reasons, the transition graph makes a poor model of a configuration space for these systems.

2.4. *Digital microfluidics*

An even better physical instantiation of the previous system arises in recent work on **digital microfluidics** (see, e.g., the work of Fair in [27, 28]). In digital microfluidics, small (appx. 1 mm diameter) droplets of fluid can be quickly and accurately manipulated in an inert oil suspension between two plates. The plates are embedded with a grid of wires. Droplet manipulation is performed via **electrowetting** — a process that exploits dynamic surface tension effects to propel a droplet. Applying a current in a particular manner through the grid drives the droplet a discrete distance along the wire grid. The goal of this is to create an efficient "lab on a chip" in which droplets of various chemicals or biological agents can be positioned, mixed, and then directed to the appropriate outputs. Using the grid, one can manipulate many droplets in parallel.

This scenario is reminiscent of manipulating robotics on a factory floor, or, rather, on a one-dimensional grid within the floor. The chief differences lie in (1) scale; (2) the discrete nature of motion; and (3) the fact that droplets are sometimes allowed to collide. Indeed, digital chemical and biochemical reactions are performed precisely by forcing two droplets to collide, then mix (by a rapid oscillatory motion), then split.

The fundamental problem of microfluidic control — how to locally manipulate an ensemble of droplets to effectuate a global result on the grid — is entirely commensurate with the motion-planning problems in robotics stated above. It is this passage from local motion rules in the presence of collision constraints which makes this (and many other contemporary problems in systems engineering) a prime candidate for configuration space methods. One feels that there is a sensible topological configuration space lurking beneath the surface of this and many other problems (see also Sec. 8).

3. Configuration Spaces of Graphs

Motivated by the applications to robot motion planning, we consider the configuration spaces of points on a graph Γ.

The most obvious difference between this problem and the problem of $\mathcal{C}^n(\mathbb{R}^2)$ is that $\mathcal{C}^n(\Gamma)$ is *not* a manifold: that is, one cannot hope that every point has a neighborhood that is locally homeomorphic to a Euclidean space. Indeed, if we ignore trivial graphs that are homeomorphic to a line segment or a circle, then the graph itself is not locally Euclidean and products of the graph still share this feature.

3.1. *Visualization*

It is best to consider some examples that can be visualized.

Example 3.1 ($\mathcal{C}^2(\mathsf{Y})$). Let Y denote the graph with three edges obtained by attaching three edges to a central vertex. The space $\mathcal{C}^2(\mathsf{Y})$ is a subset of $\mathsf{Y} \times \mathsf{Y}$, this product consisting of nine squares glued together. Of these, six correspond to configurations in which the two robots are on distinct edges of Y. Since there are three edges in Y, the remaining configurations, in which both robots are on the same edge, yield three square cells, each of which is divided by the diagonal Δ into a pair of triangular cells. Thus there are six triangular 2-cells corresponding to the configurations in which both robots are on the same edge, but at distinct locations. By enumerating the behaviors of each of these 2-cells, one can make the identifications to arrive at the space given in Fig. 1.

Exercise 3.2. Choose two distinct points on Y. Now, using two fingers, execute a motion that exchanges the two chosen locations without collision. Draw this motion as a path on the configuration space of Fig. 1.

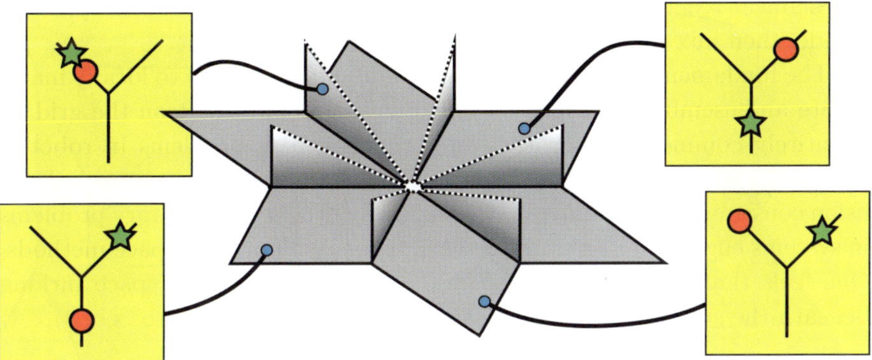

Fig. 1. The configuration space $\mathcal{C}^2(\mathsf{Y})$ of Example 3.1 embedded in $\mathbb{R}^3$. Dotted lines refer to edges that lie on the diagonal Δ. Note that the central vertex is deleted.

Example 3.3 ($\mathcal{C}^2(\mathsf{Q})$). Let Q denote the graph with three edges obtained from Y by gluing two boundary vertices together. One method of constructing $\mathcal{C}^2(\mathsf{Q})$ would be to first remove the configurations in which both robots are on the vertices to be glued. Then identify those portions of the boundary of $\mathcal{C}^2(\mathsf{Y})$ that have a robot at the vertices to be glued in Q, and glue these portions of $\mathcal{C}^2(\mathsf{Y})$ together. The result, although a very simple configuration space, is already somewhat difficult to visualize: we illustrate the space, embedded in $\mathbb{R}^3$, in Fig. 2 [left]. Each of the three "punctures" corresponds to a collision of the robots at one of the three vertices. The six dotted edges are the images of the diagonal curves from Fig. 1.

Example 3.4 ($\mathcal{C}^2(\mathsf{X})$). Increasing the incidence number of the central vertex complicates the configuration space. Consider X, a radial tree of four edges emanating from a central vertex. The visualization of $\mathcal{C}^2(\mathsf{X})$ is a bit more involved and requires some work to obtain. For the purpose of stimulating curiosity, we include this configuration space as Fig. 2 [right].

3.2. *Simplification: Deformation*

For simple graphs, though the configuration spaces illustrated earlier are visualizable, the full representation is not entirely elegant. Removal of the diagonal Δ yields a space which is not compact, and which furthermore has some topologically inessential features. In particular, many of the examples of configuration spaces we have considered are not parsimonious with respect to dimension: the configuration space can be deformed to a subset of

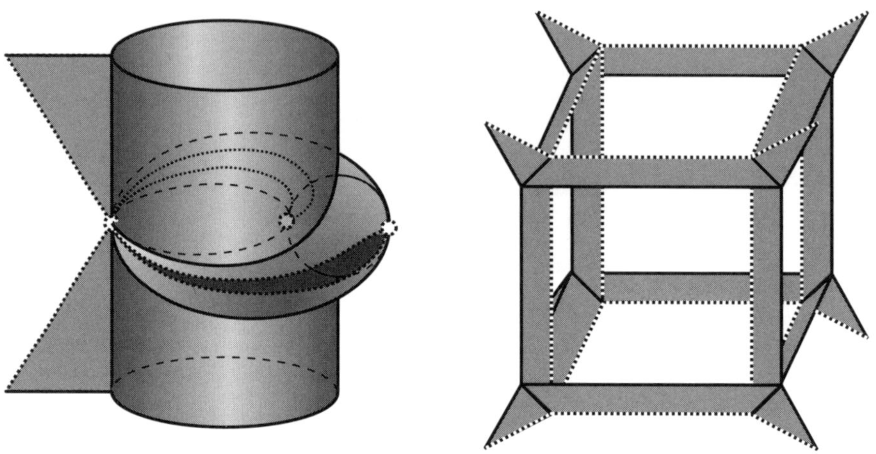

Fig. 2. The configuration spaces $\mathcal{C}^2(\mathsf{Q})$ [left] and $\mathcal{C}^2(\mathsf{X})$ [right]. Dotted lines refer to vestiges of the diagonal Δ.

strictly lower dimension. Otherwise said, the number of degrees of freedom of the system (the configuration space dimension) is not always fully needed to capture the essential features of the space.

Example 3.5. Note that the spaces illustrated in Figs. 1 and 2 can all be deformation retracted to a one-dimensional subspace: the **homotopy dimension** of these spaces is one.

The computation of this essential dimension is encapsulated in the following result:

Theorem 3.6 (Ghrist [37]). *Given a graph Γ having V vertices of valence greater than two, the space $\mathcal{C}^n(\Gamma)$ deformation retracts to a subcomplex of dimension at most V.*

The proof of this result in [37] is by inductive (and inelegant) manipulation: a recent simplification appears in [33]. We illustrate the result by continuing the some of the examples of the previous subsection. Consider for each $k > 2$ the radial k-prong tree T_k with vertices $\{v_i\}_0^k$ and edges $\{e_i\}_1^k$ attaching the central vertex v_0 to the outer vertices $\{v_i\}_1^k$. For example, $\mathsf{T}_3 = \mathsf{Y}$, and $\mathsf{T}_4 = \mathsf{X}$.

Theorem 3.6 ensures that $\mathcal{C}^n(\mathsf{T}_k)$ deformation retracts to a one-dimensional subcomplex — that is, a graph. Since the essential topological features of a graph are determined by its Euler characteristic, one need merely compute the number of vertices and edges to classify these spaces.

Proposition 3.7. *The braid group* $P_n \mathsf{T}_k = \pi_1(\mathcal{C}^n(\mathsf{T}_k))$ *is isomorphic to a free group on*

$$1 + (nk - 2n - k + 1)\frac{(n + k - 2)!}{(k - 1)!} \tag{3.1}$$

generators.

Proof. One derives a recursion relation for the Euler characteristic of $\mathcal{C}^n(\mathsf{T}_k)$ by holding fixed the point on the kth edge of T_k which is farthest from the central node. With some work, one obtains:

$$\chi(\mathcal{C}^n(\mathsf{T}_k)) = \chi(\mathcal{C}^n(\mathsf{T}_{k-1})) + n\chi(\mathcal{C}^{n-1}(\mathsf{T}_k)) - n\prod_{i=1}^{n-1}(k + i - 2). \tag{3.2}$$

The first term on the right hand side comes from the case where there is no point in the interior of the kth edge. The second term comes from fixing one of the n points in the interior of this dedicated edge. The last term comes from fixing this point at the central vertex of T_k: this subspace of the configuration space has the homotopy type of a discrete set since no other points can pass through the central vertex. Each such component contributes one edge in the deformation retracted space and hence contributes a -1 to the value of $\chi(\mathcal{C}^n(\mathsf{T}_k))$.

The seed for this recursion relation is the fact that $\mathcal{C}^1(\mathsf{T}_k) \simeq \mathsf{T}_k \simeq \{pt\}$. Solving (3.2) yields

$$\chi(\mathcal{C}^n(\mathsf{T}_k)) = -(nk - 2n - k + 1)\frac{(n + k - 2)!}{(k - 1)!},$$

which, in turn, implies that the configuration space is homotopic to a wedge of $1 - \chi$ circles. $\square$

The factorial growth of the rank in n is due to the fact that we label the n robots on T_k. If one considers the unlabeled configuration spaces, then χ is reduced by a factor of $n!$.

It is worth emphasizing that while the control problem of robots on a graph is rather intuitive for two robots, it quickly builds in complexity. Since the dimension alone makes most configuration spaces nearly impossible to visualize, Theorem 3.6 is quite helpful — the homotopy dimension of the configuration space is independent of the number of robots on the graph. For the graph T_k, Theorem 3.6 implies that there is a one-dimensional **roadmap** that gives a perfect representation of the configuration space: no topological data are lost. Since the proof of Theorem 3.6 is constructive, one can use standard algorithms for determining shortest paths on a graph to develop efficient path planning for multiple robots on T_k via the roadmap.

3.3. Simplification: Simplicial

The simplification expressed in Theorem 3.6 is not so easy to visualize. In this section, we outline a different simplification process to a topologically equivalent subset of the configuration space which has a few distinctions. First, it carries a natural simplicial structure. Second, the simplification process is easier to define explicitly and track. This simplicial model makes for easy visualization of some examples, especially radial trees.

It is easiest to explain and visualize examples in the context of unlabeled configuration spaces $\mathcal{UC}^n$. For the remainder of this subsection, we work exclusively in the unlabeled setting.

Consider again the radial k-prong tree T_k with vertices $\{v_i\}_0^k$ and edges $\{e_i\}_1^k$ attaching the central vertex v_0 to the outer vertices $\{v_i\}_1^k$. Place a Euclidean metric on each e_i of total length one. To establish convenient coordinates on $\mathcal{UC}^n(\mathsf{T}_k)$, consider any bi-indexed sequence $\mathbf{x} = \{x_j^i\}$ where $x_j^i \in [0,1]$ is the distance from the vertex v_j to the ith token on $\overline{e_j}$ closest to v_0. If there are no tokens on $\overline{e_j}$, any reference to x_j^i returns zero. It is clear that as long as (1) $x_j^i \neq x_j^{i'}$ for all $i \neq i'$; and (2) if $x_j = 1$ for some j then $x_j^1 = 1$ for all j; then, the sequence $\mathbf{x}$ determines a well defined point of $\mathcal{UC}^n(\mathsf{T}_k)$ and that all points of this configuration space can be so specified.

One can use an explicit vector field to specify a projection of $\mathcal{UC}^n(\mathsf{T}_k)$ to a simplicial complex. Consider the vector field:

$$\frac{d}{dt}x_j^i = \begin{cases} x_j^i(x_j^i - 1) & i \neq 1 \\ \Phi \cdot x_j^i(x_j^i - 1) & i = 1 \end{cases} \quad \text{where} \quad \Phi = \frac{\displaystyle\sum_{\ell=1}^{k} x_\ell^1 - 1}{1 + \left(\displaystyle\sum_{\ell=1}^{k} x_\ell^1 - 1\right)^2}. \quad (3.3)$$

Exercise 3.8. Show that this vector field determines a projection map from $\mathcal{UC}^n(\mathsf{T}_k)$ to a simplicial complex $\mathcal{B}$ whose fibers are contractible.

Example 3.9. For $n = 2$, the simplicial model $\mathcal{B}$ is always a complex of dimension one. In fact, $\mathcal{B}$ is precisely the 1-skeleton of the standard $(k-1)$-simplex. The vertices correspond to configurations with both tokens on the same edge of T_k; the edges correspond to configurations with two tokens on different edges of T_k.

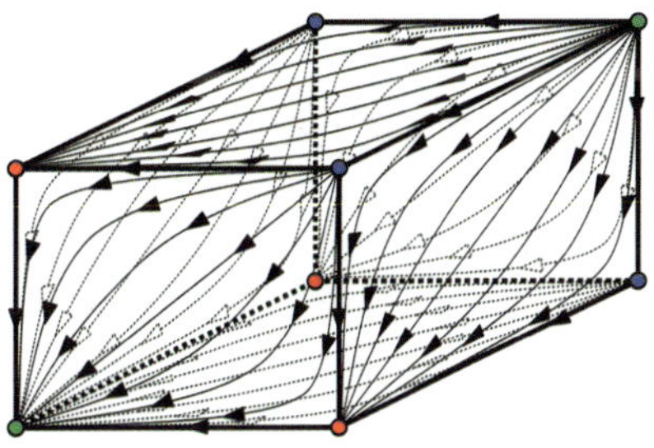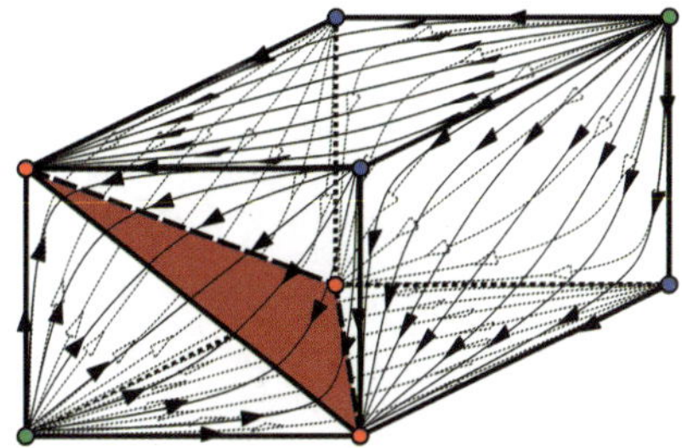

Fig. 3. [left] The vector field $x_j^1(x_j^1 - 1)$ acts on a cube of "inmost" points; [right] the rescaled field yields a deformation retraction of the cube (minus those portions of the boundary where $x_j^1 = 1$ but $x_{j'}^1 \neq 1$ for some j, j') to the unit simplex.

Example 3.10. For $k = 3$, the simplicial model $\mathcal{B}$ is a complex of dimension at most two. These complexes are pictured in Fig. 4.

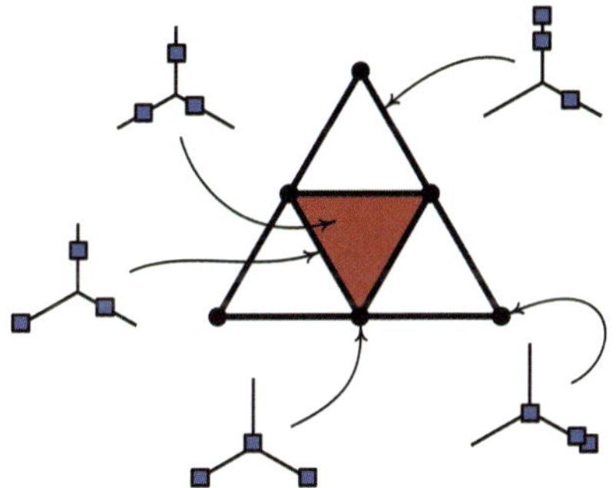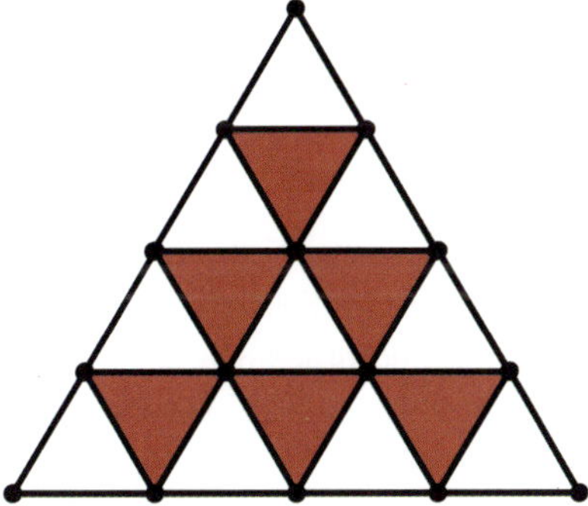

Fig. 4. Examples of $\mathcal{B}$ for $(k = 3, n = 3)$ [left], along with samples of representative configurations in the fiber; [right] $(k = 3, n = 5)$.

The general case is not difficult to discern: one inducts on n and k as in the example of Fig. 5. Note that the simplicial model $\mathcal{B}$ will have principal p-simplices for all $0 \leq p \leq \min\{n - 1, k - 1\}$. Thus, this simplification reduces dimension to a subset of dimension potentially far below that of n. This is a foreshadowing of Theorem 3.6.

More general trees can be handled by induction and a **graph of spaces** construction.

Exercise 3.11. Give a simplicial model for the configuration space of points on a tree which is the wedge product of two copies of T_k along a common boundary vertex.

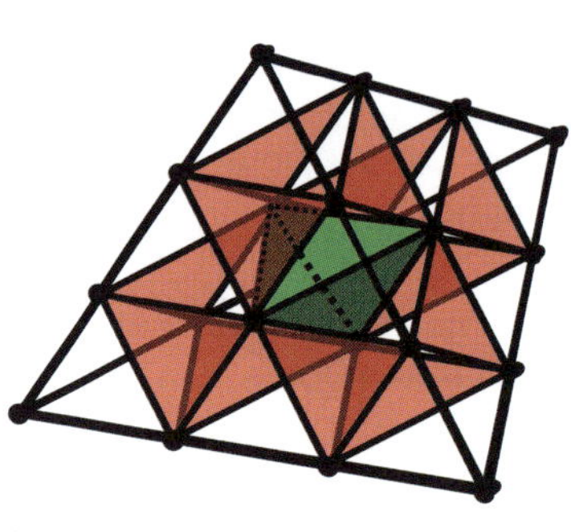 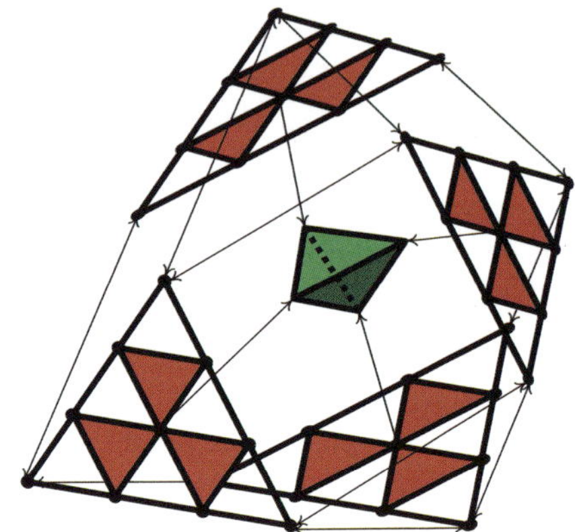

Fig. 5.   An example of $\mathcal{B}$ for $(k=4, n=4)$ [left]. There is precisely one solid 3-simplex in the center of the complex, surrounded by four identified copies of the simplicial model for $(k=3, n=4)$ [right].

Exercise 3.12. Can you draw the simplicial model associated with $\mathcal{C}^2(\mathsf{T}_k)$ based on that of $\mathcal{U}\mathcal{C}^2(\mathsf{T}_k)$?

3.4. *Simplification: Discretization*

The last example of simplification for configuration spaces is a reduction not to the smallest possible dimension subspace (as in Sec. 3.2), nor to a simplicial complex via a projection (as in Sec. 3.3), but rather to a subset that approximates the space with cubes. This simplification method has a fundamental impact in construction configuration spaces for a number of related problems in robotics: we thus relegate it to its own section.

4. Discretization

In this section, we approximate configuration spaces of graphs with **cubical complexes** — cell complexes which are built from finite-dimensional Euclidean cubes and whose gluing maps, like those of a simplicial complex, are well-behaved with respect to gluing along faces, see, e.g., [65, 8]. This **discretization** method will be seen to lead us to several interesting mathematical detours, as well as a deeper understanding of what configuration spaces are and how they arise in the applied settings of Sec. 2.

4.1. *Cubes and collisions*

Any graph Γ comes equipped with a cellular structure: 0-cells (vertices) and 1-cells (edges). The n-fold cross product of Γ with itself inherits a cell structure, each cell being a product of n (not necessarily distinct) cells in Γ. However, the configuration space does not quite have a natural cell structure, since the diagonal Δ slices through all product cells with repeated factors. Notice, however, that in several of the previous examples, these partial cells dangle "inessentially" and could be collapsed onto a more "essential" skeleton of the configuration space.

This skeleton can be specified as follows [1]. Consider the **discretized configuration space** of Γ, denoted $\mathcal{D}^n(\Gamma)$, defined as $(\Gamma \times \cdots \times \Gamma) - \tilde{\Delta}$, where $\tilde{\Delta}$ denotes the set of all product cells in $\Gamma \times \cdots \times \Gamma$ whose closures intersect the diagonal Δ. Equivalently, we can describe $\mathcal{D}^n(\Gamma)$ as the set of configurations for which, given any two robots on Γ and any path in Γ connecting them, the path contains at least one entire edge. Thus, instead of restricting robots to be at least some intrinsic distance ϵ apart (i.e., removing an ϵ neighborhood of Δ), one now restricts robots on Γ to be "at least one full edge apart". Note that $\mathcal{D}^n(\Gamma)$ is a subcomplex of the cubical complex Γ^n and a subset of $\mathcal{C}^n(\Gamma)$ (it does not contain "partial cells" that arise when cutting along the diagonal), and is, in fact, the largest subcomplex of Γ^n that does not intersect Δ.

With this natural cell structure, one can think of the vertices (0-cells) of $\mathcal{D}^n(\Gamma)$ as "discretized" configurations — arrangements of labeled tokens at the vertices of the graph. The edges of $\mathcal{D}^n(\Gamma)$, or 1-cells, tell us which discrete configurations can be connected by moving one token along an edge of Γ. Each 2-cell in $\mathcal{D}^n(\Gamma)$ represents two physically independent edges: one can move a pair of tokens independently along disjoint edges. A k-cell in $\mathcal{D}^n(\Gamma)$ likewise represents the ability to move k tokens along k disjoint edges in Γ in an independent manner.

Returning to Fig. 1, discretizing $\mathcal{C}^2(\mathsf{Y})$ removes much of the space. For example, the triangular two-dimensional cells represent configurations in which both robots are on the interior of the same edge. Since they are not "one full edge apart", these cells are deleted. The same is true of all the other two-dimensional cells that represent robots in the interior of separate edges. Which configurations of two robots on Y are separated by a full edge?

Exercise 4.1. Show that discretizing the configuration spaces of Examples 3.1 through 3.4 yields the configuration spaces of Fig. 6. How well do these spaces "approximate" the configuration spaces?

One could compute the discretization of Example 3.1 in a less direct manner that generalizes to some lovely examples to follow. For this example, simple counting reveals that the space $\mathcal{D}^2(\mathsf{Y})$ possesses twelve 0-cells (both

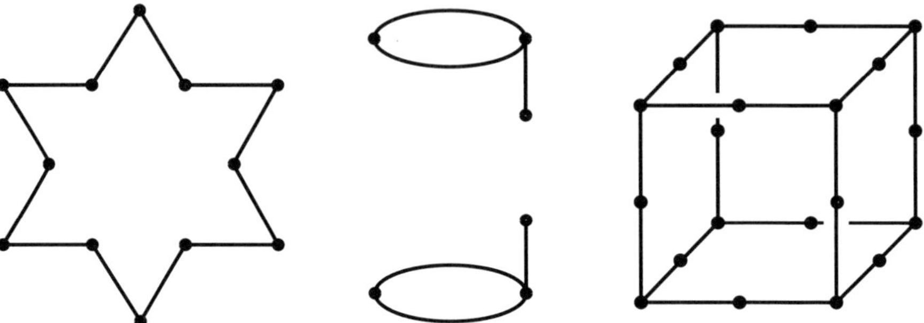

Fig. 6.   The discretizations of the configuration spaces in Examples 1-3 (left to right).

robots are at distinct vertices of Y), twelve 1-cells (one robot is at a vertex and the other is on an edge whose closure does not contain said vertex), and zero 2-cells (since any two edges intersect at the central vertex). With a little thought, one can see that $\mathcal{D}^2(\mathsf{Y})$ is a connected manifold: each zero-cell connects to exactly two 1-cells, and all of the 1-cells are joined end-to-end cyclically. It follows that $\mathcal{D}^2(\mathsf{Y})$ is a topological circle, obtained by deleting all the near-diagonal cells from $\mathcal{C}^2(\mathsf{Y})$ in Fig. 1. The discretization operation yields a homotopy-equivalent subcomplex of $\mathcal{C}^2(\mathsf{Y})$. However, this is certainly not the case for the discretization of $\mathcal{C}^2(\mathsf{Q})$, which becomes disconnected.

4.2. *Manifold examples*

Combinatorial arguments like those above can sometimes determine the discretized configuration space, even when the full configuration space is hidden from view. The following are some surprising examples of interesting spaces that arise as the discretized configuration space of non-planar graphs [1].

Example 4.2. Consider the complete graph K_5 pictured in Fig. 7 [left]. The discretized configuration space of two robots on this graph is a two-dimensional complex. A simple counting argument reveals the cell-structure.

Each 0-cell corresponds to a configuration in which the two robots are at distinct vertices. Since K_5 has five vertices, there are exactly $(5)(5-1) = 20$ such 0-cells. (There is no vertex where two edges cross in the picture; there are vertices only at the corners of the pentagon.) Each 1-cell corresponds to a configuration in which one robot is at a vertex and the other is on

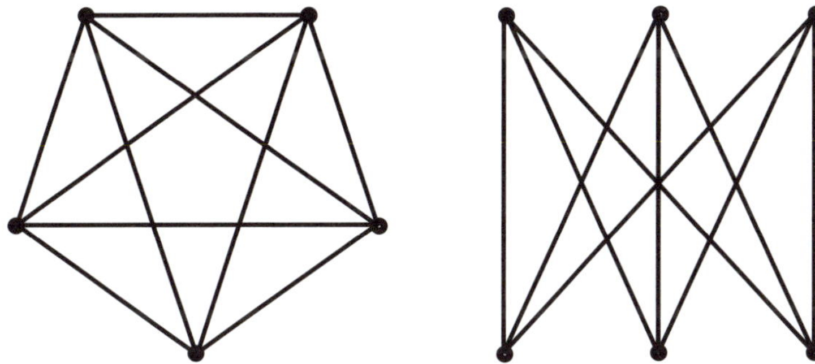

Fig. 7. The non-planar graphs K_5 [left] and $K_{3,3}$ [right].

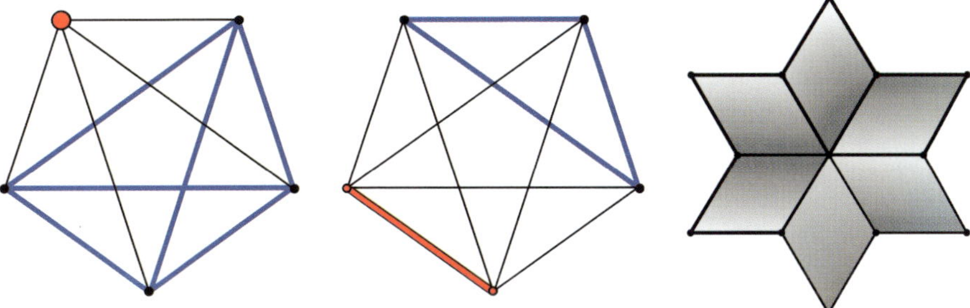

Fig. 8. [left] For every vertex in the space K_5 there are six disjoint edges. Likewise [middle] for each edge there are three totally disjoint edges. In $\mathcal{D}^2(K_5)$, these cells fit together to form a piecewise-Euclidean two-dimensional complex [right].

an edge whose endpoints do not include the vertex already occupied. From the diagram of K_5 one counts that there are $(2)(5)(6) = 60$ such 1-cells, as in Fig. 8 [left]. The factor of two is present because we label the two robots on K_5. Each 2-cell corresponds to a configuration in which the two robots occupy edges whose closures are disjoint. Again, from the diagram (and Fig. 8 [middle]) one counts that there are $(10)(3) = 30$ such 2-cells in the complex.

One then demonstrates that each edge borders a pair of 2-cells preserving an orientation and that each vertex is incident to six edges, as in Fig. 8 [right]. Also, the space $\mathcal{D}^2(K_5)$ is connected: one can move from any configuration to any other. Thus $\mathcal{D}^2(K_5)$ is a connected orientable surface, and the classification theorem for surfaces implies that the space is determined uniquely up to homeomorphism by the Euler characteristic:

$$\chi(\mathcal{D}^2(K_5)) = \#\text{faces} - \#\text{edges} + \#\text{vertices} = 30 - 60 + 20 = -10. \quad (4.1)$$

Thus, $\mathcal{D}^2(K_5)$ is a closed orientable surface of genus $g = 1 - \frac{1}{2}\chi = 6$.

It is not at all obvious that the motion of two robots on this graph should produce a genus six surface. Obtaining a manifold is surprising enough, but a manifold with genus larger than one is at odds with the notion that all of the interesting topology in these spaces is "localized" in configurations about a vertex.

Fig. 9. The space $\mathcal{D}^2(K_5)$ is homeomorphic to a closed orientable surface of genus six.

Exercise 4.3. For K_5 as above and $K_{3,3}$ the complete bipartite graph of Fig. 7 [right], show the following using combinatorics and the Euler characteristic:

(1) $\mathcal{D}^2(K_{3,3})$ is a closed orientable surface of genus four.
(2) $\mathcal{D}^3(K_5)$ is a closed orientable surface of genus 16.
(3) $\mathcal{D}^4(K_{3,3})$ is a closed orientable surface of genus 37.

For the latter two, try tracking the holes instead of the robots, and be careful about labels!

4.3. *Everything that rises must converge*

Question 4.4. When does the discretized configuration space of a graph represent an accurate approximation to the topological configuration space? That is, when is $\mathcal{D}^n(\Gamma)$ a deformation retract of $\mathcal{C}^n(\Gamma)$?

This question was answered definitively by Abrams in his 2000 thesis.

Theorem 4.5 (Abrams [1]). *For any $n > 1$ and any graph Γ with at least n vertices, the inclusion $\mathcal{D}^n(\Gamma) \hookrightarrow \mathcal{C}^n(\Gamma)$ is a homotopy equivalence if*

(1) *Each path between distinct vertices of valence not equal to two passes through at least $n + 1$ edges; and*

(2) *Each path from a vertex to itself that cannot be shrunk to a point in Γ passes through at least $n + 1$ edges.*

Thus, the braid group of two strands on K_5 is the fundamental group of the closed oriented surface of genus 6: a hyperbolic group. This is surprising, and differs starkly from the intuition of Artin braid groups.

4.4. *Curvature*

Notice that all of the interesting examples of discretized configuration spaces of graphs have the homotopy type of a graph or of a surface of nonzero genus.

Question 4.6. Are (discretized) configuration spaces of graphs ever spheres? Are they ever manifolds of dimension greater than two?

The answers to these questions require a detour from topology to geometry. The scenic route is worth the time.

The spaces $\mathcal{D}^n(\Gamma)$ can be given a natural piecewise Euclidean geometry inherited from (1) the path metric on Γ, combined with (2) the flat product metric on Γ^n. Every k-cube of $\mathcal{D}^n(\Gamma)$ thus has the geometry of a Euclidean cube.[b]

The critical concept is that of a metric space which is **nonpositively curved**. This notion of curvature for metric spaces is both old and rich, dating back to work of Alexandrov, Rauch, Hadamard, Toponogov, Cartan, and others. We modify and specialize the definitions to the class of Euclidean cubical complexes: see [8] for a more robust treatment.

Let X denote a piecewise Euclidean cubical complex. Though it may seem ironic to talk about the curvature of a space built from decidedly flat pieces, there is, nevertheless, a great deal of positive or negative curvature that can hide in the corners of the cubes.

Definition 4.7. A piecewise Euclidean cube complex X is **nonpositively curved** or **NPC** if, for any sufficiently small geodesic triangle $\triangle PQR$ in X, the sum of the angles $\angle RPQ + \angle PQR + \angle QRP$ is no greater than π.

A geodesic triangle is a set of three points $\{P, Q, R\}$ together with geodesics between them. These edges are piecewise linear (since X is piecewise Euclidean), and, consequently, the angles at the three vertices are

---

[b]It may be a rectangular prism, if the edge lengths differ. For this article, we will assume uniform lengths on all edges.

well-defined. The intuition behind this (non-standard!) definition of NPC is that on a flat Euclidean plane, the angles of a triangle sum to π, whereas on a positively curved sphere, the angle sum is greater than π, and on a negatively curved hyperbolic space, the angle sum is less than π. The *"sufficiently small"* portion of Definition 4.7 is meant to ensure that the triangle is a contractible loop in X.

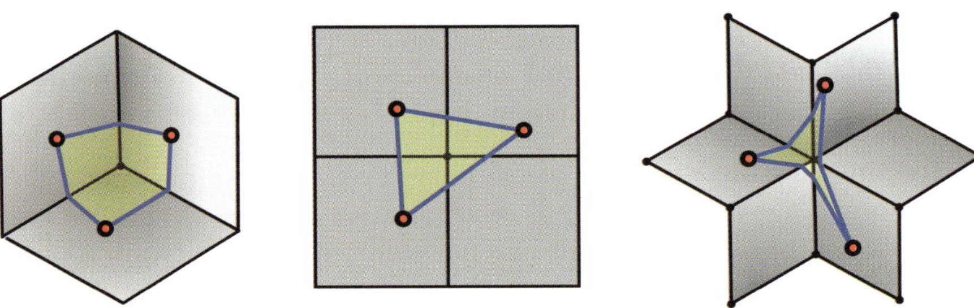

Fig. 10. Geodesic triangles in piecewise Euclidean cube complexes can detect vertices of positive [left], zero [center], or negative [right] curvature based on angle sums.

We note that the definition of NPC extends beyond piecewise Euclidean spaces to arbitrary metric spaces for which local geodesic paths exist. NPC spaces are of fundamental importance in geometric group theory: see [8] for an extensive introduction. They are local versions of **CAT(0) spaces**, another foundational class of geometric spaces.

There is a well-known combinatorial approach due to Gromov for determining when a cubical complex is nonpositively curved in terms of vertex links.

Definition 4.8. Let X denote a piecewise Euclidean cubical complex and let v denote a vertex of X. The **link** of v, LINK(v), is defined to be the abstract simplicial complex whose k-dimensional simplices are the $(k+1)$-dimensional cubes incident to v with the natural boundary relationships.

Certain global topological features of a metric cubical complex are completely determined by the local structure of the vertex links: a theorem of Gromov [42] proves that a finite-dimensional Euclidean cubical complex is NPC if and only if the link of every vertex is a flag complex without digons. Recall: a **digon** is a pair of vertices connected by two edges, and a **flag complex** is a simplicial complex which is maximal among all simplicial complexes with the same one-dimensional skeleton. Gromov's theorem permits an elementary proof of the following general result.

Theorem 4.9 (Abrams [1]; Swiatkowski [78]). *$\mathcal{D}^n(\Gamma)$ is NPC.*

Exercise 4.10. Prove it using Gromov's criterion. See Theorem 6.1 for a complete proof of a more general setting.

Thanks to the body of knowledge about NPC spaces, Theorem 4.9 combined with Theorem 4.5 yield a number of important corollaries:

Corollary 4.11. *Configuration spaces of graphs are **aspherical** (all higher homotopy groups vanish) and their fundamental groups are **torsion-free**.*

5. Reconfigurable Systems

The discretization in Sec. 4 has a far-reaching generalization which provides models for the many discretely actuated systems in robotics and beyond surveyed in Sec. 2.

5.1. *Generators and relations*

A reconfigurable system is a collection of states on a graph, where each state is thought of as a vertex labeling function. Any state can be modified by local rearrangements, these local changes being rigidly specified. We distinguish between the amount of information needed to determine the legality of an elementary move (the "support" of the move) and the precise subset on which the reconfiguration physically occurs (the "trace" of the move).

Definition 5.1. Fix $\mathcal{A}$ to be a set of labels. Fix $\mathcal{G}$ to be a graph. A **generator** ϕ for a local reconfigurable system is a collection of three objects:

(1) the **support**, $\mathrm{SUP}(\phi) \subset \mathcal{G}$, a subgraph of $\mathcal{G}$;
(2) the **trace**, $\mathrm{TR}(\phi) \subset \mathrm{SUP}(\phi)$, a subgraph of $\mathrm{SUP}(\phi)$;
(3) an *unordered* pair of **local states**

$$\mathbf{u}_0^{loc}, \mathbf{u}_1^{loc} : V(\mathrm{SUP}(\phi)) \to \mathcal{A},$$

which are labelings of the vertex set of $\mathrm{SUP}(\phi)$ by elements of $\mathcal{A}$. These local states must agree on $\mathrm{SUP}(\phi) - \mathrm{TR}(\phi)$: i.e.,

$$\mathbf{u}_0^{loc}\big|_{\mathrm{SUP}(\phi)-\mathrm{TR}(\phi)} = \mathbf{u}_1^{loc}\big|_{\mathrm{SUP}(\phi)-\mathrm{TR}(\phi)} . \tag{5.1}$$

All generators are assumed to be **nontrivial** in the sense that $\mathbf{u}_0^{loc} \neq \mathbf{u}_1^{loc}$.

Definition 5.2. A **state** is a labeling of the vertices of $\mathcal{G}$ by $\mathcal{A}$. A generator ϕ is said to be **admissible** at a state $\mathbf{u}$ if $\mathbf{u}|_{\mathrm{SUP}(\phi)} = \mathbf{u}_0^{loc}$. For such a pair $(\mathbf{u}, \phi)$, we say that the **action** of ϕ on $\mathbf{u}$ is the new state given by

$$\phi[\mathbf{u}] = \begin{cases} \mathbf{u} &: \text{ on } \mathcal{G} - \mathrm{SUP}(\phi) \\ \mathbf{u}_1^{loc} &: \text{ on } \mathrm{SUP}(\phi) \end{cases}, \tag{5.2}$$

Remark 5.3. Since the local states of each generator are unordered, it follows that any generator ϕ which is admissible at a state $\mathbf{u}$ is also admissible at the state $\phi[\mathbf{u}]$, and that $\phi[\phi[\mathbf{u}]] = \mathbf{u}$.

Definition 5.4. A **reconfigurable system** on $\mathcal{G}$ consists of a collection of generators and a collection of states closed under all possible admissible actions.

Given a pair of local moves, one can understand intuitively what it means for them to act "independently". One way to codify this in a formal reconfigurable system is as follows:

Definition 5.5. In a reconfigurable system, a collection of generators $\{\phi_{\alpha_i}\}$ is said to **commute** if

$$\mathrm{TR}(\phi_{\alpha_i}) \cap \mathrm{SUP}(\phi_{\alpha_j}) = \emptyset \quad \forall i \neq j. \tag{5.3}$$

This definition is the reason for the distinction between a local move's support and its trace. The trace is where the move "happens" and when execution can disrupt other moves trying to execute simultaneously.

Commutativity connotes physical independence.

5.2. *State complexes*

The reconfigurable systems of the previous section possess a configuration space which naturally generalizes the example of the discretized configuration space of a graph. Recall, $\mathcal{D}^n(\Gamma)$ is a cubical complex where edges denote moves (or, in our language, "generators") and cubes correspond to a commutative collection of physically independent moves.

We define the state complex of a reconfigurable system to be the cube complex with an abstract k-cube for each collection of k admissible commuting generators:

Definition 5.6. The **state complex** $\mathcal{S}$ of a local reconfigurable system is the following abstract cubical complex. Each abstract k-cube $e^{(k)}$ of $\mathcal{S}$ is an equivalence class $[\mathbf{u}; (\phi_{\alpha_i})_{i=1}^k]$ where

(1) $(\phi_{\alpha_i})_{i=1}^{k}$ is a k-tuple of commuting generators;

(2) $\mathbf{u}$ is some state for which all the generators $(\phi_{\alpha_i})_{i=1}^{k}$ are admissible; and

(3) $[\mathbf{u}_0; (\phi_{\alpha_i})_{i=1}^{k}] = [\mathbf{u}_1; (\phi_{\beta_i})_{i=1}^{k}]$ if and only if the sequence (β_i) is a permutation of (α_i) and $\mathbf{u}_0 = \mathbf{u}_1$ on the set $\mathcal{G} - \bigcup_i \mathrm{SUP}(\phi_{\alpha_i})$.

The boundary of each abstract k-cube is the collection of $2k$ faces obtained by deleting the ith generator from the list and using $\mathbf{u}$ and $\phi_{\alpha_i}[\mathbf{u}]$ as the ambient states, for $i = 1, \ldots, k$. Specifically,

$$\partial[\mathbf{u}; (\phi_{\alpha_i})_{i=1}^{k}] = \bigcup_{i=1}^{k} \left([\mathbf{u}; (\phi_{\alpha_j})_{j\neq i}] \cup [\phi_{\alpha_i}[\mathbf{u}]; (\phi_{\alpha_j})_{j\neq i}] \right) . \qquad (5.4)$$

The weak topology is used for reconfigurable systems which are not locally finite. In the locally finite case, the state complex is a locally compact cubical complex.

It follows from repeated application of Remark 5.3 that the k-cells are well-defined with respect to admissibility of actions.

Exercise 5.7. Show that the 1-skeleton of a state complex $\mathcal{S}$ is precisely the transition graph associated to the reconfigurable system.

5.3. *Examples*

The following examples come from [3, 41].

Example 5.8 (Hex-lattice metamorphic robots). This reconfigurable system is based on the first metamorphic robot system pioneered by Chirikjian [16]. It consists of a finite aggregate of planar hexagonal units locked in a hex lattice, with the ability to pivot sufficiently unobstructed units on the boundary of the aggregate.

More specifically, $\mathcal{G}$ is a graph whose vertices correspond to hex lattice points and whose edges correspond to neighboring lattice points. The alphabet is $\mathcal{A} = \{0, 1\}$ with 0 connoting an unoccupied site and 1 connoting an occupied site. There is one type of generator, represented in Fig. 11 [left], which generates a homogeneous system: this local rule can be applied to any translated or rotated position in the lattice. This generator allows for local changes in the topology of the aggregate (disconnections are possible). For physical systems in which this is undesirable — say, for power transmission purposes — one can choose a generator with larger support.

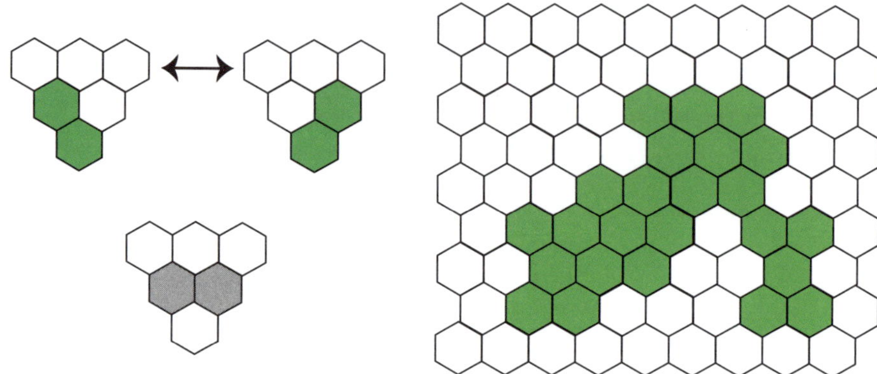

Fig. 11.   The generator for a 2-d hexagonal lattice system with pivoting locomotion. The domain is the graph dual to the hex lattice shown. Shaded cells are occupied, white are unoccupied. [left, top] The local states $\mathbf{u}_0^{loc}$ and $\mathbf{u}_1^{loc}$ are shown. [left, bottom] The support of the generator, with trace shaded. [right] A typical state in this reconfigurable system.

As an example of a state complex for this system, consider a workspace $\mathcal{G}$ consisting of three rows of lattice points with a line of occupied cells as in Fig. 12. This line of cells can "climb" on itself from the left and migrate to the right, one by one. The entire state complex is illustrated in Fig. 12 [center]. Although the transition graph appears complicated, this state complex is contractible and remains so for any length channel.

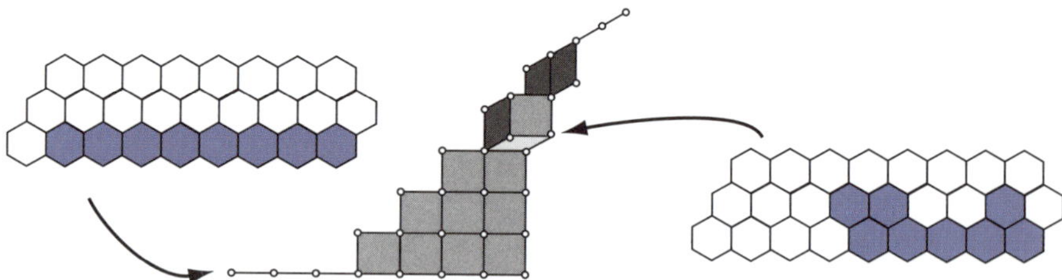

Fig. 12.   For a line of hexagons filing out of a constrained tunnel, the state complex is contractible.

Example 5.9 (2-d articulated planar robot arm). Consider as a domain $\mathcal{G}$ the lattice of edges in the first quadrant of the plane. This system consists of two types of generators, pictured in Fig. 13. The support of each generator is the union of eight edges as shown. The trace of each generator is as described in the figure caption. Beginning with a state having

n vertical edges end-to-end, the reconfigurable system models the position of an articulated robotic arm which is fixed at the origin and which can (1) rotate at the end and (2) flip corners as per the diagram. This arm is **positive** in the sense that it may extend up and to the right only.

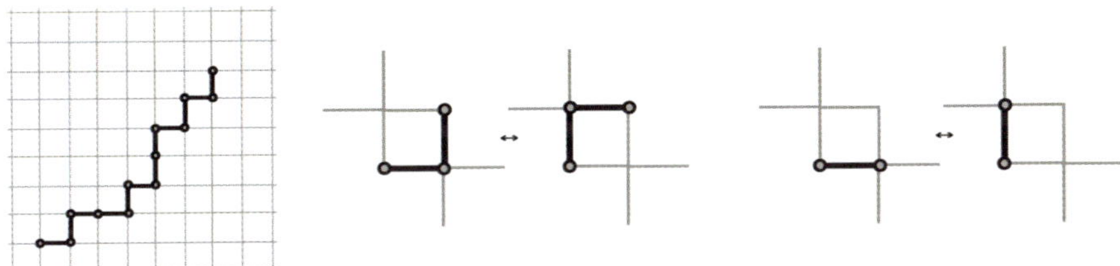

Fig. 13. A positive articulated robot arm example [left] with fixed endpoint. One generator [center] flips corners and has as its trace the central four edges. The other generator [right] rotates the end of the arm, and has trace equal to the two activated edges.

The state complex in the case $n = 5$ is illustrated in Fig. 14. Note that there can be at most three independent motions (when the arm is in a "staircase" configuration); hence the state complex has top dimension three. In this case also, although the transition graph for this system is complicated, the state complex itself is contractible.

Exercise 5.10. Show that the state complex of Example 5.9 is always contractible. (Hint: Use induction on n, the length of the arm.)

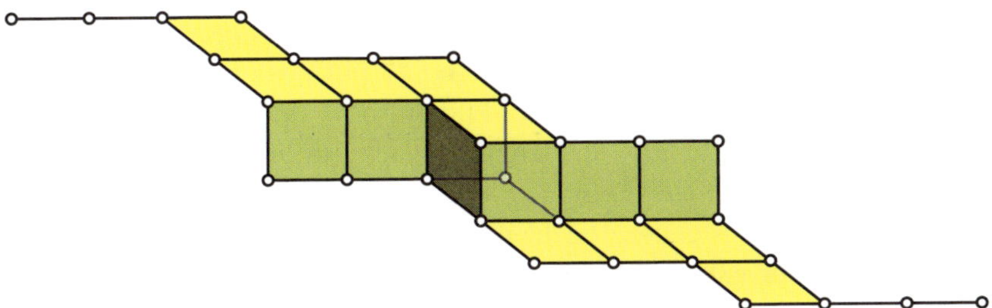

Fig. 14. The state complex of a 5-link positive arm has one cell of dimension three, along with several cells of lower dimension.

Example 5.11 (2-d expansion-compression square system). Consider a planar square lattice workspace. This system will use an alphabet of labels $\{0, 1, =, \|\}$ whose interpretation is as follows: "0" means that a

cell is unoccupied; "1" means that the cell is occupied by one module; "="
and "‖" imply that the cell is occupied by two modules compressed together in a horizontal or vertical orientation (resp.). The catalogue consists
of six generators, illustrated in Fig. 15 (the lower two generators are only
represented up to flips). The trace is equal to the support for the top two
generators illustrated; for the bottom two generators, the trace is equal to
the support minus the single square which remains unoccupied (label "0").

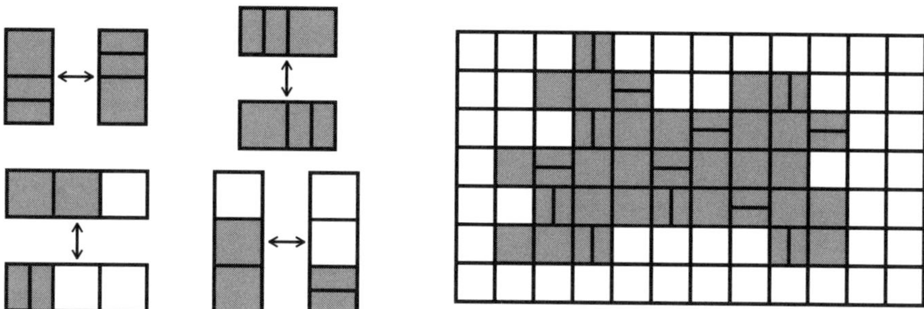

Fig. 15. Generators for a simple compression-expansion system on the square planar
lattice [left]; an example of a typical state [right].

This example is based on the Crystalline robots of Rus *et al.* [72]. Extensions to 3-d cubical systems and more elaborate motions can be accom-
modated with minor modifications.

Example 5.12 (Configuration space of points on a graph). Consider
a graph $\mathcal{G}$ and alphabet $\mathcal{A} = \{0, \ldots, n\}$ used to specify empty/occupied
vertices. There are n types of generators $\{\phi_i\}_1^n$ in this homogeneous system,
one for each nonzero element of $\mathcal{A}$. The support and trace of each ϕ_i is
precisely the closure of an (arbitrary) edge. The local states of this ϕ_i
evaluate to 0 on one of the endpoints and i on the other. The homogeneous
reconfigurable system generated from a state $\mathbf{u}$ on $\mathcal{G}$ having exactly one
vertex labeled i for each $i = 1, \ldots, n$ mimics an ensemble of n distinct noncolliding points on the graph $\mathcal{G}$. If we reduce the alphabet to $\{0, 1\}$, then
the system represents n identical agents.

Example 5.13 (Digital microfluidics). An even better physical instantiation of the previous system arises in digital microfluidics (recall Sec. 2.4).
Representing system states as marked vertices ("droplets") on a graph is
appropriate given the discrete nature of the motion by electrophoresis on a

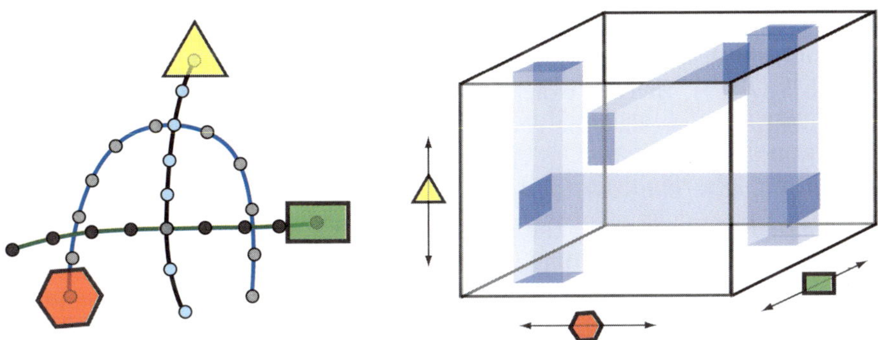

Fig. 16. [left] A coordination problem with three robots translating on (discretized, intersecting) intervals. [right] The state complex approximation to the coordination space, with collision set shaded.

graph of wires. This adds a few new ingredients to the setting of the previous example. For n different chemical agents, an alphabet of $\{0, \ldots, n\}$ is appropriate (the "0" connoting absence); however, a typical state may have many vertices with the same nonzero label (corresponding to the number of droplets of substance i in use at a given time). Furthermore, it is possible to mix droplets by merging them together, rapidly oscillating along an edge, then splitting the mixed product. This leads to a new type of generator of the form $(i \text{---} j) \Longleftrightarrow (k \text{---} k)$.

Example 5.14 (Robot coordination). There is a broad generalization of configuration spaces of graphs developed in [40, 39] which has an interpretation as a state complex. We outline a simple example. Consider a collection of n planar graphs $(\Gamma_i)_1^n$, each embedded in the plane of a common workspace (with intersections between different graphs permitted). On each Γ_i, a robot R_i with some particular fixed size/shape is free to translate along Γ_i: one thinks of the graph as being a physical groove in the floor, or perhaps an electrified overhead guidewire. The **coordination space** of this system is defined to be the space of all configurations in $\prod_i \Gamma_i$ for which there are no collisions — the robots R_i have no intersections.

We can approximate these coordination spaces by the following reconfigurable systems. Assume that each graph Γ_i has been refined by adding multiple (trivial) vertices along the interiors of edges. We will approximate the robot motion by performing discrete jumps to neighboring vertices, much as in Example 5.12.

Let the underlying graph be $\mathcal{G} = \coprod \Gamma_i$, the disjoint union of the individual graphs. The generators for this system are as follows. For each edge $\alpha \in E(\Gamma_i)$, there is exactly one generator ϕ_α. The trace is the edge itself, $\mathrm{TR}(\phi_\alpha) = \alpha$, and the generator corresponds to sliding the robot R_i from one end of the edge to the other. The support, $\mathrm{SUP}(\phi_\alpha)$ consists of the edge $\alpha \in E(\Gamma_i)$ along with any other edges β in Γ_j ($j \neq i$) for which the robot R_j sliding along the edge β can collide with R_i as it slides along α. The alphabet is $\mathcal{A} = \{0, 1\}$ and the local states for ϕ_α have zeros at all vertices of all edges in the support, except for a single 1 at the boundary vertices of α (these two boundary vertices yield the two local states, as in Example 5.12).

Any state for this reconfigurable system is one for which all vertices of each Γ_i are labeled with zeros except for one vertex with a label 1. The resulting state complex is a cubical complex which approximates the cylindrical coordination space, as pictured in Fig. 16 [40, 39]. Of course, in the case where $\Gamma_i = \Gamma$ for all i and the robots R_i are sufficiently small, this reconfigurable system is exactly that of Example 5.12.

Other concrete examples of reconfigurable systems fall outside the realm of robotics applications. These include the discrete models of protein folding considered in [75, 76], spaces of phylogenetic trees [6], and parallel computation with shared resource constraints [67]. More abstract and algebraic examples abound.

6. Curvature Again

In Sec. 4.4, a discrete curvature approach to discretized configuration spaces of graphs was seen to be extremely powerful. This theme remains true in the context of reconfigurable systems.

6.1. *State complexes are NPC*

Theorem 6.1. *The state complex of any (locally finite) reconfigurable system is NPC.*

Proof. Let $\mathbf{u}$ denote a vertex of $\mathcal{S}$. Consider the link $\mathrm{LINK}(\mathbf{u})$. The 0-simplices of the $\mathrm{LINK}(\mathbf{u})$ correspond to all edges in $\mathcal{S}^{(1)}$ incident to $\mathbf{u}$; that is, actions of generators based at $\mathbf{u}$. A k-simplex of $\mathrm{LINK}(\mathbf{u})$ is thus a commuting set of $k + 1$ of these generators based at $\mathbf{u}$.

We argue first that there are no digons in $\mathrm{LINK}(\mathbf{u})$ for any $\mathbf{u} \in \mathcal{S}$. Assume that ϕ_1 and ϕ_2 are admissible generators for the state $\mathbf{u}$, and that these two generators correspond to the vertices of a digon in LINK($\mathbf{u}$). Each edge of the digon in LINK($\mathbf{u}$) corresponds to a distinct 2-cell in $\mathcal{S}$ having a corner at $\mathbf{u}$ and edges at $\mathbf{u}$ corresponding to ϕ_1 and ϕ_2. By Definition 5.6, each such 2-cell is the equivalence class $[\mathbf{u}; (\phi_1, \phi_2)]$: the two 2-cells are therefore equivalent and not distinct.

To complete the proof, we must show that the link is a flag complex. The interpretation of the flag condition for a state complex is as follows: if at $\mathbf{u} \in \mathcal{S}$, one has a set of k generators ϕ_{α_i}, of which each pair of generators commutes, then the full set of k generators must commute. The proof follows directly from the definitions, especially from two observations from Definition 5.5: (1) commutativity of a set of actions is independent of the states implicated; and (2) any collection of pairwise commutative actions is totally commutative. $\qquad\square$

This results in informative restrictions on the topology of state complexes: e.g., the impossibility of spheres as state complexes for any system of reconfiguration. More important to applications, however, is the implication to path-planning and efficient coordination.

6.2. *Efficient state planning*

Theorem 6.1 has an important corollary for optimal planning.

Corollary 6.2. *Each homotopy class of paths connecting two given points of a state complex contains a unique shortest path.*

This is well-known for NPC spaces [8]. Figure 17 [left] gives a simple example of a 2-d cubical complex with positive curvature for which the above corollary fails.

This corollary is a key ingredient in the applications of NPC geometry to path-planning on a configuration space, since one expects geodesics on $\mathcal{S}$ to coincide with optimal solutions to the state planning problem. However, in the context of robotics applications, the goal of solving the state-planning problem is *not* necessarily coincident with the geodesic problem on the state complex. Figure 18 [left] illustrates the matter concisely. Consider a portion of a state complex $\mathcal{S}$ which is planar and two-dimensional. To get from point p to point q in $\mathcal{S}$, any edge-path which is weakly monotone increasing in the horizontal and vertical directions is of minimal length in the transition graph. The true geodesic is, of course, the straight line, which is not well-positioned with respect to the discrete cubical structure.

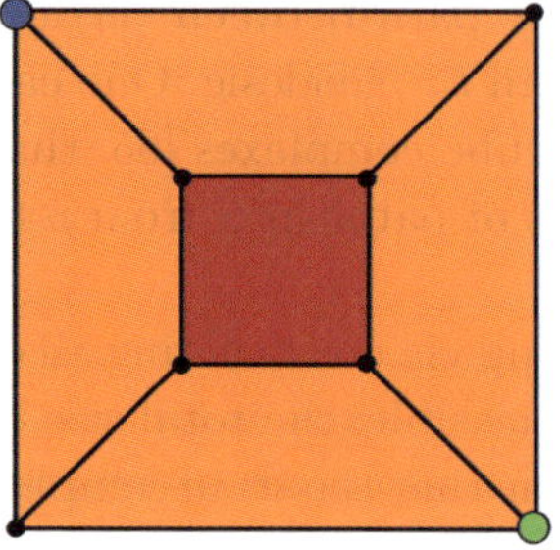 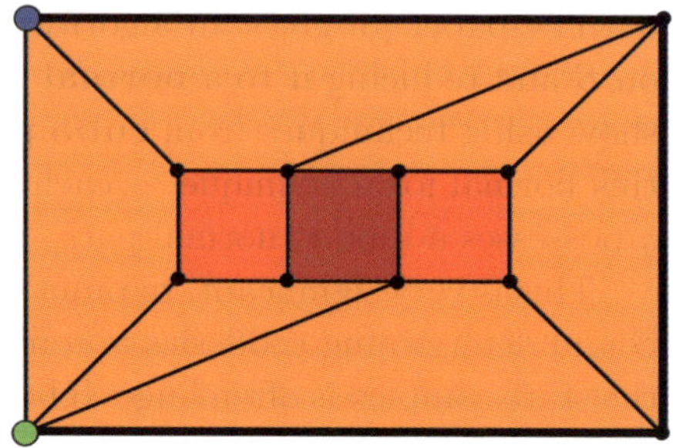

Fig. 17. [left] A 2-d cube-complex (with positive curvature) possessing two distinct homotopic shortest paths between a pair of marked points; [right] a 2-d cube-complex (with positive curvature) possessing an edge path (the thick line on the boundary) which is a locally (but not globally) shortest path. Any cube path near this path is strictly longer. Note: for both of these complexes, all 2-cells are unit-length Euclidean cubes — the figures are deformed for purposes of illustration.

Given the assumption that *each elementary move can be executed at a uniform maximum rate*, it is clear that the ℓ^2 geodesic on $\mathcal{S}$ is also **time-minimal** in the sense that the elapsed time is minimal among all paths from p to q. However, there is an envelope of non-geodesic paths which are yet time-minimizing. Indeed, the ℓ^2 geodesic in Fig. 18 [left] "slows down" some of the moves unnecessarily in order to maintain the constant slope.

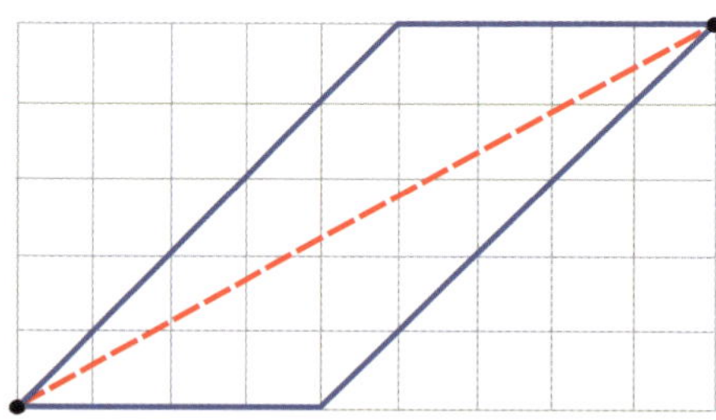 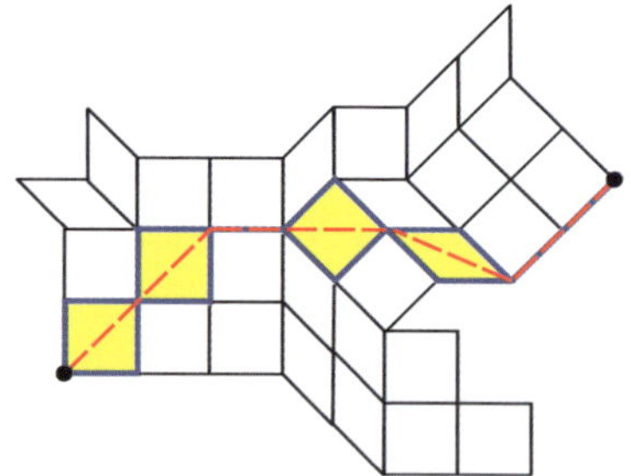

Fig. 18. [left] The true geodesic lies within an envelope of time-minimizing paths. No minimal edge-paths are time-minimizing; [right] a normal path in an NPC complex follows along the diagonals of cubes in a greedy manner.

This leads us to define a second metric on $\mathcal{S}$, one which measures elapsed time. Namely, instead of the Euclidean metric on the cells of $\mathcal{S}$, consider the space $\mathcal{S}$ with the ℓ^∞ norm on each cell. The geodesics in this geometry represent reconfiguration paths which are *time minimizing*.

The paper [3] gives an algorithm for taking any path between vertices on $\mathcal{S}$ and reducing it to a **normal form** which is an ℓ^∞ geodesic. One can show using techniques from **cube paths** in NPC cube complexes [65] that this normal form is unique — each homotopy class of paths in $\mathcal{S}$ from p to q possesses a unique normal path.

This is very significant. In practice, state-planning via constructing all of $\mathcal{S}$ and determining geodesics is computationally infeasible: the total size of the state complex is often huge. If however we assume that the state complex is unknown *but* that some path between states has been pre-computed (in the case of metamorphic robots, this done often by a decentralized planner [9, 10, 82, 83, 84, 88]), then we may optimize this trajectory. One can employ a gradient-descent curve shortening on that portion of the state complex "explored" in real-time by the path. The above results imply that any algorithm which monotonically reduces path length must converge to the shortest path (in that homotopy class) and cannot be hung up on a locally minimal path. Thus nonpositive curvature of $\mathcal{S}$ allows for a time-optimization which does not require explicit construction of $\mathcal{S}$.

The presence of local minima is a persistent problem in optimization schemes across many disciplines: nonpositive curvature is a handy antidote.

6.3. *Back to braids*

The fundamental group of a configuration space is a braid group. By analogy, the fundamental groups of state complexes ought to bear some resemblance to braid groups. This begs the question:

Question 6.3. Which groups are realizable as the fundamental group of a state complex?

Nonpositive curvature provides some elementary restrictions: e.g., the fundamental groups are torsion-free, like the Artin braid groups. The careful reader will note that not all braid groups are torsion-free — the braid groups of points on the 2-sphere S^2 are a classical example. One notes, however, that S^2 is a font of positive curvature, and one is not surprised at the algebraic reflection of this geometric fact.

Returning to state complexes, the following result yields some additional restrictions.

Theorem 6.4 (Ghrist and Peterson [41]). *The fundamental group of any finite state complex $\mathcal{S}$ embeds into the group with presentation*

$$\mathcal{A}_\mathcal{S} = \langle \phi_\alpha : [\phi_\beta, \phi_\gamma] \rangle, \tag{6.1}$$

whose generators ϕ_α are the generators of the reconfigurable system, and whose relations are commutators of pairs of generators in the reconfigurable system which commute.

Groups for which all relations are commutators of generators are called **Artin right-angled groups**. These play an important role in geometric group theory, along side Coxeter groups and similar objects. Theorem 6.4 validates the terminology of *generators* and *commuting* from Definitions 5.1 and 5.5.

The proof follows almost directly from the proof of Theorem 2 of [20], which states that the fundamental groups of $\mathcal{D}^n(\Gamma)$ embed in right-angled Artin groups.

Corollary 6.5. *Fundamental groups of state complexes are linear: they embed in $GL_n(\mathbb{R})$.*

Such embedding properties come for free from properties of Artin right-angled groups [23].

7. Last Strands

The topics touched upon in these notes are, of necessity, selective in both nature and scope. To give a glimpse of the available breadth, we list a few related results, future directions, and extensions.

7.1. *Configuration spaces*

The topology of configuration spaces goes well beyond the examples and perspectives sampled in these notes. The following remarks offer pointers to the literature for the interested reader.

7.1.1. *Linkages*

Absent from these notes is an explication of the long and detailed history of configuration spaces for mechanical linkages. Readers are perhaps familiar with the classical **Peaucellier linkage** [19] for converting circular to linear motion. This is a question about the algebraic geometry of linkage configuration spaces. Numerous papers have considered the topology and geometry of configuration spaces of closed-chain linkages [81, 46, 48, 80, 59, 44]. A universality result [47, 49] states that every orientable manifold arises as (a connected component of) the configuration space of some (planar) linkage.

7.1.2. *Algorithmic complexity*

Upon translating the problem of motion-planning in robotics to that of path-planning on a configuration space, the topologist breathes a sigh of contentment: "problem solved". The roboticist, on the other hand, finds that the problem is just beginning. A real physical environment is cluttered with obstacles to be avoided. Real robots have physical shape and appendages which must not collide. Real motion planning involves programming work, with all the difficulties of kinematics, friction, and the like.

But for the sake of maintaining the topologists' equanimity, let us assume all these problems away and return to the setting of point robots in a domain X. Even here, actually computing $\mathcal{C}^n(X)$ and performing path-planning is computationally challenging. There is an innumerable literature on the computational complexity of path planning in robotics. The books [54, 55] give an overview with references. Suffice to say that the standard complete algorithm of Canny [11] using cylindrical algebraic decompositions is doubly exponential in the geometric complexity of the workspace. Worse still, *optimal* path-planning — that is, finding a shortest path between two points in a space — was shown to be NP-complete for spaces of dimension three and above [12]. (Interestingly enough, the hardness proof requires using quite a bit of positive curvature in the space to force many geodesics between points.)

7.1.3. *Topological complexity*

A more recent reformulation of the topology of path planning on configuration spaces has been given by Farber [29]. In his work, the goal of path-planning is not to generate a single path between initial and terminal points on a configuration space, but rather to construct a **path planner**: a function from $\mathcal{C}^n(X) \times \mathcal{C}^n(X)$ to the path space $\mathcal{P}(\mathcal{C}^n(X))$. For purposes of stability, one wants this mapping to be continuous. For example, on a web-based mapping system that gives directions (e.g., `GoogleEarth`), it would be suboptimal if perturbing the start or end points on the map gave a wholly different set of directions.

However, such is often unavoidable for topological reasons. Farber observes that a continuous path planner is possible if and only if $\mathcal{C}^n(X)$ is a contractible space. He then transforms the question into one of complexity: what is the smallest number of continuous path-planners on subsets of $\mathcal{C}^n(X) \times \mathcal{C}^n(X)$ which cover all possible initial and final locations? Farber relates this **topological complexity**, $TC(\mathcal{C}^n(X))$, to Schwarz genus and the algebraic topology of $\mathcal{C}^n(X)$. This topological measure of the difficulty of path planning is the subject of much investigation. For example, [31] shows that the topological complexity of $\mathcal{C}^n(\mathbb{R}^2)$ and $\mathcal{C}^n(\mathbb{R}^3)$ is linear in n. In contrast, the topological complexity of motion planning on graphs has an upper bound independent of n [30], cf. Theorem 3.6.

7.2. Graph braid groups

There has been a flurry of recent work in understanding configuration spaces of graphs and their braid groups. The following are examples:

7.2.1. Embeddings

Crisp and Wiest showed [20] that braid groups of graphs embed in Artin right-angled groups. The embedding of Theorem 6.4 agrees with their embedding; indeed, the proof of [20] is the basis of the more general result in [41]. A dual result of that cited above has appeared recently in the thesis of Sabalka [73]. Namely, any right-angled Artin group embeds into the braid group of some number of points on a graph. (In combination with Theorem 6.4, we see that any state complex group embeds into a graph braid group.)

7.2.2. Right-angled groups

Determining which groups arise as graph braid groups has received a great deal of recent attention. It was conjectured in [37] that all graph braid groups were right-angled Artin groups. This conjecture has been repeatedly disproved: see [1, 78]. A cohomological argument of Farley and Sabalka shows that $\mathcal{UC}^n(\mathsf{T})$ is Artin right-angled if and only if $n < 4$ or there exists an embedded arc in T passing through all vertices of T [34].

7.2.3. π_1 and H^*

Farley and Sabalka have recently [33] given a presentation for $\mathcal{UC}^n(\mathsf{T})$ for any tree T. Their methods stem from examining the discretized configuration space and using Forman's discrete Morse theory to determine critical cells. These methods have led also to an understanding of the cohomology ring [34], which has surprising connections to face algebras and Stanley-Reisner rings.

7.2.4. *Rigidity*

Sabalka [74] gives a rigidity result for tree braid groups. A simple version of the result is the following: if one is given the isomorphism type of $\pi_1(\mathcal{U}\mathcal{C}^4(\mathsf{T}))$ — the 4-strand braid group, and no other information — then one can determine the combinatorial type of T. This is an incredibly surprising and elegant result.

7.3. *State complexes*

There are a number of interesting questions concerning state complexes, many of which are open.

7.3.1. *Manifolds*

Which manifolds can arise as state complexes of a reconfigurable system? Thanks to Theorem 6.1, only aspherical manifolds are within the realm of possibility. There is a simple proof in [41] that given any simplicial complex L which is flag, there exists a reconfigurable system whose state complex has link L at each state. Thus, many different manifolds best described as *hyperbolic* may arise. Combining this result with Theorem 6.4, we see that there are hyperbolic 3-manifold groups which embed in Artin right-angled groups.

7.3.2. *Realization*

It is currently unknown which NPC cubical complexes can arise as state complexes for some reconfigurable system. The thesis of Peterson (in progress) gives some surprising examples of NPC cubical complexes which cannot be realized as state complexes: see Fig. 19.

7.3.3. *Limits*

To some extent, state complexes are generalizations of the discretized configuration spaces $\mathcal{D}^n(\Gamma)$ and thus "approximate" a topological configuration space. It is tempting to guess that all state complexes are "discretizations" of some topological configuration space. There are certain examples of reconfigurable systems for which it makes sense to refine the underlying graph and obtain a sequence of reconfigurable systems. In such examples, one can ask whether the sequence of state complexes enjoys any sort of convergence properties. A canonical example of such refinement occurs in the statement

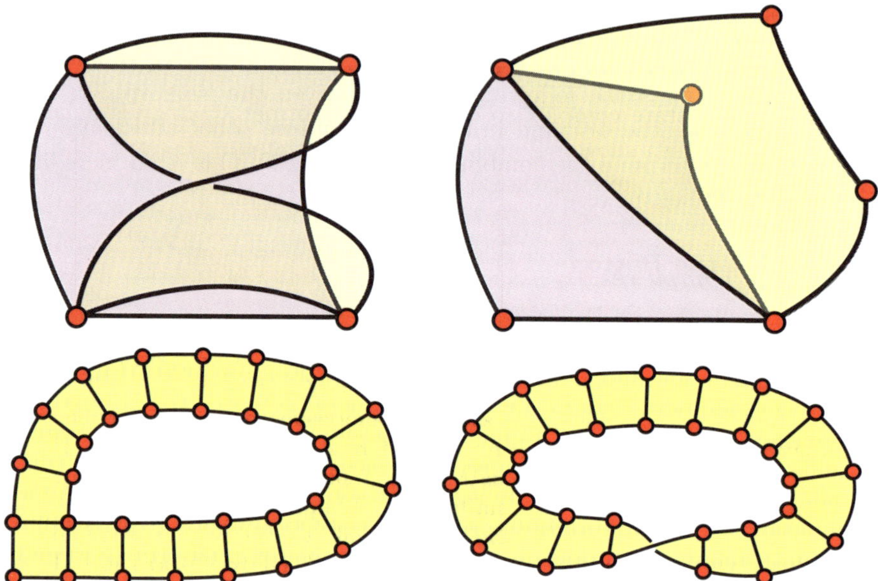

Fig. 19. None of these NPC cube complexes can be realized as a state complex of a reconfigurable system.

of Theorem 4.5. The refinement of the underlying edges of Γ which inserts additional vertices with the zero label has the effect of enlarging the state complex. Let $\Gamma^{(i)}$ denote the ith refinement of Γ (each edge of the original Γ now has 2^i subedges). Under the canonical inclusions of $\mathcal{D}^n(\Gamma)$ into $\mathcal{C}^n(\Gamma)$, the sequence $\mathcal{D}^n(\Gamma^{(i)})$ can be discussed in terms of Hausdorff and Gromov-Hausdorff limits. It follows from Theorem 4.5 that the state complexes $\mathcal{D}^n(\Gamma^{(i)})$ stabilizes in homotopy type as $i \to \infty$ and that this stable homotopy type is precisely that of $\mathcal{C}^n(\Gamma)$.

It is intriguing to consider similar refinement and limit questions for general state complexes. One can certainly construct examples for which this type of refinement does not lead to state complexes which stabilize in homotopy type. However, it may be that there is a notion of refinement for which convergence in the Gromov-Hausdorff sense works.

7.3.4. *Embedding*

An attractive conjecture is that all reconfigurable systems are isomorphic to a configuration space of points on graphs, with constraints more general than the standard non-collision avoidance of the diagonal. For example, one might hope that every state complex embeds in a product of graphs. This is true for simply-connected state complexes [41], but not in general. The best one can hope for is that either some finite cover, or perhaps a cubical refinement of a state complex so embeds. The former strategy is delicate: it has recently been shown (Peterson, thesis in progress) that not all finite covers of state complexes are state complexes.

8. Hope at the Bottom of the Box

There are numerous settings across many scientific disciplines in which a configuration space seems to be lurking beneath the surface. These notes have detailed a few examples, mostly inspired by robotics applications and mostly amenable to techniques from the topology and geometry of cube complexes.

Some important examples not covered here include the following.

8.1. *Protein folding*

Perhaps the largest scientific challenge of our day is the need for fast, accurate models of protein folding that can be used to assist drug design. The sheer enormity of degrees of freedom (read, "dimension of the configuration space") frustrates analysts. Putting this problem on the "Moore's Law Credit Card"[c] is futile.

It is perhaps not unreasonable to hope that understanding better the configuration space associated to a protein will carry explanatory power in a biological setting. One famous enigma concerning the folding of proteins, is the **Levinthal paradox** [56]: although the number of possible conformations of a polypeptide chain is too large to be sampled exhaustively, protein sequences do fold into unique native states in seconds. The paradox, posed by Cyrus Levinthal in 1969, is still not completely resolved. Since in order to decipher and control cellular processes one needs to master the mechanisms regulating the configurational changes of nucleic acids, the demand for algorithms that accurately predict the paths of these changes is high.

A speculative geometric explanation for the Levinthal paradox is that the typical protein configuration space has a residue of hyperbolic geometry. Recall that in hyperbolic space, the volume of a ball grows exponentially as a function of the radius: hence a configuration space with lots of negative (sectional) curvature would encompass a large volume (and thus, a large

---

[c]That is, waiting for processor speed to catch up to the complexity of the problem.

number of discretized states) within a set of small diameter (as measured by time-to-goal under local folding rules).

8.2. *Self-assembly*

The collective behavior of mechanical devices at micro- and nano- scales increasingly resembles that of chemical or biological reactions, in which individual particles endowed with intrinsic geometric data (the shape of the particle) and algorithmic data (reactivity, bonding proclivities, repulsions) yield assemblies of complex devices through purely local means. The challenge of being able to control these processes for purposes of design in the mechanical realm is two-fold. First is the *physical* challenge: to be able to understand and exploit the behavior of mechanical devices at micro- (or smaller) scales. Second is the *algorithmic* challenge: how does one "program" a collection of parts under physical constraints to assemble into a desired device. At small scales, one exchanges the ability to fabricate very intricate components for the ability to produce huge numbers of simple components. A significant algorithmic question is therefore which complex devices can be built out of simple pieces with simple assembly rules?

Not surprisingly, experimentalists have been able to build objects at small scales which have complete regularity with respect to the local geometry: planar hex and square lattices [85], cylindrical sheets [45], and spheres [5]. These have the property that the component "tiles" are regularly shaped and the assembly "rules" are determined by passive physical means, e.g., via surface tension effects on a liquid surface. Such geometrically regular assemblies can and are having significant impact on manufacturing at small scales [53, 60].

With respect to "programmable" self-assembly of mechanical devices at small scales, a fairly recent NASA-IAC report on *Kinematic Cellular Automata* [79] contains extensive information on small-scale programmable mechanical assembly, mostly in the context of self-replication. We note in particular the claim presented there that self-assembly requires "a new mathematics" to be effective in programming devices.

One possible avenue of exploration is that of modeling the appropriate variables as topological spaces, with a *parts space*, a *design space*, an *assembly products space* and other constructs. All of these would be examples of systems with very many degrees-of-freedom and potentially intricate features. The paper [50] gives a grammatical approach to some of these problems.

8.3. *Genetics*

One area in which a great deal of attention has been recently focused is on configuration spaces associated to **phylogenetic trees** — data structures for organizing and comparing genetic sequencing of organisms. Here, we refer the interested reader to the seminal paper [6], which build configuration spaces for rooted labeled trees and uses the NPC geometry of these spaces to, e.g., give a well-defined notion of averages for phylogenetic data. A more recent survey article [77] gives a broader treatment of applications of geometric and topological combinatorics to a broad class of problems in phylogenetics. In all contexts, the challenge of finding the correct space for representing phylogenetic data is central. It is not a coincidence that, as in the case of configuration spaces for robot coordination, the geometry as well as the topology of these spaces are central.

8.4. *"O donna in cui la mia speranza vige"*

In all the above cases, whatever configuration spaces regulate the phenomena, they are not of the "traditional" form of (1.1). Some of these examples are being addressed by mathematicians in creative and compelling ways.

A mathematician should be optimistic in the face of what seems to scientists insurmountable challenges. After all, classifying spaces, sheaf theory, K-theory, Floer theory, and the entire Grothendieck programme are but a few examples of the tools topologists have invented to manage the complexities of topological spaces, and most of these tools are not merely comfortable with but wholly dependent upon the use of infinite-dimensional spaces. The fact that there has been so little historical communication between the topologists who have invented the tools and the scientists who may need the tools is, counter-intuitively, a reason to rejoice. For it may come to pass that the solutions to these significant scientific challenges rely upon a smile from the Queen of Sciences.

Acknowledgments

The writing of these notes is supported in part by DARPA # HR0011-07-1-0002 and by NSF-DMS 0337713.

References

1. A. Abrams, *Configuration spaces and braid groups of graphs*, Ph.D. thesis, UC Berkeley (2000).

2. A. Abrams and R. Ghrist, Finding topology in a factory: Configuration spaces, *Amer. Math. Monthly* 109 (2002) 140–150.

3. A. Abrams and R. Ghrist, State complexes for metamorphic robot systems, *Int. J. Robotics Research* 23 (2004) 809–824.

4. S. Alexander, R. Bishop, and R. Ghrist, Pursuit and evasion on non-convex domains of arbitrary dimensions, in *Proc. Robotics: Systems and Science* (2006).

5. B. Berger, P. Shor, L. Tucker-Kellogg, and J. King, Local rule-based theory of virus shell assembly, *Proc. National Academy of Science, USA* 91(6) (1994) 7732–7736.

6. L. Billera, S. Holmes, and K. Vogtmann, Geometry of the space of phylogenetic trees, *Adv. Applied Math.* 27 (2001) 733–767.

7. J. Birman, *Braids, Links, and Mapping Class Groups*, Princeton University Press, Princeton, N.J. (1974).

8. M. Bridson and A. Haefliger, *Metric Spaces of Nonpositive Curvature*, Springer-Verlag, Berlin (1999).

9. Z. Butler, S. Byrnes, and D. Rus, Distributed motion planning for modular robots with unit-compressible modules, in *Proc. IROS* (2001).

10. Z. Butler, K. Kotay, D. Rus, and K. Tomita, Cellular automata for decentralized control of self-reconfigurable robots, in *Proc. IEEE ICRA Workshop on Modular Robots* (2001).

11. J. Canny, *The Complexity of Robot Motion Planning*, MIT Press, Cambridge, MA (1988).

12. J. Canny and J. Reif, Lower bounds for shortest path and related problems, in *Proc. 28th Ann. IEEE Symp. Found. Comp. Sci.* (1987) 49–60.

13. A. Castano, W. M. Shen, and P. Will, CONRO: Towards deployable robots with inter-robots metamorphic capabilities, *Autonomous Robots* 8(3) (2000) 309–324.

14. G. Castleberry, *The AGV Handbook*, Braun-Brumfield, Ann Arbor, MI (1991).

15. R. Charney, The Tits conjecture for locally reducible Artin groups, *Internat. J. Algebra Comput.* 10 (2000) 783–797.

16. G. Chirikjian, Kinematics of a metamorphic robotic system, in *Proc. IEEE ICRA* (1994).

17. G. Chirikjian and A. Pamecha, Bounds for self-reconfiguration of metamorphic robots, in *Proc. IEEE ICRA* (1996).

18. G. Chirikjian, A. Pamecha, and I. Ebert-Uphoff, Evaluating efficiency of self-reconfiguration in a class of modular robots, *J. Robotics Systems* 13(5) (1996) 317–338.

19. R. Courant and H. Robbins, *What is Mathematics?*, Oxford University Press (1941).

20. J. Crisp and B. Wiest, Embeddings of graph braid groups and surface groups in right-angled Artin groups and braids groups, *Alg. & Geom. Top.* 4 (2004) 439–472.

21. P. Csorba and F. Lutz, Graph coloring manifolds, *Contemp. Math.* 423 (2006) 51–69.

22. M. Davis, Groups generated by reflections and aspherical manifolds not covered by Euclidean space, *Ann. Math.* (2) 117 (1983) 293–324.

23. M. Davis and T. Januszkiewicz, Right angled Artin groups are commensurable with right angled Coxeter groups, *J. Pure and Appl. Algebra* 153 (2000) 229–235.

24. D. Epstein *et al.*, *Word Processing in Groups*, Jones & Bartlett Publishers, Boston MA (1992).

25. M. Erdmann and T. Lozano-Perez, On multiple moving objects, in *Proc. IEEE ICRA* (1986).

26. E. Fadell and S. Husseini, *Geometry and Topology of Configuration Spaces*, Springer, Berlin (2001).

27. R. Fair, V. Srinivasan, H. Ren, P. Paik, V. Pamula, and M. Pollack, Electrowetting based on chip sample processing for integrated microfluidics, in *Proc. IEEE Inter. Electron Devices Meeting (IEDM)* (2003).

28. R. Fair, A. Khlystov, V. Srinivasan, V. Pamula, and K. Weaver, Integrated chemical/biochemical sample collection, pre-concentration, and analysis on a digital microfluidic lab-on-a-chip platform, in *Lab-on-a-Chip: Platforms, Devices, and Applications, Conf. 5591, SPIE Optics East* (2004) 25–28.

29. M. Farber, Topological complexity of motion planning, *Discrete and Computational Geometry* 29 (2003) 211–221.

30. M. Farber, Collision free motion planning on graphs, in *Algorithmic Foundations of Robotics IV*, M. Erdmann, D. Hsu, M. Overmars, A. F. van der Stappen, eds., Springer (2005) 123–138.

31. M. Farber and S. Yuzvinsky, Topological robotics: Subspace arrangements and collision free motion planning, *Transl. of Amer. Math. Soc.* 212 (2004) 145–156.

32. D. Farley, Finiteness and CAT(0) properties of diagram groups, *Topology* 42 (2003) 1065–1082.

33. D. Farley and L. Sabalka, Discrete Morse theory and graph braid groups, *Algebraic & Geometric Topology* 5 (2005) 1075–1109.

34. D. Farley and L. Sabalka, On the cohomology rings of tree braid groups, *J. Pure and Applied Algebra* 212(1) (2008) 53–71.

35. P. Gaucher, About the globular homology of higher dimensional automata, *Cahiers de Top. et Geom. Diff. Categoriques* 43(2) (2002) 107–156.

36. R. Ghrist, Shape complexes for metamorphic robot systems, in *Algorithmic Foundations of Robotics V, STAR 7* (2004) 185–201.

37. R. Ghrist, Configuration spaces and braid groups on graphs in robotics, in *AMS/IP Studies in Mathematics* 24 (2001) 29–40.

38. R. Ghrist and D. Koditschek, Safe, cooperative robot dynamics on graphs, *SIAM J. Control Optim.* 40 (2002) 1556–1575.

39. R. Ghrist and S. LaValle, Nonpositive curvature and Pareto optimal motion planning, *SIAM J. Control Optim.* 45(5) (2006) 1697–1713.

40. R. Ghrist, J. O'Kane, and S. M. LaValle, Computing Pareto optimal coordinations on roadmaps, *Int. J. Robotics Research* 12 (2006) 997–1012.

41. R. Ghrist and V. Peterson, The geometry and topology of reconfiguration, *Adv. Appl. Math.* 38 (2007) 302–323.

42. M. Gromov, Hyperbolic groups, in *Essays in Group Theory*, MSRI Publ. 8, Springer-Verlag (1987).

43. F. Haglund and D. Wise, Special cube complexes, *Geom. and Func. Anal.* 17(5) (2008) 1551–1620.

44. J.-C. Hausmann and A. Knutson, Polygon spaces and Grassmannians, *Enseign. Math. (2)* 43(1-2) (1997) 173–198.

45. H. Jacobs, A. Tao, A. Schwartz, D. Gracias, and G. Whitesides, Fabrication of a cylindrical display by patterned assembly, *Science* 296 (2002) 4763–4768.

46. M. Kapovich and J. Millson, Moduli spaces of linkages and arrangements, in *Advances in Geometry*, Progr. Math. 172, Birkhäuser, Boston, MA (1999) 237–270.

47. M. Kapovich and J. Millson, Universality theorems for configuration spaces of planar linkages, *Topology* 41(6) (2002) 1051–1107.

48. M. Kapovich and J. Millson, On the moduli space of polygons in the Euclidean plane, *J. Diff. Geom.* 42 (1995) 133–164.

49. H. King, Planar linkages and algebraic sets, *Turkish J. Math.* 23(1) (1999) 33–56.

50. E. Klavins, R. Ghrist, and D. Lipsky, The graph grammatical approach to self-organizing robotic systems, *IEEE Trans. Automatic Controls* 51(6) (2006) 949–962.

51. D. Koditschek and E. Rimon, Robot navigation functions on manifolds with boundary, *Adv. in Appl. Math.* 11 (1990) 412–442.

52. K. Kotay and D. Rus, The self-reconfiguring robotic molecule: Design and control algorithms, in *Proc. Workshop Alg. Found. Robotics* (1998).

53. D. Lammers, Motorola speeds the move to nanocrystal flash, *EE Times* (8 December 2003).

54. J.-C. Latombe, *Robot Motion Planning*, Kluwer Academic Press, Boston, MA (1991).

55. S. LaValle, *Planning Algorithms*, Cambridge University Press (2006).

56. C. Levinthal, in *Mössbauer Spectroscopy in Biological Systems, Proceedings of a Meeting held at Allerton House, Monticello, Illinois*, J. T. P. DeBrunner and E. Munck, eds., University of Illinois Press, Illinois (1969) 22–24.

57. L. Lovász, Kneser's conjecture, chromatic number, and homotopy, *J. Combin. Theory Ser. A* 25 (1978) 319–324.

58. C. McGray and D. Rus, Self-reconfigurable molecule robots as 3-d metamorphic robots, in *Proc. Intl. Conf. Intelligent Robots & Design* (2000).

59. R. Milgram and J. Trinkle, The geometry of configuration spaces of closed chains in two and three dimensions, *Homology, Homotopy, and Applications* 6(1) (2004) 237–267.

60. N. Mokhoff, Crystals line up in IBM flash chip, *EE Times* (8 December 2003).

61. S. Murata, H. Kurokawa, and S. Kokaji, Self-assembling machine, in *Proc. IEEE ICRA* (1994).

62. S. Murata, H. Kurokawa, E. Yoshida, K. Tomita, and S. Kokaji, A 3-d self-reconfigurable structure, in *Proc. IEEE ICRA* (1998).

63. S. Murata, E. Yoshida, A. Kamikura, H. Kurokawa, K. Tomita, and S. Kokaji, M-TRAN: Self-reconfigurable modular robotic system, *IEEE-ASME Trans. on Mechatronics* 7(4) (2002) 431–441.

64. A. Nguyen, L. Guibas, and M. Yim, Controlled module density helps reconfiguration planning, in *Proc. Workshop on Algorithmic Foundations of Robotics* (2000).

65. G. Niblo and L. Reeves, The geometry of cube complexes and the complexity of their fundamental groups, *Topology* 37 (1998) 621–633.

66. A. Pamecha, I. Ebert-Uphoff, and G. Chirikjian, Useful metric for modular robot motion planning, *IEEE Trans. Robotics & Automation* 13(4) (1997) 531–545.

67. V. Pratt, Modelling concurrency with geometry, in *Proc. Symp. on Principles of Programming Languages* (1991).

68. M. Raussen, State spaces and dipaths up to dihomotopy, *Homotopy, Homology, & Appl.* 5 (2003) 257–280.

69. M. Raussen, On the classification of dipaths in geometric models of concurrency, *Math. Structures Comp. Sci.* 10 (2000) 427–457.

70. L. Reeves, Rational subgroups of cubed 3-manifold groups, *Michigan Math. J.* 42 (1995) 109–126.

71. C. Reutenauer, *The Mathematics of Petri Nets*, Prentice-Hall (1990).

72. D. Rus and M. Vona, Crystalline robots: Self-reconfiguration with unit-compressible modules, *Autonomous Robots* 10(1) (2001) 107–124.

73. L. Sabalka, Embedding right-angled Artin groups into graph braid groups, *Geom. Dedicata* 124(10) (2007) 191–198.

74. L. Sabalka, On rigidity and the isomorphism problem for tree braid groups, Preprint (2007).

75. A. Sali, E. Shakhnovich, and M. Karplus, How does a protein fold? *Nature* 369 (1994) 248–251.

76. A. Sali, E. Shakhnovich, and M. Karplus, Kinetics of protein folding, *J. Mol. Biol.* 235 (1994) 1614–1636.

77. B. Sturmfels and L. Pachter, The mathematics of phylogenomics, *SIAM Review* 49 (2007) 3–31.

78. J. Swiatkowski, Estimates for homological dimension of configuration spaces of graphs, *Colloq. Math.* 89 (2001) 69–79.

79. T. Toth-Fejel, Modeling Kinematic Cellular Automata, NASA Institute for Advanced Concepts Phase I: CP-02-02, General Dynamics Advanced Information Systems Contract # P03-0984, 30 April (2004).

80. J. Trinkle and R. Milgram, Complete path planning for closed kinematic chains with spherical joints, *Int. J. Robotics Research* 21(9) (2002) 773–789.

81. K. Walker, *Configuration Spaces of Linkages*, Undergraduate thesis, Princeton University (1985).

82. J. Walter, J. Welch, and N. Amato, Distributed reconfiguration of metamorphic robot chains, in *Proc. ACM Symp. on Distributed Computing* (2000).

83. J. Walter, E. Tsai, and N. Amato, Choosing good paths for fast distributed reconfiguration of hexagonal metamorphic robots, in *Proc. IEEE ICRA* (2002).

84. J. Walter, J. Welch, and N. Amato, Concurrent metamorphosis of hexagonal robot chains into simple connected configurations, *IEEE Trans. Robotics & Automation* 15 (1999) 1035–1045.

85. G. Whitesides and B. Grzybowski, Self assembly at all scales, *Science* 295 (2002) 2418–2421.

86. M. Yim, A reconfigurable robot with many modes of locomotion, in *Proc. Intl. Conf. Adv. Mechatronics* (1993).

87. M. Yim, J. Lamping, E. Mao, and J. Chase, Rhombic dodecahedron shape for self-assembling robots, Xerox PARC Tech. Rept. P9710777 (1997).

88. M. Yim, Y. Zhang, J. Lamping, and E. Mao, Distributed control for 3-d metamorphosis, *Autonomous Robots* 10 (2001) 41–56.

89. E. Yoshida, S. Murata, K. Tomita, H. Kurokawa, and S. Kokaji, Distributed formation control of a modular mechanical system, in *Proc. Intl. Conf. Intelligent Robots & Sys.* (1997).

BRAIDS AND MAGNETIC FIELDS

Mitchell A. Berger

Mathematics, University of Exeter
North Park Road, Exeter EX4 4QF, United Kingdom
E-mail: m.berger@exeter.ac.uk

Magnetic field lines can exhibit a wide range of topological structures, including braid structure. This chapter describes the importance of braided magnetic fields in the atmosphere of the sun. Braided coronal loops store vast amounts of energy; this energy can power small flares and heat the solar atmosphere to millions of degrees. A simple model for self-organization of braided loops will be presented. The basic theory of magnetized fluids will be briefly reviewed, emphasizing topological quantities.

1. Introduction: Braids in Nature

Wherever we find a collection of partially aligned curves, we will see braid structure. The set of curves may be integral curves (*field lines*) for a vector field in three dimensions — for example the lines of force of a magnetic field, or the streamlines and vortex lines in a moving fluid. If there is a dominant component of the field or flow, then the lines will all travel in one direction, with meanderings about that direction (for example in a rotating fluid there will usually be a dominant component of vorticity). Even though the field is continuous, we can derive information about the field structure by choosing some number N individual field lines, and examining the braid structure of these lines.

In dynamical systems theory, there are several contexts where braids become visible. One involves a set of points moving in a plane (e.g. [24]). The space-time curves $(x(t), y(t), t)$ will in general be braided. Even motion in one dimension has braid structure. Consider a dynamical system with a set of solutions $x(a, t)$ where a is some parameter (e.g. $a = x(t = 0)$).

Fig. 1. *The Tuscan Straw Plaiter* (1868) by the pre-Raphaelite painter William Holman Hunt. Plaiting (braiding) straw was an important rural industry in the 19th century. Often different regions would use their own special braid patterns. Courtesy of *The Rossetti Archive.*

Let the system evolve for a time T. Then for each solution labeled by a, the set of points $(x, \mathrm{d}x/\mathrm{d}t, t)$ for $0 \leq t \leq T$ is a curve in three dimensions (essentially the *phase curve* with an extra dimension along t added). Several of these curves form a braid (in fact a positive braid) [27]. Braids have also been of great practical importance in many industries (see Fig. 1).

Section 2 describes a particularly interesting natural environment for braiding: the solar atmosphere. Magnetized clouds form long thin loops which can become mutually entangled. The stresses caused by this entangling can be released in violent storms called *solar flares*. While these braided loops form randomly, they may acquire a more coherent structure through selective reconnection (surgery). Section 3 gives a review of magnetic field theory and topology.

2. Solar X-Ray Loops

2.1. *Basic description*

The sun (see Figs. 2 and 3) has an atmosphere (*the corona*) full of interesting structures and violent activity (e.g. [2, 34]). While the temperature at the surface measures at a relatively cool 5500 degrees Kelvin, most of the

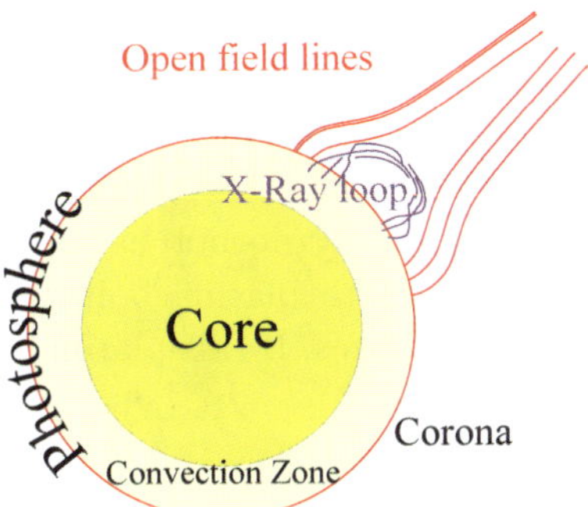

Fig. 2.   A cross-section showing different regions of the sun.

Fig. 3.   Clouds of hot plasma in the solar atmosphere (TRACE image). The plasma traces the magnetic field lines. Braided field lines within the cloud store magnetic energy which is released in numerous small flares, heating the plasma to 1–2 million degrees.

atmosphere above is heated to over 1 million degrees. The nature of this heating is still puzzling, but the most prominent theories suggest conversion of magnetic energy to heat. In essence, while the Earth's atmosphere has electrical storms, the solar atmosphere has magnetic storms, called *flares*.

When looking at images like Fig. 3, bear in mind that there may be just as strong a magnetic field outside the visible loops as inside. What you do not see may be invisible because the plasma is too hot or too cold (and hence radiates at the wrong wavelengths for the telescope's detector). There can also be invisible loops which simply do not have enough plasma to emit much radiation. (Loops fill up with plasma boiled off of the surface just after they have been heated by flares. If this plasma drains away, they may disappear.)

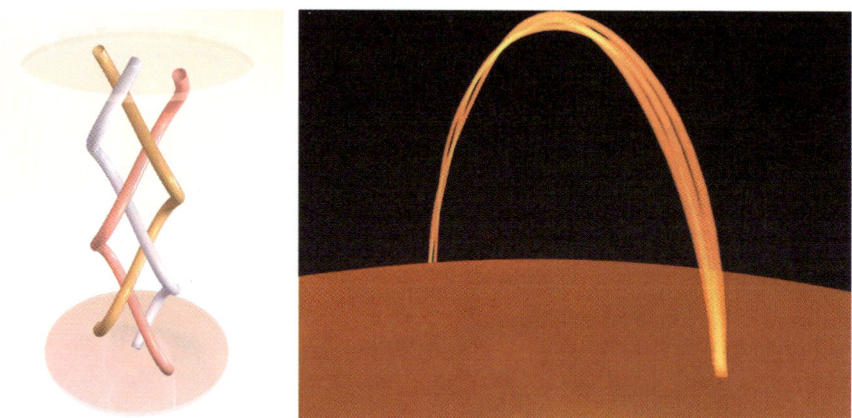

Fig. 4. On the left, a standard pigtail braid. On the right, the tubes in the pigtail have been compressed together, arched into loop shapes, and drawn as clear tubes with glowing gas in the interior. This makes the braid picture somewhat closer in visualization to the coronal images. However, some additional effects are missing (e.g. background radiation, distortion of the tubes from circular cross-sections, artifacts of the observation process ...).

The structure of the solar atmosphere can be seen by telescopes sensitive to radiation emitted by hot plasma — for example in extreme ultraviolet or x-ray wavelengths (Fig. 3). The images (in false color) show clouds of plasma forming long loops anchored in the surface (*photosphere*). These loops follow the magnetic field lines, just as iron filings trace out a magnetic field. Loops have a typical length of about 10^5 km, and a typical radius of about $R = 10^3$ km. Parker [31, 32] has suggested that the field lines inside a loop are highly braided. Braided fields carry more energy than smooth fields; this energy can be released if the field lines are cut and reconnected, simplifying the topology.

Note that the loops begin and end on the same surface, rather than traveling between parallel planes as in a conventional mathematical braid. However, apart from the very small loops which form a carpet-like floor to the corona, the two endpoints of a loop are well separated. Thus the arch-like geometry of loops can usually be ignored when discussing their topology.

The physics of the magnetic forces will be discussed in the next section (including the relation between magnetic energy and braid word length). For now, however, we will ask two questions: first, can braid structure be seen, even at the relatively large scales detectable by present instruments? Secondly, what sort of braids should we expect to develop inside coronal loops?

2.2. *Are solar magnetic fields braided?*

At first sight, the simple pictures of braids found either in braid diagrams or three-dimensional plots (Fig. 4, left) bear little resemblance to the solar plasma. However, drawing the braids with closely packed strings, of the same radius and length as seen in the solar images, makes the resemblance more plausible (Fig. 4, right). There are two major difficulties in deducing braid structure from present observations, however. First, we only see a two-dimensional projection, making it more difficult to properly deduce three-dimensional structure. The recent launch of the *Stereo* satellites may help here. These are two satellites in Earth orbit but observing the sun at different angles, thus allowing a stereoscopic view. Secondly, the coronal plasma is optically thin, i.e. almost transparent. Thus when two loops cross each other in the images, it is difficult to decide which one is in front. Thus it seems plausible that braid structure (of several crossings) exists at the length scales of current observations — but unambiguous delineation of the braid types will be difficult.

What is the interior structure of a coronal loop? (Here we are calling a *loop* one of the 10^3 km radius arches seen in images like Fig. 3; thinner structures will be called *magnetic flux elements*.) The two ends of a loop at the solar surface are called *footpoints*. At small scales inside the loop, there are strong theoretical indications that the field is highly braided. First, the field must store extra energy somewhere to power flares and coronal heating. If the magnetic flux within the loop is untwisted and aligned parallel to the axis, then there will be little free energy. Suppose instead that all the flux within a loop is twisted about a central axis. An integral invariant of the field called *magnetic helicity* measures this net twist (see next section).

This invariant is approximately conserved even during flares [4, 17]. Helicity conservation reduces the amount of energy released during reconnection.

Much more energy can be stored, and released, if the internal field inside a loop has more small scale structure. If we add small scale structure to a set of parallel curves, then they will become braided about each other. We can think about the braid structure in a few ways. First, we can simply choose (randomly or otherwise) a set of N field lines within the loop and ask what their braid structure is. A second approach is more qualitative. We can divide the interior of the loop into subsets of field lines: i.e. sub-volumes which all stretch from one footpoint to the other, where field lines do not cross boundaries (except at the footpoints). The subvolumes should be chosen (if possible!) so that their boundaries do not branch, and the magnetic lines within stay together coherently from one footpoint to the other. Call each subvolume a *magnetic flux element*. In total, the loop will then contain N of these flux elements which will generally be braided. At this point, we should note that the footpoints of a loop in the surface of the sun (the *photosphere*) are not single objects, but sets of discrete intense concentrations of magnetic flux. The coherent coronal flux elements may connect to one or more of these photospheric flux concentrations.

A second reason to believe in braiding comes from consideration of the boundary motions. The photospheric flux concentrations are transported by the ≈ 0.5 km/sec. turbulent flows beneath the photosphere. If an individual flux concentration at the photosphere spins about its own axis, then the magnetic flux connecting to it will become twisted. If several flux concentrations move about each other, then the flux elements connecting to them will become more braided. Further complications can be seen in detailed observations of the surface: flux concentrations can disappear and reform, possibly with the flux from several concentrations merging. This process can further tangle the overlying field. Also the photosphere is covered by a fine *magnetic carpet* of small arched field lines. The large coronal field lines we have been discussing interact with this carpet, further enhancing the boundary motion.

One may also consider how magnetic flux is created. The magnetic fields in the corona are first generated deep in the turbulent convection zone. The *dynamo* generation process is likely to create fields which are twisted and tangled well before they emerge into the corona (for observations of the effects of surface motions and emergence on the photospheric field, see [38]).

What are the possibilities for the interior structure of a loop? One possibility might be that each flux element twists about a central axis, as in

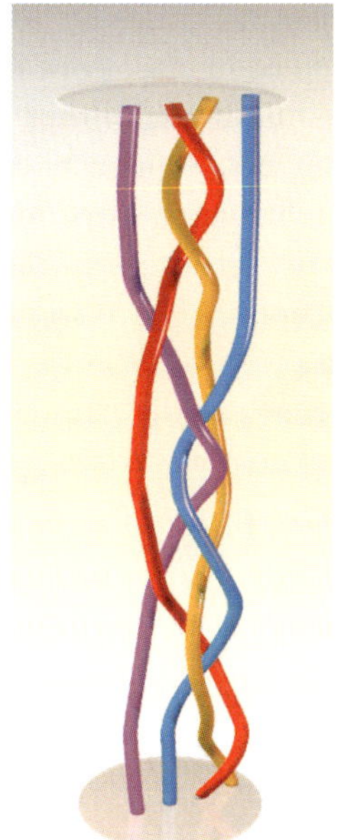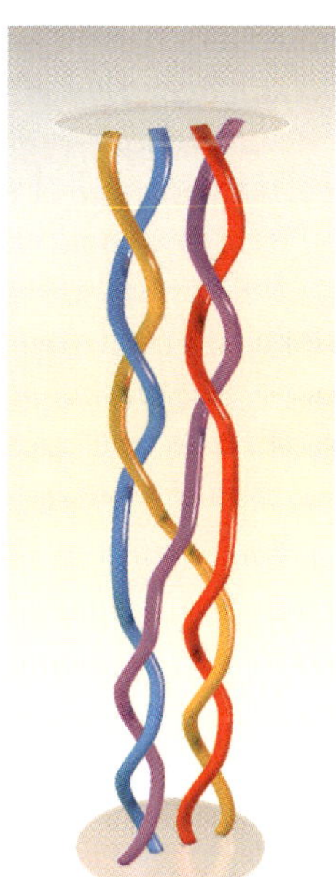

Fig. 5. Some possibilities for the interior structure of a coronal loop. Left: A loop uniformly twisted through one turn. Middle: A randomly generated braid. Right: A highly coherent braid. Apart from the "interchange" crossing in the middle, the lower and upper parts of the braid are quite regular.

Fig. 5 (left) (analogous to a periodic braid). A second possibility is a set of randomly braided flux elements within each loop (Fig. 5 (middle)). Such a loop structure can be modeled by first generating a random braid (for example, from a random Artin word, or from a random walk of the endpoints), and then relaxing the geometry of the braid to a state of minimum magnetic energy [3]. A third example is a more well-organized braid like Fig. 5 (right), which would be reducible to the trivial braid with reconnection at one central crossing.

Numerical simulations [16] suggest that flares (reconnection) are triggered when neighboring flux elements are misaligned by more than a certain angle μ (typically $\mu \approx 30^\circ\text{--}45^\circ$). Let us (crudely!) estimate what this implies for the braid structure. Consider a set of N braided strings, radius R, stretching between two planes separated by a height L. The strings are assumed to be closely packed — each string is touching at least one other string at all heights. If the number of strings is small, then each crossing requires two neighboring strings to make a half turn about each other (for more than 5 strings or so, distant strings will often be seen to cross in projection, even though they are not in close contact). If two circles of radius R touching at one point rotate about each other by an angle of π, then their centres each travel a distance πR. Thus a braid with C crossings requires its strings to travel in the transverse direction a total distance of about $\ell_{total} = 2\pi RC$. Suppose for our crude estimation that over the braid as a whole the net transverse distance is shared equally amongst the strings, so each string has a length projected onto the $x-y$ plane of $\ell = \ell_{total}/N$. Also we assume this transverse length is distributed uniformly in height. Then the typical angle with respect to the vertical of a string is

$$\theta \approx \tan^{-1}\left(\frac{\ell}{L}\right) \approx \tan^{-1}\left(\frac{2\pi RC}{NL}\right). \tag{2.1}$$

Two crossing flux elements with angles with respect to the vertical $\theta = \mu/2$ are liable to flare. This occurs when the number of crossings reaches about

$$C_{\text{critical}} \approx \frac{NL}{2\pi R}\tan(\mu/2). \tag{2.2}$$

For flux elements with aspect ratio $L/R = 200$, $N = 4$ and $\mu = \pi/6$, we find $C_{\text{critical}} \approx 32$.

2.3. *Self-organized braids*

Figure 6 demonstrates the basic process behind a small flare. The left diagram shows a 3-braid with Artin word $B = (\sigma_1^4\,\sigma_2^{-4})$ (reading upwards with right-handed crossings positive). In the middle two diagrams a reconnection changes the connectivity; the braid now becomes $B' = (\sigma_1^4\,\sigma_2^{-1}\,\sigma_1^{-1}\,\sigma_2^{-3})$. One might note that this *adds* a crossing (and hence a letter to the word). This is what we see from the front, of course. As seen from the side, a crossing disappears — so simplifying a geometric braid by reconnection does not necessarily simplify the algebraic braid (at least not immediately). The new braid can now be reduced in both word length and magnetic energy — B' is in fact equivalent to $(\sigma_1\,\sigma_2^{-1}\sigma_1^{-1})$.

This example (and the third example in Fig. 5) have been set up so that one reconnection leads to a considerable reduction of the braid. Are such examples likely or highly artificial? We can investigate this question most easily with braids on 3 strings, where exact algorithms exist for minimizing the word length [9, 36, 26].

Consider colored braids with 3 strings colored red, blue, and green. Suppose these braids are described in the following way: first two of the

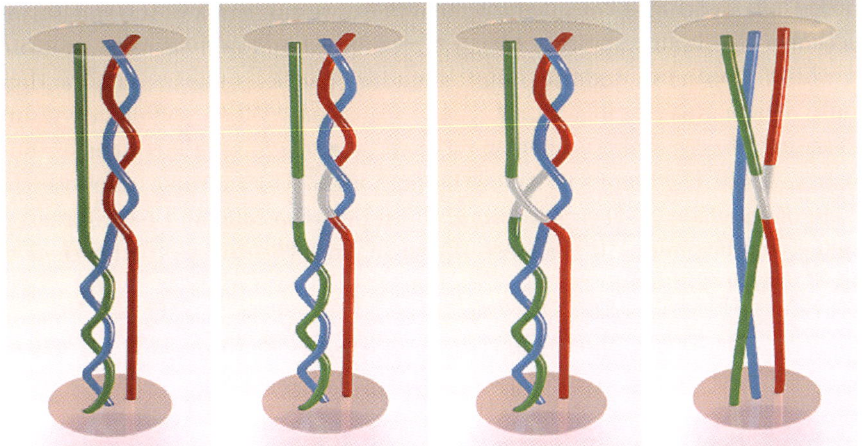

Fig. 6.   Reconnection can reduce topological complexity and liberate energy.

strings (say blue and red) turn about each other T_1 times, with the green string vertical, then a different pair (say green and red) turn T_2 times, and so on. We will ignore uniform rotations of all three strings. We could notate this braid

$$C = br(T_1)gr(T_2)\ldots. \tag{2.3}$$

Thus the first braid in Fig. 6 has word $C = bg(2)br(-2)$ (we define $T = 1$ to be a right-handed turn through 2π). Each piece of the braid (here $br(T_1)$ and $gr(T_2)$) will be called a *segment*.

We can now define the distribution of turns for a colored braid $n(T)$ as the number of times T appears in the word. For example, the word

$$C = br(3)bg(-2)gr(3)bg(1/2) \tag{2.4}$$

has

$$n(-2) = n(1/2) = 1, \qquad n(3) = 2. \tag{2.5}$$

A given method for generating braids, when sampled over many instances (or over a very long braid) will give us a probability distribution of the frequencies $n(T)$. Suppose, for example, we generate the braid by a Poisson process: first randomly choose a half-turn amongst the 6 possibilities (positive or negative bg, br, and gr). Next repeat the same half-turn with probability p or change to a different set of two strings with probability $q = 1 - p$. Then $n(T) = qp^{2T-1}$, i.e. an exponential decrease in T.

For braided strings or magnetic field lines, we can think of the generation as occurring over time. Start with a trivial braid and add turns by motions at one (or both) boundaries. The boundary motions might be a random walk, or perhaps motions in a turbulent medium. When sufficiently many crossings are created, the field may then start to flare. Flares correspond to changes in the connectivity of the field lines. For example, suppose the braid of equation (2.4) sees a reconnection between blue and red right after the $bg(-2)$ segment. In this case (if we rename the curves to the *right* of the reconnection site)

$$br(3)bg(-2)gr(3)bg(1/2) \rightarrow br(3)bg(-2)bg(3)gr(1/2) \tag{2.6}$$

$$= br(3)bg(1)gr(1/2). \tag{2.7}$$

A cancellation of four turns (two positive, two negative) has occurred. If the two segments on either side of the reconnection were both of the same sign they would simply merge rather than cancel (e.g. $bg(2)bg(3) \rightarrow bg(5)$).

The distribution $n(T)$ has astronomical implications. Large cancellations create large flares; but this will rarely happen if $n(T)$ decreases exponentially with T. Observations show that the energy distribution of flares obeys a power law over several orders of magnitude [14] (roughly $n(E) \sim E^{-1.8}$ for energies 10^{25} ergs $< E < 10^{32}$ ergs). Will the reconnection process create a power law distribution of twists?

A simple model suggests that the answer is yes. Consider a colored braid consisting of a large number N segments. We change the braid one step at a time. At each time step, we randomly choose two neighboring segments. Reconnection then merges or partially cancels these two segments. Also, we introduce a new segment at the left side, so that N stays constant. The twist of the new segment is distributed with probability function $p(T)$.

At each time step the distribution function $n(T)$ changes by $\delta n(T)$: first, there is a probability of $p(T)$ that the new segment will add to $n(T)$. Next, if the segment to the left of the reconnection has twist T, then the total number with twist T, $Nn(T)$ reduces by one. Similarly considerations apply to the segment to the right. Finally, if the left segment has twist w and the right segment has twist $T - w$, a new T-segment will be created. We simplify the model to assume T is continuous rather than half-integer.

$$N\delta n(T) = p(T) - 2n(T) + \int_{-\infty}^{\infty} n(w)\mathrm{d}w \int_{-\infty}^{\infty} n(u)\mathrm{d}(u)\delta(T - (u+w)) \tag{2.8}$$

$$= p(T) - 2n(T) + \int_{-\infty}^{\infty} n(w)n(T - w)\mathrm{d}w. \tag{2.9}$$

In a steady state, the left-hand side vanishes. Thus

$$p(T) - 2n(T) + \sqrt{2\pi}(n * n)(T) = 0, \tag{2.10}$$

where $f * g$ is the Fourier convolution. To solve this, we take the Fourier transform,

$$\tilde{n}^2(k) - 2\,\tilde{n}(k) + \tilde{p}(k) = 0. \tag{2.11}$$

This has solution

$$\tilde{n}(k) = \left(1 \pm \sqrt{1 - \tilde{p}(k)}\right). \tag{2.12}$$

Note that we must choose the negative square root in order to insure that $\tilde{n}(k) \to 0$ as $k \to \infty$.

Say the input is a Poisson process, so that for some λ,

$$p(T) = \frac{\lambda}{2} e^{-\lambda|T|}. \tag{2.13}$$

Then

$$\tilde{p}(k) = \frac{\lambda^2}{\lambda^2 + k^2}, \tag{2.14}$$

$$\tilde{n}(k) = \left(1 - \frac{|k|}{\sqrt{\lambda^2 + k^2}}\right). \tag{2.15}$$

We can solve the inverse transform using standard integrals:

$$n(T) = \frac{\pi\lambda}{2}\left(I_1(\lambda T) - L_{-1}(\lambda T)\right) \tag{2.16}$$

where I_0 is a Bessel–I function and L_0 is a Struve–L function (see Fig. 7). This function falls asymptotically as T^{-2}.

3. Magnetohydrodynamics

3.1. *The Maxwell equations*

The Maxwell equations (e.g. [21]) govern the structure of electromagnetic fields as well as their generation by matter.

First, let the space-time metric be Minkowski $\eta = \mathrm{diag}(1, -1, -1, -1)$. Let the Faraday tensor F collect the components of the electric field $\vec{E}$ and magnetic field $\vec{B}$. There are 3 components each; 6 terms fit nicely into an antisymmetric 2nd rank tensor:

$$F^{ab} = \begin{pmatrix} 0 & E_x & E_y & E_z \\ -E_x & 0 & B_z & -B_y \\ -E_y & -B_z & 0 & B_x \\ -E_z & B_y & -B_x & 0 \end{pmatrix}. \tag{3.1}$$

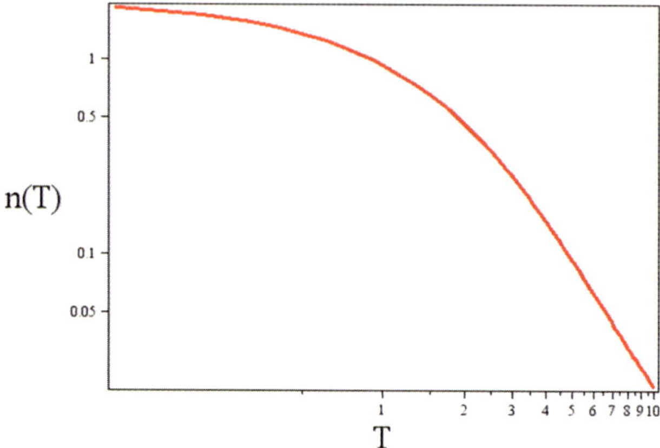

Fig. 7.   The distribution $n(T)$ of braid segment lengths given by equation (2.16), shown with a log-log plot. The asymptotic slope is -2.

There is also the dual Faraday tensor

$$^{*}F^{ab} = \begin{pmatrix} 0 & +B_x & +B_y & +B_z \\ -B_x & 0 & -E_z & E_y \\ -B_y & E_z & 0 & -E_x \\ -B_z & -E_y & E_x & 0 \end{pmatrix}. \tag{3.2}$$

The *Internal Maxwell Equation* structures the fields:

$$\partial_b \, {}^{*}F^{ab} = 0 \tag{3.3}$$

(with summation on repeated indices). This equation tells us that the magnetic part of the field is divergence-free, and time-changes of magnetic flux generate electric forces.

The *Source Maxwell Equations* show how the fields are generated by the distribution and motion of electric charges (Fig. 8). Let $j^a = (\rho_e, j^1, j^2, j^3)$ be a four-vector representing the charge density ρ_e and the three components of electric current. The divergence of F equals this four-vector:

$$\partial_b F^{ab} = j^a. \tag{3.4}$$

In three-vector notation (in SI units with permittivity ϵ_0 and permeability μ_0) the Maxwell equations read

$$\nabla \cdot \vec{B} = 0, \qquad \nabla \times \vec{E} + \frac{\partial \vec{B}}{\partial t} = 0. \tag{3.5}$$

$$\nabla \cdot \epsilon_0 \vec{E} = \rho_e, \qquad \nabla \times \frac{1}{\mu_0}\vec{B} - \epsilon_0 \frac{\partial \vec{E}}{\partial t} = \vec{J}. \tag{3.6}$$

Fig. 8.   Electric fields pour out of electric charges, whereas magnetic fields wrap around electric currents.

For sources moving at non-relativistic speeds $\partial_t \vec{E}$ is small, so we have to an excellent approximation

$$\nabla \times \vec{B} \approx \mu_0 \vec{J} \qquad (3.7)$$

so magnetic fields wrap around electrical currents (see Fig. 8).

3.2. *Magnetohydrodynamics*

Magnetohydrodynamics is a rather long word meaning the study of magnetized fluids. A fluid can carry a magnetic field if it is highly conducting. Examples include molten metal, for example the liquid core of the Earth. Ionized plasmas (where electrons have been stripped off of atoms) can also carry magnetic fields. Most of the the universe is ionized, including the interior and atmosphere of stars, and the magnetosphere of the Earth.

In a highly conducting medium the negative charges (electrons) do not stray far away from the positive charges (ions). Thus we can usually assume charge neutrality at each point in space, i.e. $\rho = 0$ (in fact in a plasma there are small oscillations about this state, see e.g. [15]). Can there be an electric field if there is no source charge present? The answer is yes; in fact there are two ways electrical fields are generated. First, we are considering the macrophysics of the medium, i.e. we average over the microphysics of many interacting particles. In addition to their mean motion giving the bulk fluid velocity $\vec{V}$, these particles have random thermal motions and sometimes collide with each other (and with plasma oscillations). The collisions resist the electrical current; as a result an electrical field is generated. In ordinary circumstances this field is to an excellent approximation parallel to and proportional to the current. This approximation is called *Ohm's law*: for a material at rest $\vec{E} = \eta \vec{J}$, where η is the *resistivity*. Secondly, if the fluid moves with a velocity field $\vec{V}(\vec{x})$, then the rest frame electric field must be transformed to a moving frame. The result is

$$\vec{E} = \eta \vec{J} + \vec{B} \times \vec{V}. \tag{3.8}$$

Combined with equation (3.5), we obtain the *magnetic induction equation*:

$$\frac{\partial \vec{B}}{\partial t} = \nabla \times (\vec{V} \times \vec{B}) - \nabla \times (\eta \vec{J}) \tag{3.9}$$

$$= \nabla \times (\vec{V} \times \vec{B}) + \eta \nabla^2 \vec{B} \tag{3.10}$$

(the latter equation holds for uniform η).

While we now have an equation for the evolution of $\vec{B}$, we still need to think about the velocity field $\vec{V}$. First, by conservation of mass,

$$\frac{d\rho}{dt} + \nabla \cdot \rho \vec{V} = 0. \tag{3.11}$$

This needs to be combined with the Navier-Stokes equation for momentum conservation:

$$\rho \left(\frac{d\vec{V}}{dt} + \vec{V} \cdot \nabla \vec{V} \right) = -\nabla p + \vec{J} \times \vec{B} + \nu \nabla^2 \vec{V} \tag{3.12}$$

where p and ρ are the pressure and density of the flow, and ν is the kinematic viscosity. To complete this set of equations some sort of equation of state is needed to determine ρ and p. (For example, consider an incompressible fluid with $\nabla \cdot \vec{V} = 0$ and $\rho = $ constant. Pressure p can be found by taking the divergence of the Navier-Stokes equation.)

We now have a system of equations (3.10), (3.11), and (3.12) involving the vector fields $\vec{V}$ and $\vec{B}$. Note that the electric field $\vec{E}$ is no longer mentioned explicitly. This is fine — we no longer need $\vec{E}$. In fact, as the careful reader might notice, Ohm's law in a moving frame (3.8) is not quite consistent with $\nabla \cdot \vec{E} = 0$ in a neutral medium. But we have already violated strict adherence to Maxwell with the non-relativistic approximation in equation (3.7). The equations involving $\vec{V}$ and $\vec{B}$ are perfectly consistent on their own, and describe a wide range of physical phenomena. If we wish to include additional physics on small time and length scales, then we must bring back $\vec{E}$ and consider the behavior of individual particles.

In the solar atmosphere, the Earth's magnetosphere, and in large plasma fusion devices, the resistivity η is very small. More precisely, given a physical system with a characteristic size L and characteristic fluid velocity V, a dimensionless number called the *magnetic Reynolds number* can be formed:

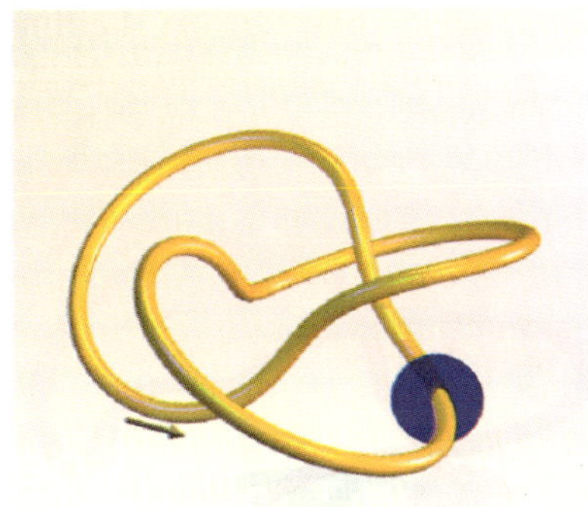 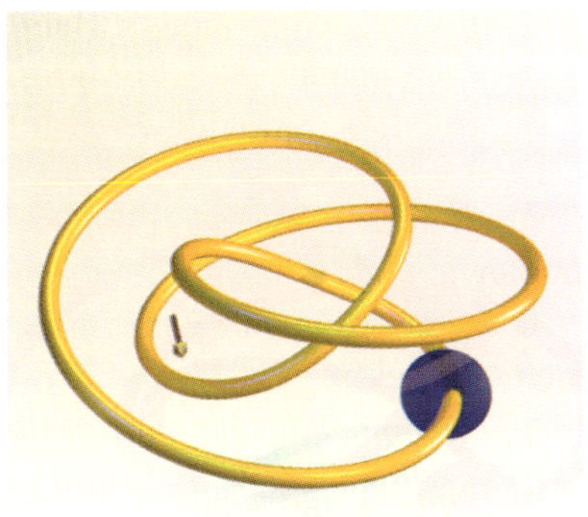

Fig. 9. A magnetic flux tube in the shape of a trefoil knot. Ideal MHD preserves the topological structure of the field. The figures display a transformation from a 2-3 torus knot to a 3-2 torus knot. The axial flux crossing any cross-sectional surface S is a constant and preserved by the motion.

$R_m = VL/\eta$. Note that the choice of V and L is not always obvious (for a sphere do we choose radius or diameter for L?) — but once these have been agreed upon, we can compute R_m. We can then get a feel for the relative importance of the first and second times on the right hand side of the magnetic induction equation, (3.10). Large R_m means that the second term is small and unimportant — unless, of course, $\nabla^2 \vec{B}$ is large! In the solar atmosphere we might have $\eta \approx 1\,\mathrm{m}^2/\sec$, $L \approx 10^6\,\mathrm{m}$, and $V \approx 10^6\,\mathrm{m}/\sec$. This gives $R_m \approx 10^{12}$. The resistive term only becomes important if there are extremely large gradients in the magnetic field direction, i.e. regions of intense electrical current. These regions are the ones susceptible to reconnection and flaring.

In between flares, we can for most practical purposes ignore the resistive term. Equation (3.10) upgrades its name to the *ideal* magnetic induction equation, which now reads

$$\frac{\partial \vec{B}}{\partial t} = \nabla \times (\vec{V} \times \vec{B}) = -\mathcal{L}_V \vec{B} \tag{3.13}$$

where $\mathcal{L}_V \vec{B}$ is a Lie derivative. Thus the field is *Lie dragged* by the fluid flow (also called *frozen in* to the flow). Individual field lines move like thin infinitely elastic strings. This means that a closed field line in the shape of a trefoil knot will preserve that topology no mater how it is distorted (see Fig. 9). One immediate invariant is magnetic flux. Let Φ be the magnetic flux through a co-moving surface S with boundary curve C:

$$\Phi = \int_S \vec{B} \cdot \hat{n}\,\mathrm{d}^3x = \oint_C \vec{A} \cdot \mathrm{d}\vec{\ell} \tag{3.14}$$

where $\vec{A}$ is a *vector potential* for $\vec{B}$, $\nabla \times \vec{A} = \vec{B}$. Then Φ is conserved by the flow.

The energy inside a magnetized fluid is quadratic in both velocity and magnetic field:

$$E = \frac{1}{2} \int \rho V^2 \,\mathrm{d}^3x + \frac{1}{2\mu_0} \int B^2 \,\mathrm{d}^3x. \tag{3.15}$$

Fluid motions exchange energy between the kinetic and magnetic terms. Viscosity and resistivity decrease E, converting it to thermal energy.

3.3. *Magnetostatics*

Of special interest are static solutions to the ideal MHD equations. These solutions have implications for knot theory [29]: consider a knotted tube filled with magnetic field. Start with the tube in some arbitrary initial geometrical configuration. In general, this configuration will have net forces, so the tube will accelerate and move. But suppose the fluid medium has substantial viscosity (like a jar of honey). Then energy will be steadily depleted and the magnetic flux tube will relax to some energy minimum with no net forces. The minimum energy will be a function of the knot type and the internal twist of the field lines within the tube (analogous to the *framing* of a tube). Thus one can extract knot invariants from this procedure, as well as generating aesthetically pleasing *ideal knots* for a given knot type [37].

Here $\vec{V} = 0$ and $\partial_t = 0$. This implies that

$$\vec{J} \times \vec{B} = \nabla p. \tag{3.16}$$

By taking the scalar product with either $\vec{J}$ or $\vec{B}$ we see that gas pressure p is constant along both $\vec{B}$ lines and $\vec{J}$ lines. So (as long as $\nabla p \neq 0$) the vector fields $\vec{J}$ and $\vec{B}$ determine level surfaces of p.

On the other hand, in a region of space where $\nabla p = 0$ the current and field will be parallel. This implies

$$\vec{J} = \lambda(\vec{x})\vec{B}. \tag{3.17}$$

This is called a *force free field*, because the Lorentz force $\vec{J} \times \vec{B} = 0$. A field with uniform λ is called a *linear* force free field. It is linear because λ is the

Fig. 10.   A static magnetic field inside a torus. We can describe the topology of the field by describing how much each surface twists about the central axis. Suppose a (rational) field line closes upon itself after encircling the torus n times the short way around and m times the long way around. Then it has a net twist over one axial circuit of $T = n/m$. One can, of course, have flux surfaces where T is irrational.

eigenvalue of a linear equation $\nabla \times \vec{B} = \lambda \vec{B}$. Note that for nonlinear force free fields $(\nabla \lambda \neq 0)$ $\lambda(\vec{x})$ is constant along field lines (and hence current lines).

Consider the structure of a static field inside a compact volume, bounded by a magnetic surface (where the normal component of the field vanishes, $\vec{B} \cdot \hat{n} = 0$). In other words, we have a finite volume where the magnetic lines do not cross the boundary. For both the pressure balanced equilibria and non-linear force free equilibria we can readily deduce a qualitative picture of the structure of the field. The volume must foliate into the level surfaces of either p or λ (see Fig. 10). For example, the Hill's vortex solution inside a sphere consists of nested torii. The outermost torus is singular, consisting of the spherical boundary itself plus a line segment inside (e.g. the axis from North to South poles). When λ is uniform it is possible for field lines to be ergodic within a compact volume, leading to very interesting topological behavior [19].

3.4. *Magnetic helicity*

Magnetic helicity is an ideal MHD invariant (and is approximately conserved in non-ideal MHD) [28, 5, 4]. A vector field $\vec{B}$ will be called a *closed field* in a volume $\mathcal{D}$ if $\vec{B} \cdot \hat{n}|_{\partial\mathcal{D}} = 0$ on the boundary $\partial\mathcal{D}$. Also $\partial\mathcal{D}$ will be called a *closed field region* for $\vec{B}$. We can define helicity integrals for both closed and open regions (as well as all of $\mathbb{R}^3$). The closed region definition is simplest.

Helicity integrals are vector field generalizations of the Gauss linking integral. In general, magnetic field lines within $\mathcal{D}$ will not be closed curves, even if it is a closed field (field lines can be ergodic within a subvolume, for example). However, we can consider approximations of the field consisting of N closed flux tubes, containing fluxes $\delta\Phi$ (for details of this approximation, see [1]). Let $\vec{x}(\sigma)$ be a point on tube i of $\vec{V}$, and $\vec{y}(\tau)$ be a point on tube j of $\vec{W}$, with $\vec{r} = \vec{y} - \vec{x}$. The linking integral between tube i and tube j is

$$L_{ij} = \frac{1}{4\pi} \oint_i \oint_j \frac{d\vec{x}}{d\sigma} \cdot \frac{\vec{r}}{r^3} \times \frac{d\vec{y}}{d\tau} \, d\tau \, d\sigma. \tag{3.18}$$

We now define

$$H_N = \sum_{i=1}^{N} \sum_{j=1}^{N} L_{ij} \, \delta\Phi^2 . \tag{3.19}$$

Taking the limit as $N \to \infty$ and $\delta\Phi \to 0$ gives [28, 1]

$$H \equiv \frac{1}{4\pi} \int_{\mathcal{D}} \int_{\mathcal{D}} \vec{B}(\vec{x}) \cdot \frac{\vec{r}}{r^3} \times \vec{B}(\vec{y}) \, d^3x \, d^3y \qquad \text{(closed field region)}. \tag{3.20}$$

Note that we can make $\mathcal{D}$ be infinite space provided $|\vec{B}|$ decreases faster than d^{-2} where d is distance from the origin.

For a pair of closed magnetic flux tubes of flux Φ_1 and Φ_2, we can express the helicity in terms of the interlinking L_{12} between the tubes, plus the internal linking of field lines within each tube. The latter can be expressed, via the Călugăreanu theorem, as sums of *Twist numbers* T_w and Writhing numbers W_r ([13, 12]):

$$H = (T_{w1} + W_{r1})\Phi_1^2 + (T_{w2} + W_{r2})\Phi_2^2 + 2L_{12}\Phi_1\Phi_2. \tag{3.21}$$

Next, suppose we divide our closed field region $\mathcal{D}$ into two subvolumes $\mathcal{D}_1$ and $\mathcal{D}_2$. We can define helicity integrals in each subvolume even though the subvolumes themselves are not closed [5]. We do this by measuring relative to a *vacuum reference field* (for example, see Fig. 11). For a magnetic field in $\vec{B}_1$ in $\mathcal{D}_1$, the vacuum field $\vec{P}_1$ is defined by

$$\nabla \cdot \vec{P}_1 = 0; \tag{3.22}$$

$$\nabla \times \vec{P}_1 = 0; \tag{3.23}$$

$$\vec{P}_1 \cdot \hat{n}|_{\partial\mathcal{D}_1} = \vec{B}_1 \cdot \hat{n}|_{\partial\mathcal{D}_1}. \tag{3.24}$$

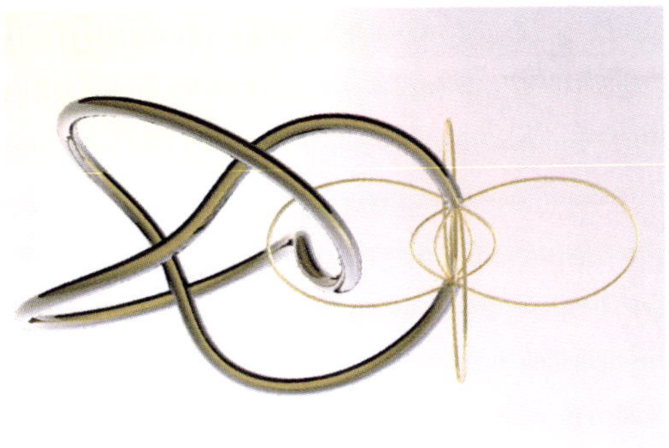

Fig. 11.   A volume $\mathcal{D}_1$ in the shape of a thin tube. The axis is an open curve which forms part of the 2-3 trefoil seen in Fig. 9. In this example, magnetic flux only crosses the boundary at the ends of the tube. The vacuum lines in the complement of $\mathcal{D}_1$ spread out away from the end regions. A vacuum field inside $\mathcal{D}_1$ would thread the tube roughly parallel to its axis; the *Twist number* of vacuum field lines with respect to the axis is zero.

The last condition says that the normal flux at the boundary of $\mathcal{D}_1$ is the same for $\vec{P}_1$ as for the original magnetic field $\vec{B}_1$ (as $\mathcal{D}$ is a closed region, the only normal flux will be at the boundary between $\mathcal{D}_1$ and $\mathcal{D}_2$).

Suppose $\vec{B} = \vec{B}_1$ in $\mathcal{D}_1$ and $\vec{B} = \vec{B}_2$ in $\mathcal{D}_2$. Then we can call the total helicity $H_{\mathcal{D}}(\vec{B}_1, \vec{B}_2)$. We define the *relative helicity* in $\mathcal{D}_1$ to be

$$H_{\mathcal{D}1}(B_1) = H_{\mathcal{D}}(\vec{B}_1, \vec{B}_2) - H_{\mathcal{D}}(\vec{P}_1, \vec{B}_2). \tag{3.25}$$

One can prove that with this definition $H_{\mathcal{D}1}(B_1)$ is independent of the field $\vec{B}_2$ [5] (in particular, one can express $H_{\mathcal{D}1}(B_1)$ as an integral over $\mathcal{D}_1$ alone). We also note that one can show

$$H_{\mathcal{D}}(\vec{B}_1, \vec{B}_2) = H_{\mathcal{D}1}(B_1) + H_{\mathcal{D}2}(B_2) + H_{\mathcal{D}}(\vec{P}_1, \vec{P}_2). \tag{3.26}$$

What is the helicity of a braided magnetic field? Let $\mathcal{D}_1$ be the space between two parallel planes. Suppose that the field $\vec{B}_1$ consists of two flux tubes of flux Φ stretching between the planes (see Fig. 12). We can compute a *winding number* w_{12} between the two curves as the number of turns one curve makes about the other in going from bottom to top plane. We suppose that the field lines within tube 1 wind by a uniform number of turns T_1 about the axis of tube 1 (and similarly for tube 2). Then

$$H_{\mathcal{D}1}(B_1) = (T_1 + T_2 + 2w_{12})\Phi^2. \tag{3.27}$$

With more than two tubes, we would sum the winding numbers (weighted by flux squared) over all pairs of tubes.

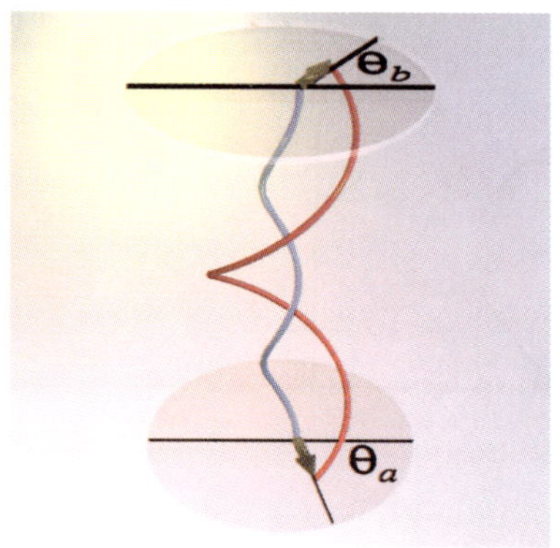

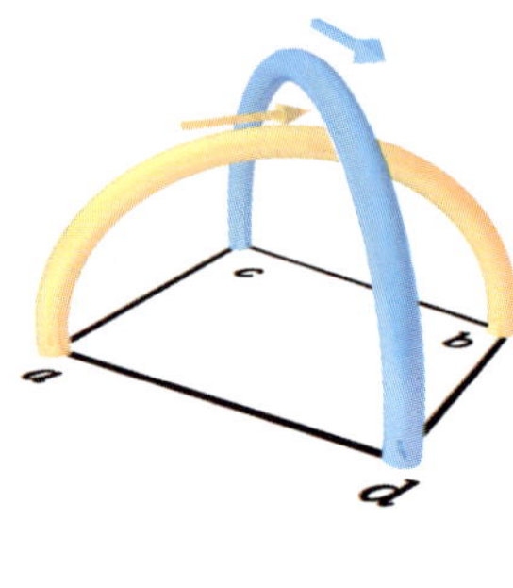

Fig. 12.   Left: Two flux tubes between parallel planes. Their winding number is $w_{12}/(2\pi)$ where $w_{12} = \theta_b - \theta_a + 2\pi n$ for some integer n. Right: Two arches. Their winding number is related to a cross ratio involving the four points a, b, c, and d (equation (3.28)).

For a magnetic field consisting of arched flux tubes, computing the magnetic helicity is a bit more complicated. For details, see [33]. Here is one result: given two arched tubes (see Fig. 12), we can assign the four end points coordinates in the complex plane a, b, c, and d, where tube 1 connects a to b. Then one can show (for $T_1 = T_2 = 0$)

$$\frac{H_{\mathcal{D}1}(B_1)}{\Phi^2} \quad \mathrm{mod}\ 2 = \frac{1}{\pi}\mathcal{I}m \log \frac{(a-c)(b-d)}{(a-d)(b-c)}. \tag{3.28}$$

3.5. *Higher order invariants*

Just as magnetic helicity is based on the Gauss linking number, we can base other magnetic invariants on higher order linking numbers [30, 6, 7, 35, 20]. These numbers are related to Massey invariants and can be computed from Kontsevich integrals [39, 10]. For example, consider three braided flux tubes between parallel planes. Let the axes of the tubes follow the curves $a(z)$, $b(z)$, and $c(z)$ where $0 \le z \le 1$ gives height between the planes, and $a, b, c \in \mathbb{C}$. Let

$$\lambda_{ab}(z) \equiv \frac{1}{2\pi i} \int_0^z \frac{\mathrm{d}a - \mathrm{d}b}{a - b} \tag{3.29}$$

and similarly for λ_{bc} and λ_{ca}. The real part of $\lambda_{ab}(z)$ gives the winding number of the curves a and b between the lower plane and z. We now construct a closed form

$$\psi_{abc} \equiv \frac{1}{2}\left((\lambda_{ab} - \lambda_{bc})\mathrm{d}\lambda_{ca} + (\lambda_{bc} - \lambda_{ca})\mathrm{d}\lambda_{ab} + (\lambda_{ca} - \lambda_{ab})\mathrm{d}\lambda_{bc}\right). \tag{3.30}$$

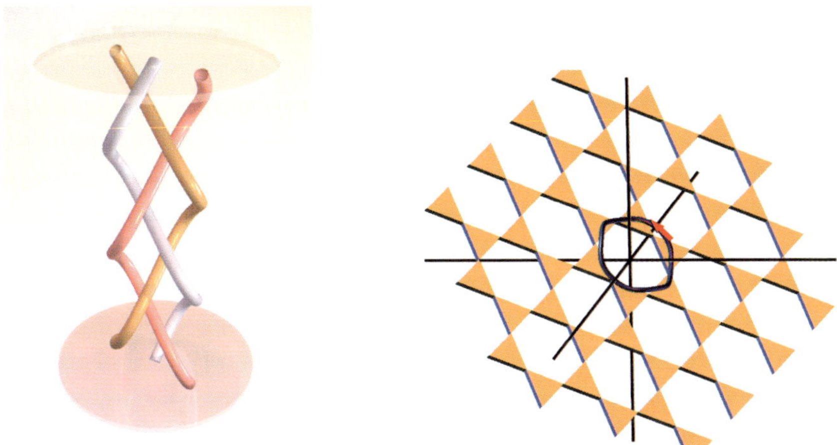

Fig. 13. A three-dimensional space with axes the three winding numbers $\mathcal{R}e(\lambda_{ab})$, $\mathcal{R}e(\lambda_{bc})$, and $\mathcal{R}e(\lambda_{ca})$ (equation (3.29)). The hexagonal regions correspond to forbidden combinations of winding numbers. The pigtail braid on the left corresponds to a curve in the winding space which encircles one of the hexagons.

The integral of this form

$$\Psi_{abc} = \int_0^1 \psi_{abc}(z) \tag{3.31}$$

is then a conserved quantity. A geometric interpretation of Ψ_{abc} can be seen in Fig. 13 [9]. This figure shows a three-dimensional space whose axes give the three winding numbers $\mathcal{R}e(\lambda_{ab})$, $\mathcal{R}e(\lambda_{bc})$, and $\mathcal{R}e(\lambda_{ca})$. A three braid determines a curve in this winding space. The three winding numbers cannot be chosen completely independently, however: their projections in the $(1,1,1)$ direction must stay within the triangular regions seen in the figure (in other words the curves stay within prisms extending along $(1,1,1)$). If the curve in the winding space is closed, then we can ask how many of the hexagonal forbidden regions are enclosed. In particular, the second order quantity Ψ_{abc} gives the signed number of hexagons linked by the winding curve. An interpretation in terms of Hamiltonian dynamics is given in [11].

It can be shown [17] that magnetic helicity is the only knot-like invariant which is resilient, i.e. the only invariant which will be guaranteed to survive reconnection with only small losses (there are small upper bounds on these losses [4]). For example, a sequence of a few reconnections can transform the Borromean rings (say) into three unlinked rings. In a similar manner, for any set of knotted and linked flux tubes, higher order invariants can be removed quickly with an appropriate set of reconnections. Why not magnetic helicity? The answer is that helicity can hide *inside* the tubes, in the form of internal twist.

3.6. *Crossing numbers*

Crossing numbers can be defined for vector fields as well as knots. If we have a finite set of closed curves then we can project them onto a plane and count (unsigned) crossings; we can then average over all projection angles. If we approximate a vector field as a set of N thin closed flux tubes then we can count the crossings of the tubes. Freedman and He (1991) employed a technique similar to Arnold's analysis of magnetic helicity [1] by finding a suitable limit for the number of crossings as $N \rightarrow \infty$, calling this the *asymptotic crossing number*. They then found lower bounds on the energy given this number.

Asymptotic crossing numbers can also be computed for braided magnetic fields stretching between parallel planes [8]. Here field lines are projected onto vertical planes only, so averaging is over a single angle rather than a sphere. By projecting a geometrical braid onto a plane, we determine an Artin braid — the crossing number in fact equals the Artin word length for that projection angle. For a field confined to a cylindrical region, with uniform B_z, one can find lower bounds on the energy given the asymptotic crossing number. Of course, crossing number is not a topological or ideal invariant. It has, however, a positive minimum given the field topology just like magnetic energy. Methods for finding minima (at least local minima) of the asymptotic crossing number for a braided magnetic field are given in [3].

References

1. V. I. Arnold and B. A. Khesin, *Topological Methods in Hydrodynamics*, New York: Springer, 1998.
2. M. J. Aschwanden, *Physics of the Solar Corona. An Introduction*, New York: Springer, 2004.
3. P. D. Bangert, M. A. Berger, and R. Prandi, "In search of minimal random braid configurations", *J. Physics A: Mathematical and General*, **35** (2002), 43–59.
4. M. A. Berger, "Rigorous new limits on magnetic helicity dissipation in the solar corona", *Geophysical and Astrophysical Fluid Dynamics*, **30** (1984), 79–104.
5. M. A. Berger and G. B. Field, "The topological properties of magnetic helicity", *J. Fluid Mechanics*, **147** (1984), 133–148.

6. M. A. Berger, "Third order link integrals", *J. Physics A: Mathematical and General*, **23** (1990), 2787–2793.

7. M. A. Berger, "Third order braid invariants", *J. Physics A: Mathematical and General*, **24** (1991), 4027–4036.

8. M. A. Berger, "Energy-crossing number relations for braided magnetic fields", *Phys. Rev. Lett.*, **70** (1993), 705–708.

9. M. A. Berger, "Minimum crossing numbers for three-braids", *J. Physics A: Mathematical and General*, **27** (1994), 6205–6213.

10. M. A. Berger, "Topological invariants in braid theory", *Letters in Math. Physics*, **55** (2001), 181–192.

11. M. A. Berger, "Hamiltonian dynamics generated by Vassiliev invariants", *J. Physics A: Mathematical and General*, **34** (2001), 1363–1374.

12. M. A. Berger and C. Prior, "The writhe of open and closed curve", *J. Phys. A: Math. Gen.*, **39** (2006), 8321–8348.

13. Călugăreanu, "On isotopy classes of three dimensional knots and their invariants", *Czechoslovak Math. J.*, **T11** (1961), 588.

14. P. Charbonneau, S. W. McIntosh, H. L. Liu, and T. J. Bogdan, "Avalanche models for solar flares", *Solar Phys.*, **203** (2001), 321–353.

15. A. R. Choudhuri, *The Physics of Fluids and Plasmas: An Introduction for Astrophysicists*, Cambridge: Cambridge University Press, 1998.

16. R. D. Dahlburg, J. A. Klimchuk, and S. K. Antiochos, "An explanation for the "switch-on" nature of magnetic energy release and its application to coronal heating", *Astrophysical J.*, **622** (2005), 1191–1201.

17. M. Freedman and M. A. Berger, "Combinatorial relaxation", *Geophysical and Astrophysical Fluid Dynamics*, **73** (1993), 91–96.

18. M. H. Freedman and Z. X. He, "Divergence-free fields — energy and asymptotic crossing number", *Ann. Math.*, **134** (1991), 189.

19. R. Ghrist and R. Komendarczyk, "Overtwisted energy-minimizing curl eigenfields", *Nonlinearity*, **19** (2006), 41–52.

20. G. Hornig and C. Mayer, "Towards a third-order topological invariant for magnetic fields", *J. Physics A: Mathematical and General*, **35** (2002), 3945–3959.

21. J. D. Jackson, *Classical Electrodynamics*, Wiley, New York, 1998.

22. J. A. Klimchuck, "On solving the coronal heating problem", *Solar Phys.*, **234** (2006), 41–77.

23. M. Kontsevich, "Vassiliev's knot invariants", *Adv. Soviet Math.*, **16** (1993), 137.

24. K. D. Kristiansen, G. Helgesen, and A. T. Skjeltorp, "Braid theory and Zipf-Mandelbrot relation used in microparticle dynamics", *European Physical J.*, **B 51** (2006), 363–371.

25. P. Laurence and E. Stredulinsky, "Asymptotic Massey products, induced currents and Borromean torus links", *J. Math. Physics*, **41** (2000), 3170.

26. J. Mairesse and F. Mathéus, "Randomly growing braid on three strands and the manta ray", *Ann. Appl. Prob.*, **17** (2007), 502–536.

27. F. A. McRobie and J. M. T. Thompson, "Braids and knots in driven oscillators", *International J. of Bifurcation and Chaos*, **3** (1993), 1343–1361.

28. H. K. Moffatt, "The degree of knottedness of tangled vortex lines", *J. Fluid Mech.*, **35** (1969), 117–129.

29. H. K. Moffatt, "The energy spectrum of knots and links", *Nature*, **347** (1990), 367–369.

30. M. I. Monastrysky and P. V. Sasarov, "Topological invariants in magnetic hydrodynamics", *Sov. Physics JETP*, **93** (1987), 1210.

31. E. N. Parker, "Magnetic neutral sheets in evolving fields. II — Formation of the solar corona", *Astrophysical J.*, **264** (1983), 642–647.

32. E. N. Parker, "Tangential discontinuities in untidy magnetic topologies", *Phys. Plasmas*, **11** (2004), 2328–2332.

33. E. Pariat, P. Démoulin, and M. A. Berger, "Basic properties of mutual magnetic helicity", *Solar Physics*, **233** (2006), 3–27.

34. E. R. Priest, *Solar Magnetohydrodynamics*, Dordrecht: Reidel, 1982.

35. A. Ruzmaikin and P. Akhmetiev, "Topological invariants of magnetic fields, and the effect of reconnections", *Physics of Plasmas*, **1** (1994), 331.

36. H. Şimşek, M. Bayram, and I. Can, "Automatic calculation of minimum crossing numbers of 3-braids", *Appl. Math. Comp.*, **144** (2003), 507–516.

37. A. Stasiak, V. Katritch and L. H. Kauffman (eds.), *Ideal knots*, Vol. 19 Series on knots and everything, Singapore: World Scientific, 1998.

38. L. van Driel-Gesztelyi, P. Démoulin, and C. H. Mandrini, "Observations of magnetic helicity", *Advances in Space Research*, **32** (1993), 1855–1866.

39. S. Willerton, "On the Vassiliev invariants for knots and for pure braids", Ph.D. Thesis, University of Edinburgh, 1997.

BRAID GROUP CRYPTOGRAPHY

David Garber

Department of Applied Mathematics
Faculty of Sciences
Holon Institute of Technology
52 Golomb Street, PO Box 305
58102 Holon, Israel
E-mail: garber@hit.ac.il

In the last decade, a number of public-key cryptosystems based on combinatorial group theoretic problems in braid groups have been proposed. We survey these cryptosystems and some known attacks on them.

This survey includes: Basic facts on braid groups and on Garside's normal form of its elements, some known algorithms for solving the word problem in the braid group, the major public-key cryptosystems based on the braid group, and some of the known attacks on these cryptosystems. We conclude with a discussion of future directions (which includes also a description of some cryptosystems which are based on other noncommutative groups).

1. Introduction

In many situations, we need to transfer data in a secure way: credit cards information, health data, security uses, etc. The idea of public-key cryptography in general is to make it possible for two parties to agree on a shared secret key, which they can use to transfer data in a secure way (see [80]).

There are several known public-key cryptosystems that are based on the *discrete logarithm problem*, which is the problem of finding x in the equation $g^x = h$ where g, h are given, and on the *factorization problem*, which is the problem of factoring a number into its prime factors: Diffie-Hellman [39] and RSA [113]. These schemes are used in most of the present-day applications using public-key cryptography.

There are several problems with this situation:

- **Subexponential attacks on the current cryptosystems' underlying problems:** Diffie-Hellman and RSA are breakable in time that is subexponential (i.e. less than an exponential amount of time) in the size of the secret key [2]. The current length of secure keys is at least 1000 bits. Thus, the length of the key should be increased every few years. This makes the encryption and decryption algorithms very heavy.
- **Quantum computers:** If quantum computers are in future implemented in a satisfactory way, then RSA will not be secure anymore, since there are polynomial (in $\log(n)$) run-time algorithms of Shor [117] that solve the factorization problem and the discrete logarithm problem. Hence, they solve the problems that RSA and Diffie-Hellman are based on (for more information, see for example [3]).
- **Too much secure data is transferred by the same method:** It is not healthy that most of the secure data in the world be transferred by the same method, since in case this method is broken, too much secure data will be revealed.

Hence, for solving these problems, one should look for a new public-key cryptosystem that on the one hand will be efficient for implementation and use, and on the other hand will be based on a problem that is different from the discrete logarithm problem and the factorization problem. Moreover, the problem should have no subexponential algorithm for solving it, and it is preferable that it has no known attacks by quantum computers.

Combinatorial group theory is a fertile ground for finding hard problems which can serve as a base for a cryptosystem. The braid group defined by Artin [7] is a very interesting group from many aspects: it has many equivalent presentations in entirely different disciplines; its word problem (to determine whether two elements are equal in the group) is relatively easy to solve, but some other problems (such as the conjugacy problem, decomposition problem, and more) seem to be hard to solve.

Based on the braid group and its problems, two cryptosystems were suggested about a decade ago: by Anshel, Anshel and Goldfeld in 1999 [5] and by Ko, Lee, Cheon, Han, Kang and Park in 2000 [79]. These cryptosystems initiated a wide discussion about the possibilities of cryptography in the braid group especially, and in groups in general.

An interesting point which should be mentioned here is that the conjugacy problem in the braid group attracted people even before the cryptosystems on the braid groups were suggested (see, for example, [44, 54]).

After the cryptosystems were suggested, some probabilistic solutions were given [51, 52, 70], but it gave a great push for the efforts to solve the conjugacy problem theoretically in polynomial time (see [14, 15, 16, 56, 57, 58, 59, 85, 86, 87] and many more).

The potential use of braid groups in cryptography led to additional proposals of cryptosystems that are based on apparently hard problems in braid groups (Decomposition problem [120], Triple Decomposition problem [82], Shifted Conjugacy Search problem [31], and more) and in other groups, like Thompson groups [119], polycyclic groups [42] and more. For more information, see the new book of Myasnikov, Shpilrain and Ushakov [105].

In these notes, we try to survey this fascinating subject. Section 2 deals with some different presentations of the braid group. In Sec. 3, we describe two normal forms for elements in the braid groups. In Sec. 4, we give several solutions for the word problem in the braid group. Section 5 introduces the notion of public-key cryptography. In Sec. 6, the first cryptosystems that are based on the braid group are presented. Section 7 is devoted to the theoretical solution to the conjugacy search problem, using the different variants of Summit Sets. In Sec. 8, we describe some more attacks on the conjugacy search problem. In Sec. 9, we discuss some further suggestions for cryptosystems based on the braid group and their cryptanalyses. Section 10 deals with the option of changing the distribution for choosing a key. In Sec. 11, we deal with some suggestions for cryptosystems that are based on other non-commutative groups.

2. The Braid Group

2.1. *Basic definitions*

The braid groups were introduced by Artin [7]. There are several definitions for these groups (see [13, 114]), and we need two of them for our purposes.

2.1.1. *Algebraic presentation*

Definition 2.1. For $n \geq 2$, the *braid group* B_n is defined by the presentation:

$$\left\langle \sigma_1, \ldots, \sigma_{n-1} \ \middle| \ \begin{array}{c} \sigma_i \sigma_j = \sigma_j \sigma_i \text{ for } |i - j| \geq 2 \\ \sigma_i \sigma_{i+1} \sigma_i = \sigma_{i+1} \sigma_i \sigma_{i+1} \text{ for } |i - j| = 1 \end{array} \right\rangle. \tag{2.1}$$

This presentation is called *Artin's presentation* and the generators are called *Artin's generators*.

An element of B_n will be called an *n-braid*. For each n, the identity mapping on $\{\sigma_1, \ldots, \sigma_{n-1}\}$ induces an embedding of B_n into B_{n+1}, so that we can consider an n-braid as a particular $(n+1)$-braid. Using this, one can define the limit group B_∞.

Note that B_2 is an infinite cyclic group, and hence it is isomorphic to the group of integers $\mathbb{Z}$. For $n \geq 3$, the group B_n is not commutative and its center is an infinite cyclic subgroup [26].

When a group is specified using a presentation, each element of the group is an equivalence class of words with respect to the congruence generated by the relations of the presentation. Hence, every n-braid is an equivalence class of n-braid words under the congruence generated by the relations in Presentation (2.1).

2.1.2. *Geometric interpretation*

The elements of B_n can be interpreted as geometric braids with n strands. One can associate with every braid the planar diagram obtained by concatenating the elementary diagrams of Fig. 1 corresponding to successive letters.

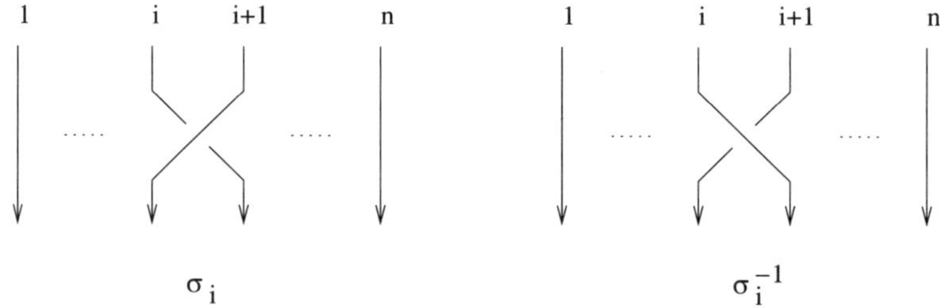

Fig. 1.   The geometric Artin's generators.

Note that some authors draw the geometric braids in the opposite way (i.e. σ_i is drawn as σ_i^{-1}, and vice versa).

A braid diagram can be seen as induced by a three-dimensional figure consisting of n disjoint curves connecting the points $(1, 0, 0), \ldots, (n, 0, 0)$ to the points $(1, 0, 1), \ldots, (n, 0, 1)$ in $\mathbb{R}^3$ (see Fig. 2).

Then the relations in Presentation (2.1) correspond to ambient isotopy, that is: to continuously moving the curves without moving their ends and without allowing them to intersect (see Figs. 3 and 4); the converse implication, i.e. the fact that the projections of isotopic three-dimensional

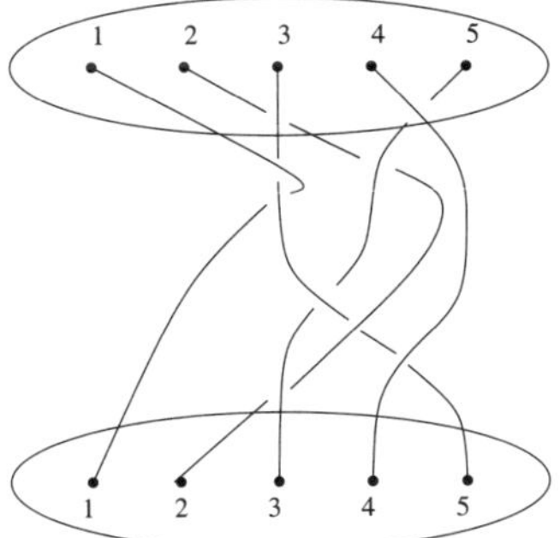

Fig. 2. An example of a braid in B_5.

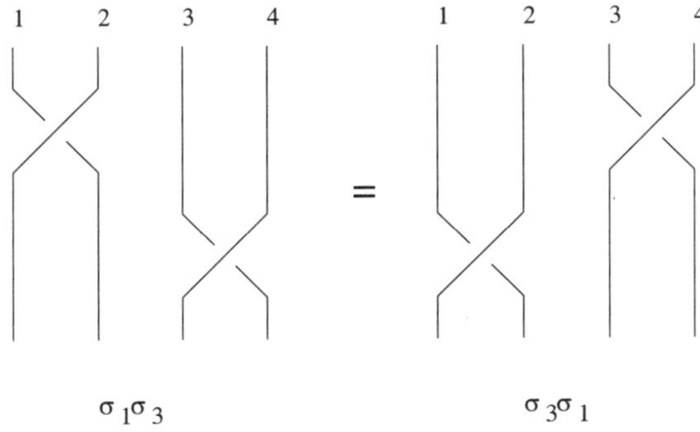

$\sigma_1\sigma_3$ $\qquad\qquad$ $\sigma_3\sigma_1$

Fig. 3. The commutative relation for geometric Artin's generators.

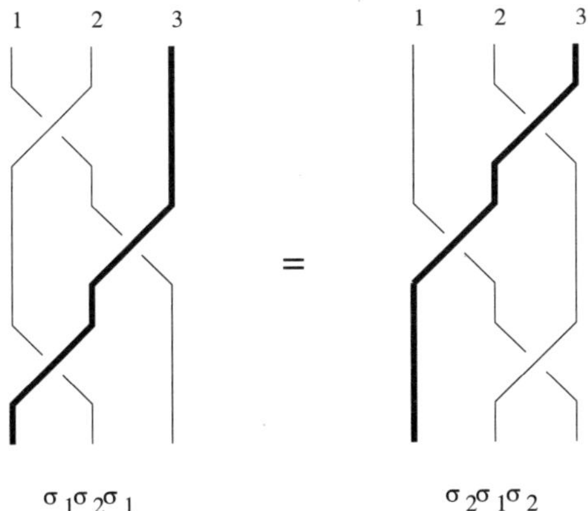

$\sigma_1\sigma_2\sigma_1$ $\qquad\qquad$ $\sigma_2\sigma_1\sigma_2$

Fig. 4. The triple relation for geometric Artin's generators.

figures can always be encoded in words connected by Presentation (2.1), was proved by Artin in [7]. Hence, the word problem in the braid group for Presentation (2.1) is also the *braid isotopy problem*, and thus is closely related to the much more difficult knot isotopy problem.

The multiplication of two geometric braids is done by concatenating the braid diagrams.

2.2. *The Birman-Ko-Lee's presentation*

Like Artin's generators, the generators of Birman-Ko-Lee [17] are braids in which exactly one pair of strands crosses. The difference is that Birman-Ko-Lee's generators includes arbitrary transpositions of strands (i, j) instead of only adjacent transpositions $(i, i + 1)$ in Artin's generators. For each t, s with $1 \leq s < t \leq n$, define the following element of B_n:

$$a_{ts} = (\sigma_{t-1}\sigma_{t-2} \cdots \sigma_{s+1})\sigma_s(\sigma_{s+1}^{-1} \cdots \sigma_{t-2}^{-1}\sigma_{t-1}^{-1}).$$

See Fig. 5 for an example (note that the braid a_{ts} is an elementary interchange of the tth and sth strands, with all other strands held fixed, and with the convention that the strands being interchanged pass in front of all intervening strands). Such an element is called a *band generator*.

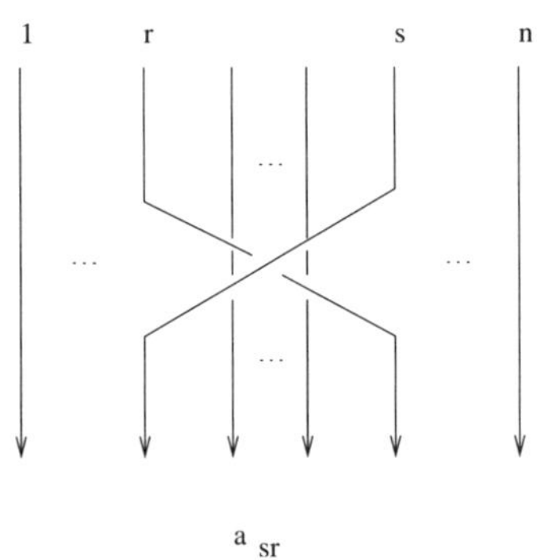

Fig. 5.   A band generator.

Note that the usual Artin's generator σ_t is the band generator $a_{t+1,t}$.

This set of generators satisfies the following relations (see Proposition 2.1 in [17] for a proof):

$$a_{ts}a_{rq} = a_{rq}a_{ts} \qquad \text{if } [s, t] \cap [q, r] = \emptyset.$$
$$a_{ts}a_{sr} = a_{tr}a_{ts} = a_{sr}a_{tr} \quad \text{for } 1 \leq r < s < t \leq n.$$

For a geometric interpretation of the second relation, see Fig. 6.

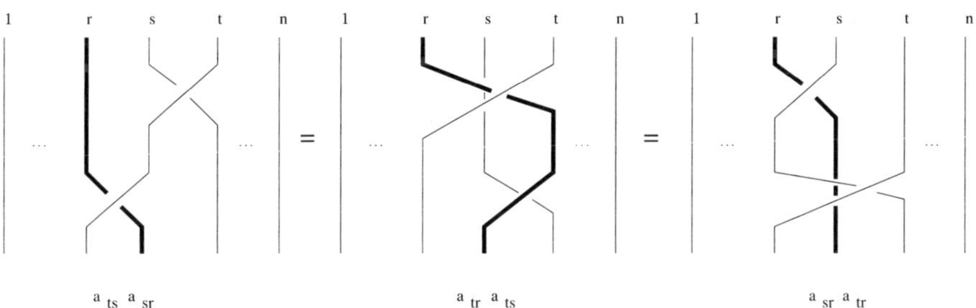

Fig. 6.   The second relation of the Birman-Ko-Lee presentation.

2.2.1. *A geometric viewpoint on the difference between presentations*

A different viewpoint on the relation between the two presentations is as follows: one can think on the braid group as the isotopy classes of boundary-fixing homeomorphisms on the closed disk $D \subset \mathbb{C}^2$ centered at 0 with n punctures [7]. Let K be the set of n punctures.

For presenting Artin's generators in this viewpoint, we have first to define the concept of a *half-twist*. Let σ be a simple path such that $\sigma \subseteq (D - \partial D - K) \cup \{a, b\}$, σ connects a with b $(a, b \in K)$. Choose now a small regular neighbourhood U of σ, and an orientation preserving diffeomorphism $f : \mathbb{R}^2 \to \mathbb{C}$ ($\mathbb{C}$ is taken with the usual "complex" orientation) such that $f(\sigma) = [-1, 1]$, $f(U) = \{z \in \mathbb{C} \mid |z| < 2\}$. Let $\alpha(x)$ be any real smooth monotone function such that

$$\alpha(x) = \begin{cases} 1 & x \in \left[0, \dfrac{3}{2}\right] \\[2mm] 0 & x \geq 2 \,. \end{cases}$$

With this function, we define a diffeomorphism $h : \mathbb{C} \to \mathbb{C}$ as follows: for any $z = re^{i\varphi} \in \mathbb{C}$, we define: $h(z) = re^{i(\varphi + \alpha(r)\pi)}$. It is clear that for all z satisfying $|z| \leq \frac{3}{2}$, $h(z)$ is a positive rotation on $180°$ and $h(z) = \mathrm{Id}$ for all z satisfying $|z| \geq 2$.

The *half-twist defined by* σ, denoted by $H(\sigma)$, is the braid defined by $(f^{-1} \cdot h \cdot f)|_D$.

In this viewpoint, we locate the punctures on the real line, and Artin's generator σ_i is the half-twist $H(\sigma)$ where the path σ is the real segment between the points i and $i + 1$ (see Fig. 7).

On the other hand, for illustrating the generators a_{ts} of the Birman-Ko-Lee presentation, let us take the punctures organized as the vertices of an n-gon contained in the disk D. Now, the generator a_{ts} is the half-twist

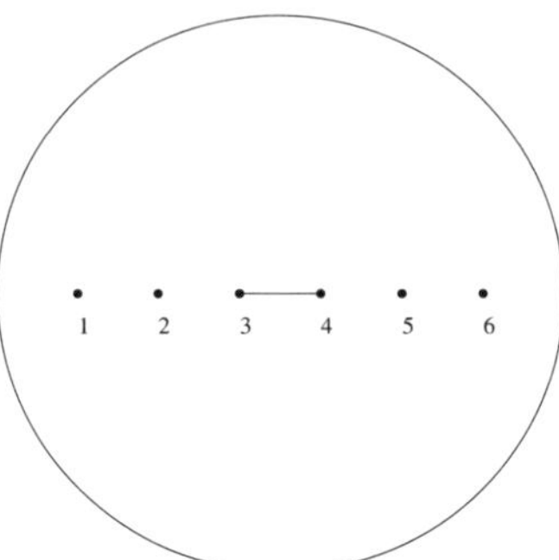

Fig. 7.   Artin's generator σ_3.

$H(\sigma)$ where the path σ is the chord which connects the points t and s (see Fig. 8).

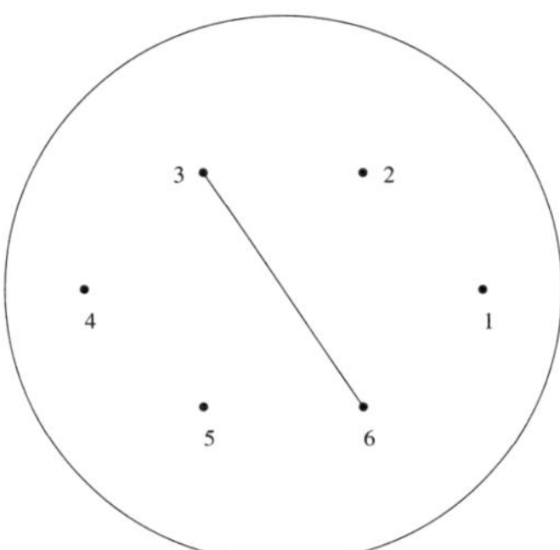

Fig. 8.   The Birman-Ko-Lee generator a_{63}.

For more information, see [9, 19].

3. Normal Forms of Elements in the Braid Group

Fix a presentation for the braid group B_n, i.e. a short exact sequence:

$$1 \to R \to F \to B_n \to 1,$$

where F is a free group, and R is generated by the relations of B_n.

A *normal form* of an element in a group is a unique presentation to each element in the group. In other words, the normal form gives a canonical representative of each equivalence class.

Having a normal form for elements in the group is very useful, since it lets us compare two elements, so it gives a solution for the *word problem*:

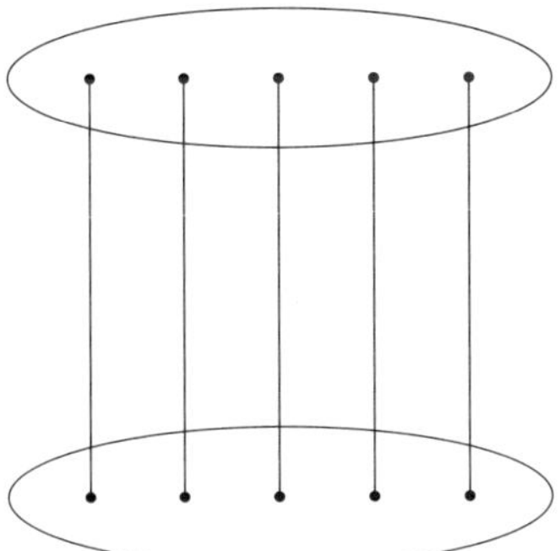

Fig. 9.   The unit braid $\varepsilon \in B_5$.

Problem 3.1. *Given a braid word $w \in F$, does w represent the unit braid ε (see Fig. 9)?*

Since B_n is a group, the above problem is equivalent to the following problem:

Problem 3.2. *Given two braids w, w', do w and w' represent the same braid?*

Indeed, $w = w'$ is equivalent to $w^{-1}w' = \varepsilon$, where w^{-1} is the word obtained from w by reversing the order of the letters and exchanging σ_i and σ_i^{-1} everywhere.

We present here two well-known normal forms of elements in the braid group. For more normal forms, see [20, 32, 41, 48].

3.1. *Garside's normal form*

Garside's normal form is initiated in the work of Garside [54], and several variants have been described in partly independent papers [1, 38, 44, 45, 128].

We start by defining a *positive braid* which is a braid that can be written as a product of positive powers of Artin's generators. We denote the set of positive braids by B_n^+. This set has the structure of a monoid under the operation of braid concatenation.

An important example of a positive braid, which has a central role in Garside's normal form, is the *fundamental braid* $\Delta_n \in B_n^+$:

$$\Delta_n = (\sigma_1 \cdots \sigma_{n-1})(\sigma_1 \cdots \sigma_{n-2}) \cdots \sigma_1 \, .$$

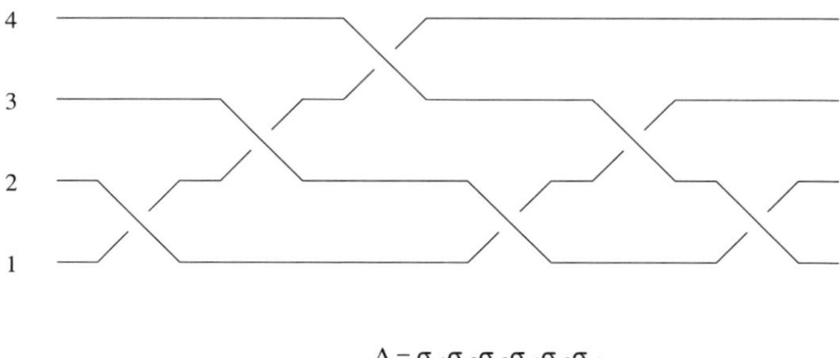

$$\Delta_4 = \sigma_1 \sigma_2 \sigma_3 \sigma_1 \sigma_2 \sigma_1$$

Fig. 10. The fundamental braid Δ_4.

Geometrically, Δ_n is the braid on n strands, where any two strands cross positively, exactly once (see Fig. 10). Later, if it is clear from the context, we will omit the index n.

The fundamental braid has several important properties:

(1) For any generator σ_i, we can write $\Delta_n = \sigma_i A = B \sigma_i$ where A, B are positive braids.
(2) For any generator σ_i, the following holds:

$$\tau(\sigma_i) := \Delta_n^{-1} \sigma_i \Delta_n = \sigma_{n-i}$$

(the inner automorphism τ on B_n is called the *shift map*).
(3) Δ_n^2 is the generator of the center of B_n.

Now, we introduce *permutation braids*. One can define a partial order on the elements of B_n: for $A, B \in B_n$, we say that A is a *prefix* of B and write $A \preceq B$ if $B = AC$ for some C in B_n^+. Its simple properties are:

(1) $B \in B_n^+ \Leftrightarrow \varepsilon \preceq B$
(2) $A \preceq B \Leftrightarrow B^{-1} \preceq A^{-1}$.

$P \in B_n$ is a *permutation braid* (or a *simple braid*) if it satisfies: $\varepsilon \preceq P \preceq \Delta_n$. Its name comes from the fact that there is a bijection between the set of permutation braids in B_n and the symmetric group S_n (there is a natural surjective map from B_n to S_n defined by sending i to the ending place of the strand which starts at position i, and if we restrict ourselves to permutation braids, this map is a bijection). Hence, we have $n!$ permutation braids.

Geometrically, a permutation braid is a braid on n strands, where any two strands cross positively, *at most* once.

Given a permutation braid P, one can define a *starting set* $S(P)$ and a *finishing set* $F(P)$ as follows:

$$S(P) = \{i | P = \sigma_i P' \text{ for some } P' \in B_n^+\}$$

$$F(P) = \{i | P = P' \sigma_i \text{ for some } P' \in B_n^+\}.$$

The starting set is the indices of the generators which can start a presentation of P. The finishing set is defined similarly. For example, by the first property of Δ_n, $S(\Delta_n) = F(\Delta_n) = \{1, \ldots, n-1\}$ since Δ_n can start and end with any of Artin's generator.

A *left-weighted decomposition* of a positive braid $A \in B_n^+$ into a sequence of permutation braids is:

$$A = P_1 P_2 \cdots P_k,$$

where P_i are permutation braids, and $S(P_{i+1}) \subseteq F(P_i)$, i.e. any adjoining of a generator from P_{i+1} to P_i, will convert P_i into a braid that is not a permutation braid.

Example 3.3. The following braid is left-weighted:

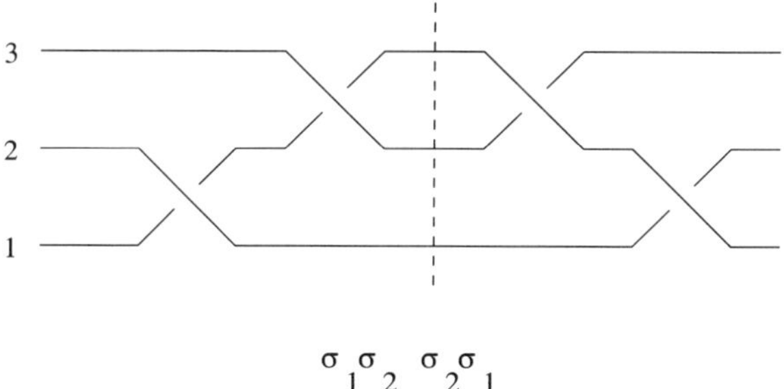

$$\sigma_1 \sigma_2 \quad \sigma_2 \sigma_1$$

The following braid is not left-weighted, due to the circled crossing which can be moved to the first permutation braid:

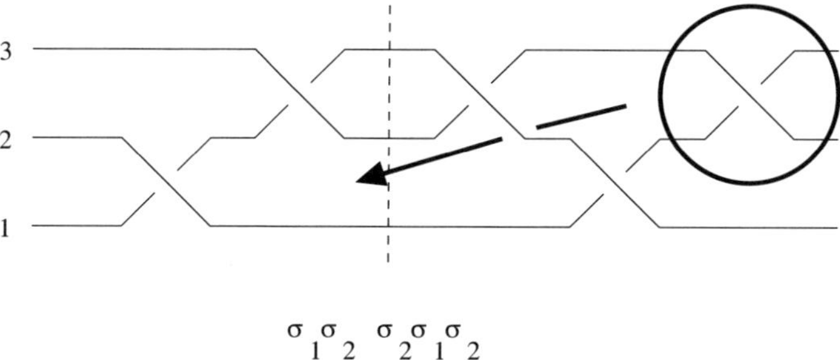

$$\sigma_1 \sigma_2 \quad \sigma_2 \sigma_1 \sigma_2$$

Now, we show it algebraically:

$$\sigma_1\sigma_2 \cdot \sigma_2\sigma_1\sigma_2 = \sigma_1\sigma_2 \cdot \sigma_1\sigma_2\sigma_1 = \sigma_1\sigma_2\sigma_1 \cdot \sigma_2\sigma_1 \,.$$

The following theorem introduces *Garside's normal form* (or *left normal form* or *greedy normal form*) and states its uniqueness:

Theorem 3.4. *For every braid $w \in B_n$, there is a unique presentation given by:*

$$w = \Delta_n^r P_1 P_2 \cdots P_k$$

where $r \in \mathbb{Z}$ is maximal, P_i are permutation braids, $P_k \neq \varepsilon$ and $P_1 P_2 \cdots P_k$ is a left-weighted decomposition.

For converting a given braid w into its Garside's normal form we have to perform the following steps:

(1) For any negative power of a generator, replace σ_i^{-1} by $\Delta_n^{-1} B_i$ where B_i is a permutation braid.
(2) Move any appearance of Δ_n to the left using the relation: $\tau(\sigma_i) = \Delta_n^{-1}\sigma_i\Delta_n = \sigma_{n-i}$. So we get: $w = \Delta_n^{r'} A$ where A is a positive braid.
(3) Write A as a left-weighted decomposition of permutation braids. The way to do this is as follows: Take A, and break it into permutation braids (i.e. take the longest possible sequences of generators that are still permutation braids). Then we get: $A = Q_1 Q_2 \cdots Q_j$ where each Q_i is a permutation braid. For each i, we compute the finishing set $F(Q_i)$ and the starting set $S(Q_{i+1})$. In case the starting set is not contained in the finishing set, we take a generator $\sigma \in S(Q_{i+1}) \setminus F(Q_i)$, and using the relations of the braid group we move it from Q_{i+1} to Q_i. Then, we get the decomposition $A = Q_1 Q_2 \cdots Q_i' Q_{i+1}' \cdots Q_j$. We continue this process till we have $S(Q_{i+1}) \subseteq F(Q_i)$ for every i, and then we have a left-weighted decomposition as needed. For more details, see [44] and Proposition 4.2 of [59] (in the latter reference, this is done based on their new idea of local slidings, see Sec. 7.5 below).

Example 3.5. Let us present the braid $w = \sigma_1\sigma_3^{-1}\sigma_2 \in B_4$ in Garside's normal form. First, we should replace σ_3^{-1} by: $\Delta_4^{-1}\sigma_3\sigma_2\sigma_1\sigma_3\sigma_2$, so we get:

$$w = \sigma_1 \cdot \Delta_4^{-1}\sigma_3\sigma_2\sigma_1\sigma_3\sigma_2 \cdot \sigma_2 \,.$$

Now, moving Δ_4 to the left yields:

$$w = \Delta_4^{-1} \cdot \sigma_3\sigma_3\sigma_2\sigma_1\sigma_3\sigma_2\sigma_2 \,.$$

Decomposing the positive part into a left-weighted decomposition, we get:

$$w = \Delta_4^{-1} \cdot \sigma_2\sigma_1\sigma_3\sigma_2\sigma_1 \cdot \sigma_1\sigma_2 \,.$$

The algorithmic complexity of transforming a word into a normal form with respect to Artin's presentation is $O(|w|^2 n \log n)$, where $|w|$ is the number of letters in a word w in F (see Sec. 9.5 in [45]).

In a similar way, one can define a *right normal form*. A *right-weighted decomposition* of a positive braid $A \in B_n^+$ into a sequence of permutation braids is:

$$A = P_k \cdots P_2 P_1$$

where P_i are permutation braids, and $F(P_{i+1}) \subseteq S(P_i)$, i.e. any addition of a generator from P_{i+1} to P_i, will convert P_i into a braid that is not a permutation braid.

Now, one has the following theorem about the *right normal form* and its uniqueness:

Theorem 3.6. *For every braid $w \in B_n$, there is a unique presentation given by:*

$$w = P_k \cdots P_2 P_1 \Delta_n^r$$

where $r \in \mathbb{Z}$, P_i are permutation braids, and $P_k \cdots P_2 P_1$ is a right-weighted decomposition.

For converting a given braid w into its right normal form we have to follow three steps, similar to those of Garside's normal form: We first replace σ_i^{-1} by $B_i\Delta_n^{-1}$, where B_i is a permutation braid. Then, we move any appearance of Δ_n to the right side. Then, we get: $w = A\Delta_n^{r'}$ where A is a positive braid. The last step is to write A as a right-weighted decomposition of permutation braids.

Now we define the *infimum* and the *supremum* of a braid w: For $w \in B_n$, set $\inf(w) = \max\{r : \Delta_n^r \preceq w\}$ and $\sup(w) = \min\{s : w \preceq \Delta_n^s\}$.

One can easily see that if $w = \Delta_n^m P_1 P_2 \cdots P_k$ is Garside's normal form of w, then: $\inf(w) = m$, $\sup(w) = m + k$.

The *canonical length of w* (or *complexity of w*), denoted by $\ell(w)$, is given by $\ell(w) = \sup(w) - \inf(w)$. Hence, if w is given in its normal form, the canonical length is the number of permutation braids in the form.

3.2. *The Birman-Ko-Lee canonical form*

Based on Presentation (2.2), Birman, Ko and Lee [17] give a new canonical form for elements in the braid group.

They define a new fundamental word:

$$\delta_n = a_{n,n-1}a_{n-1,n-2}\cdots a_{2,1} = \sigma_{n-1}\sigma_{n-2}\cdots\sigma_1 .$$

See Fig. 11 for an example for $n = 4$.

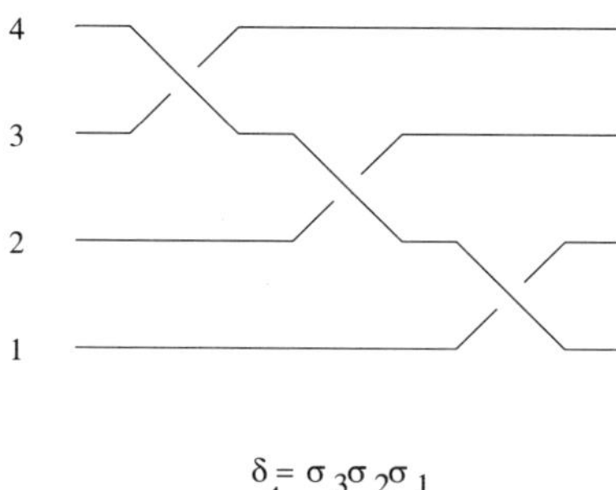

Fig. 11.　The fundamental braid δ_4.

One can easily see the connection between the new fundamental word and Garside's fundamental word Δ_n:

$$\Delta_n^2 = \delta_n^n .$$

The new fundamental word δ_n has important properties, similar to Δ_n:

(1) For any generator a_{sr}, we can write $\delta_n = a_{sr}A = Ba_{sr}$ where A, B are positive braids (with respect to the Birman-Ko-Lee generators).
(2) For any generator a_{sr}, the following holds: $a_{sr}\delta_n = \delta_n a_{s+1,r+1}$.

Similar to Garside's normal form of braids, each element of B_n has the following unique form in terms of the band generators:

$$w = \delta_n^j A_1 A_2 \cdots A_k,$$

where $A = A_1 A_2 \cdots A_k$ is positive, j is maximal and k is minimal for all such representations, also the A_i's are positive braids which are determined uniquely by their associated permutations (see [17, Lemma 3.1]). We will refer to Garside's braids P_i as *permutation braids*, and to the Birman-Ko-Lee braids A_i as *canonical factors*. Note that not every permutation corresponds to a canonical factor.

Note that there are $C_n = \frac{(2n)!}{n!(n+1)!}$ (the nth Catalan number) different canonical factors for the band-generators presentation (see [17, Corollary 3.5]), whereas there are $n!$ different permutation braids for Artin's presentation. Since C_n is much smaller than $n!$, it is sometimes computationally easier to work with the band-generators presentation than Artin's presentation (see also Sec. 8.3.2 below).

As in Garside's normal form, there is an algorithmic way to convert any braid to this canonical form: we first convert any negative power of a generator to $\delta_n^{-1} A$ where A is positive. Then, we move all the δ_n to the left, and finally we organize the positive word in a left-weighted decomposition of canonical factors.

The complexity of transforming a word into a canonical form with respect to the Birman-Ko-Lee presentation is $O(|w|^2 n)$, where $|w|$ is the length of the word w in F [17].

As in Garside's normal form, one can define infimum, supremum and canonical length for the canonical form of the Birman-Ko-Lee presentation.

4. Algorithms for Solving the Word Problem in Braid Group

Using ε for the unit word (see Fig. 9), recall that the *word problem* is the following algorithmic problem: *Given a braid word w, does $w = \varepsilon$ hold, i.e. does w represent the unit braid ε?*

In this section, we will concentrate on some solutions for the word problem in the braid group.

4.1. *Dehornoy's handles reduction*

The process of *handle reduction* has been introduced by Dehornoy [29], and one can see it as an extension of the free reduction process for free groups. Free reduction consists of iteratively deleting all patterns of the form xx^{-1} or $x^{-1}x$: starting with an arbitrary word w of length m, and no matter how the reductions are performed, one finishes in at most $m/2$ steps with a unique reduced word, i.e., a word that contains no xx^{-1} or $x^{-1}x$.

Free reduction is possible for any group presentation, and in particular for B_n, but it does not solve the word problem: there exist words that represent $\varepsilon \in B_n$, but do not freely reduce to the unit word. For example, the word $\sigma_1 \sigma_2 \sigma_1 \sigma_2^{-1} \sigma_1^{-1} \sigma_2^{-1}$ represents the unit word, but free reductions cannot reduce it any further.

The handle reduction process generalizes free reduction and involves not only patterns of the form xx^{-1} or $x^{-1}x$, but also more general patterns of the form $\sigma_i \cdots \sigma_i^{-1}$ or $\sigma_i^{-1} \cdots \sigma_i$:

Definition 4.1. *A σ_i-handle is a braid word of the form*

$$w = \sigma_i^e w_0 \sigma_{i+1}^d w_1 \sigma_{i+1}^d \cdots \sigma_{i+1}^d w_m \sigma_i^{-e},$$

with $e, d = \pm 1, m \geq 0$, and $w_0, \ldots, w_m$ containing no $\sigma_j^{\pm 1}$ with $j \leq i + 1$.

The *reduction of w* is defined as follows:

$$w' = w_0 \sigma_{i+1}^{-e} \sigma_i^d \sigma_{i+1}^e w_1 \sigma_{i+1}^{-e} \sigma_i^d \sigma_{i+1}^e \cdots \sigma_{i+1}^{-e} \sigma_i^d \sigma_{i+1}^e w_m,$$

i.e. we delete the initial and final letters $\sigma_i^{\pm 1}$, and we replace each letter $\sigma_{i+1}^{\pm 1}$ with $\sigma_{i+1}^{-e} \sigma_i^{\pm 1} \sigma_{i+1}^e$ (see Fig. 12, taken from [30]).

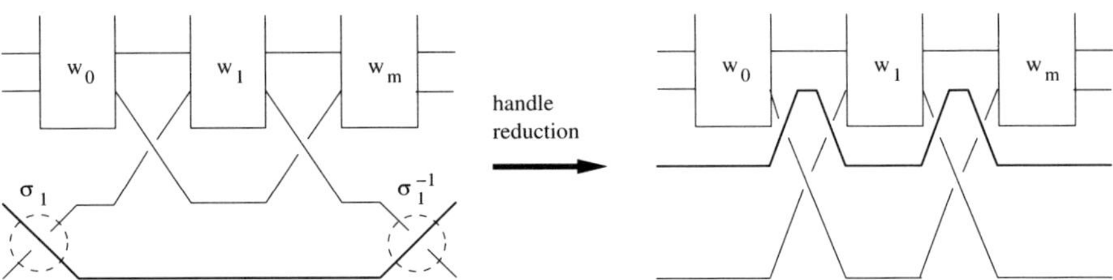

Fig. 12.   An example for a handle reduction (for σ_1). The two circled crossings in the left side are the start and the end of the handle.

Note that a braid of the form $\sigma_i \sigma_i^{-1}$ or $\sigma_i^{-1} \sigma_i$ is a handle, and hence we see that the handle reduction process generalizes the free reduction process.

Reducing a braid yields an equivalent braid: as illustrated in Fig. 12, the $(i+1)$th strand in a σ_i-handle forms a sort of handle, and the reduction consists of pushing that strand so that it passes above the next crossings instead of below. So, as in the case of a free reduction, if there is a reduction sequence from a braid w to ε, i.e., a sequence $w = w_0 \to w_1 \to \cdots \to w_N = \varepsilon$ such that, for each k, w_{k+1} is obtained from w_k by replacing some handle of w_k by its reduction, then w is equivalent to ε, i.e., it represents the unit word ε.

The following result of Dehornoy [29] shows the converse implication and the termination of the process of handle reductions:

Proposition 4.2. *Assume that $w \in B_n$ has a length m. Then every reduction sequence starting from w leads in at most $2^{m^4 n}$ steps to an irreducible* *braid (with respect to Dehornoy's reductions). Moreover, the unit word ε is the only irreducible word in its equivalence class, hence w represents the unit braid if and only if any reduction sequence starting from w finishes with the unit word.*

A braid may contain many handles, so building an actual algorithm requires one to fix a strategy prescribing in which order the handles will be reduced. Several variants have been considered; as can be expected, the most efficient ones use a "Divide and Conquer" trick.

For our current purpose, the important fact is that, although the proved complexity upper bound of the above proposition is very high, handle reduction is extremely efficient in practice, even more than the reduction to a normal form, see [30].

Remark 4.3. In [34], Dehornoy gives an alternative proof for the convergence of the handle reduction algorithm of braids which is both more simple and more precise than the one in his original paper on handle reductions [29].

4.2. *Action on the fundamental group*

As we have pointed out at Sec. 2.2.1, the braid group can be thought of as the isotopy classes of boundary-fixing homeomorphisms on the closed disk $D_n \subset \mathbb{C}^2$ centered at 0 with n punctures $p_1, \ldots, p_n$ [7]. It means that two elements are the same if their actions on $\pi_1(D_n \setminus \{p_1, \ldots, p_n\}, u)$ are equal.

In [50], we propose the following solution for the word problem: we start with a geometric base for $\pi_1(D_n \setminus \{p_1, \ldots, p_n\}, u)$ presented in Fig. 13.

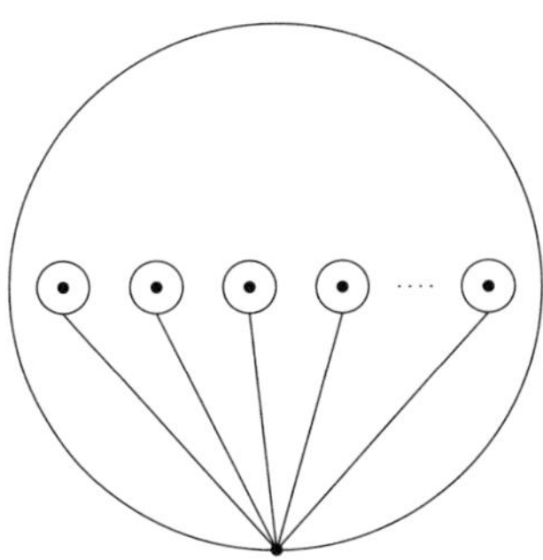

Fig. 13.   A geometric base.

Now, we apply the two braids on this initial geometric base. If the resulting bases are the same up to isotopy, it means that the braids are equal, otherwise they are different.

In Fig. 14, there is a simple example of two equal braids that result in the same base.

$\sigma_1\sigma_2\sigma_1$:

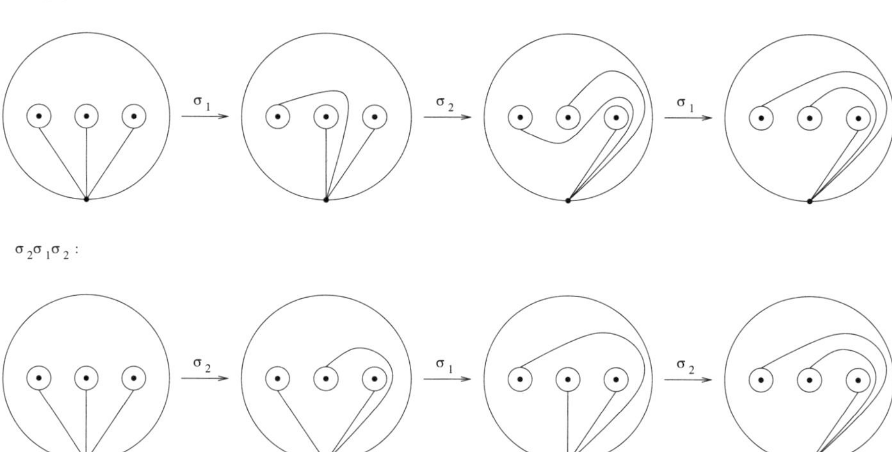

$\sigma_2\sigma_1\sigma_2$:

Fig. 14. An example of applications of two equal braids $\sigma_1\sigma_2\sigma_1 = \sigma_2\sigma_1\sigma_2$ on the initial geometric base.

This algorithm is very quick and efficient for short words, but its worst case is exponential. For more details on its implementation, see [50].

For more solutions for the word problem in the braid groups, see [40] and [48].

5. What is Public-Key Cryptography?

The idea of Public-Key Cryptography (PKC) was invented by Diffie and Hellman [39]. At the heart of this concept is the idea of using a one-way function for encryption (see the survey paper of Koblitz and Menezes [80]).

The functions used for encryption belong to a special class of *one-way functions* that remain one-way only if some information (the decryption key) is kept secret. If we use informal terminology, we can define a *public-key encryption function* as a map from plain text (that is, unencrypted) message units to ciphertext message units that can be feasibly computed by anyone having the public key, but whose inverse function (which deciphers the ciphertext message units) cannot be computed in a reasonable amount of time without some additional information, called the *private key*.

This means that everyone can send a message to a given person using the same enciphering key, which can simply be looked up in a public directory whose contents can be authenticated by some means. There is no need for the sender to have made any secret arrangement with the recipient; indeed, the recipient need never have had any prior contact with the sender at all.

Some of the purposes for which public-key cryptography has been applied are:

- **Confidential message transmission:** Two people want to exchange messages in the open airwaves, in such a way that an intruder observing the communication cannot understand the messages.
- **Key exchange** or **key agreement:** Two people using the open airwaves want to agree upon a secret key for use in some symmetric-key cryptosystem. The agreement should be in such a way that an intruder observing the communication cannot deduce any useful information about their shared secret.
- **Authentication:** The prover wishes to convince the verifier that he knows the private key without enabling an intruder watching the communication to deduce anything about his private key.
- **Signature:** The sender of the message has to send the receiver a (clear or ciphered) message together with a signature proving the origin of the message. Each signature scheme may lead to an authentication scheme: in order to authenticate the sender, the receiver can send a message to the sender, and require that the sender signs this message.

Now, we give some examples of the most famous and well-known public-key cryptosystems.

5.1. *Diffie-Hellman*

In 1976, Diffie and Hellman [39] introduced a key-exchange protocol which is based on the apparent difficulty of computing logarithms over a finite field $\mathbb{F}_q$ with q elements and on some commutative property of the exponent.

Their key-exchange protocol works as follows:

Protocol 5.1.

> *Public keys:* q and a primitive element α.
> *Private keys:* Alice: X_i; Bob: X_j.

Alice: Sends Bob $Y_i = \alpha^{X_i}$ (mod q).

Bob: Sends Alice $Y_j = \alpha^{X_j}$ (mod q).

Shared secret key: $K_{ij} = \alpha^{X_i X_j}$ (mod q).

K_{ij} is indeed a shared key since Alice can compute $K_{ij} = Y_j^{X_i}$ (mod q) and Bob can compute $K_{ij} = Y_i^{X_j}$ (mod q).

This method is secure due to the apparent hardness of the Discrete Logarithm Problem.

5.2. *RSA*

Rivest, Shamir and Adleman [113] introduced one of the most famous and common cryptosystems, which is called RSA. This method is widely used in commerce.

Find two large prime numbers p and q, each about 100 decimal digits long. Let $n = pq$ and $\phi = \phi(n) = (p-1)(q-1)$ (the Euler number). Choose a random integer E between 3 and ϕ that has no common factors with ϕ. It is easy to find an integer D that is the "inverse" of E modulo ϕ, that is, $D \cdot E$ differs from 1 by a multiple of ϕ.

Alice makes E and n public. All the other quantities here are kept secret.

The encryption is done as follows: Bob, who wants to send a plain text message P to Alice that is an integer between 0 and $n-1$, computes the ciphertext integer $C = P^E$ (mod n). (In other words, raise P to the power E, divide the result by n, and C is the remainder.) Then, Bob sends C to Alice.

For decrypting the message, Alice uses the secret decryption number D for finding the plain text P by computing: $P = C^D$ (mod n).

This method is currently secure, since in order to determine the secret decryption key D (for decrypting the message), the intruder should factor the 200 or so digits number n, which is a very hard task.

6. First Cryptosystems that are Based on the Braid Groups

In this section, we describe the first cryptosystems that are based on the braid groups. We start with the definition of some apparently hard problems which the cryptosystems are based on. After that, we describe the first two key-exchange protocols that are based on the braid group. We finish the section with some more cryptosystems based on the braid group.

6.1. *Underlying problems for cryptosystems in the braid group*

We list here several apparently hard problems in the braid group that are the base of many cryptosystems in the braid group.

- **Conjugacy Decision Problem:** Given $u, w \in B_n$, determine whether they are conjugate, i.e., there exists $v \in B_n$ such that

$$w = v^{-1}uv.$$

- **Conjugacy Search Problem:** Given conjugate elements $u, w \in B_n$, find $v \in B_n$ such that

$$w = v^{-1}uv.$$

- **Multiple Simultaneous Conjugacy Search Problem:** Given m pairs of conjugate elements $(u_1, w_1), \ldots, (u_m, w_m) \in B_n$ that are all conjugated by the same element, find $v \in B_n$ such that

$$w_i = v^{-1}u_iv, \quad \text{for all } i \in \{1, \ldots, m\}.$$

- **Decomposition Problem:** Let $w \in B_n$, $u \notin G \leq B_n$. Find $x, y \in G$ such that $w = xuy$.

6.2. *Key-exchange protocols based on the braid group*

In this section, we present two key-exchange protocols which are based on apparently hard problems in the braid group. After the transmitter and receiver agree on a shared secret key, they can use a symmetric protocol for transmitting messages in the insecure channel.

6.2.1. *The Anshel-Anshel-Goldfeld key-exchange protocol*

The following scheme was proposed theoretically by Anshel, Anshel and Goldfeld [5], and implemented in the braid group by Anshel, Anshel, Fisher and Goldfeld [4].

This scheme assumes that the Conjugacy Search Problem is difficult enough (so this scheme, as well as the other schemes described below, would keep its interest, even if it turned out that braid groups are not relevant, since it might be implemented in other groups).

Let G be a subgroup of B_n:

$$G = \langle g_1, \ldots, g_m \rangle, \qquad g_i \in B_n.$$

The secret keys of Alice and Bob are words $a \in G$ and $b \in G$ respectively.

The key-exchange protocol is as follows:

Protocol 6.1.

Public keys: $\{g_1, \ldots, g_m\} \subset B_n$.
Private keys: Alice: a; Bob: b.

Alice: Sends Bob publicly the conjugates: $ag_1a^{-1}, \ldots, ag_ma^{-1}$.
Bob: Sends Alice publicly the conjugates: $bg_1b^{-1}, \ldots, bg_mb^{-1}$.

Shared secret key: $K = aba^{-1}b^{-1}$.

K is indeed a shared key, since if $a = x_1 \cdots x_k$ where $x_i = g_j^{\pm 1}$ for some j, then Alice can compute $ba^{-1}b^{-1} = (bx_k^{-1}b^{-1}) \cdots (bx_1^{-1}b^{-1})$ and hence Alice knows $K = a(ba^{-1}b^{-1})$. Similarly, Bob can compute aba^{-1}, and hence he knows $K = (aba^{-1})b^{-1}$.

The security is based on the difficulty of a variant to the Conjugacy Search Problem in B_n, namely the *Multiple Simultaneous Conjugacy Search Problem* described above. Note that this variant may be easier than the original Conjugacy Search Problem.

In [4], it is suggested to work in B_{80} with $m = 20$ and short initial braids g_i of length 5 or 10 Artin's generators.

Remark 6.2. We simplified a bit the protocol given by Anshel-Anshel-Goldfeld, but the principle remains the same. Moreover, in their protocol, they used not the braids themselves, but their images under the colored Burau representation of the braid group defined by Morton [102] (see Sec. 8.4.1 below).

6.2.2. *Diffie-Hellman-type key-exchange protocol*

Following the commutative idea for achieving a shared secret key of Diffie-Hellman, Ko *et al.* [79] propose a key-exchange protocol based on the braid group and some commutative property of some of its elements. Although braid groups are not commutative, we can find large subgroups such that each element of the first subgroup commutes with each element of the second. Indeed, braids involving disjoint sets of strands commute. Similar approach appears also in the Algebraic Eraser Scheme (see [6] and Sec. 9.4 here).

Note that this scheme was proposed independently in [125] in the context of a general, unspecified noncommutative semigroup with difficult conjugacy problem, but the braid groups were not mentioned there explicitly.

Denote by LB_n (resp. UB_n) the subgroup of B_n generated by $\sigma_1, \ldots, \sigma_{m-1}$ (resp. $\sigma_{m+1}, \ldots, \sigma_{n-1}$) with $m = \lfloor \frac{n}{2} \rfloor$. Then, every braid in LB_n commutes with every braid in UB_n.

Here is Ko *et al.* key-exchange protocol:

Protocol 6.3.

> *Public key:* One braid p in B_n.
> *Private keys:* Alice: $s \in LB_n$; Bob: $r \in UB_n$.
>
> *Alice:* Sends Bob $p' = sps^{-1}$.
> *Bob:* Sends Alice $p'' = rpr^{-1}$.
>
> *Shared secret key:* $K = srpr^{-1}s^{-1}$.

K is a shared key since Alice can compute $K = sp''s^{-1}$ and Bob can compute $K = rp'r^{-1}$, and both are equal to K since s and r commute.

The security is based on the difficulty of the Conjugacy Search Problem in B_n, or, more exactly, on the difficulty of the following variant, which can be called the Diffie-Hellman-like Conjugacy Problem:

Problem 6.4. *Given a braid p in B_n, and the braids $p' = sps^{-1}$ and $p'' = rpr^{-1}$, where $s \in LB_n$ and $r \in UB_n$, find the braid $rp'r^{-1}$, which is also $sp''s^{-1}$.*

The suggested parameters are $n = 80$, i.e. to work in B_{80}, with braids specified using (normal) sequences of length 12, i.e. sequences of 12 permutation braids (see [23]).

6.3. *Encryption and decryption*

The following scheme is proposed by Ko *et al.* [79]. We continue with the same notation of Ko *et al.* Assume that h is a public collision-free one-way hash function of B_n to $\{0,1\}^{\mathbb{N}}$, i.e. a computable function such that the probability of having $h(b_2) = h(b_1)$ for $b_2 \neq b_1$ is negligible (collision-free), and retrieving b from $h(b)$ is infeasible (one-way) (for some examples, see Dehornoy [30, Sec. 4.4] and Myasnikov [106]).

We start with $p \in B_n$ and $s \in LB_n$. Alice's public key is the pair (p, p') with $p' = sps^{-1}$, where s is Alice's private key. For sending the message m_B, which we assume lies in $\{0,1\}^{\mathbb{N}}$, Bob chooses a random braid $r \in UB_n$

and he sends the encrypted text $m_B'' = m_B \oplus h(rp'r^{-1})$ (using $\oplus$ for the Boolean operation "exclusive-or", i.e. the sum in $\mathbb{Z}/2\mathbb{Z}$), together with the additional datum $p'' = rpr^{-1}$. Now, Alice computes $m_A = m'' \oplus h(sp''s^{-1})$, and we have $m_A = m_B$, which means that Alice retrieves Bob's original message.

Indeed, because the braids r and s commute, we have (as before):

$$sp''s^{-1} = srpr^{-1}s^{-1} = rsps^{-1}r^{-1} = rp'r^{-1},$$

and, therefore, $m_A = m_B \oplus h(rp'r^{-1}) \oplus h(rp'r^{-1}) = m_B$.

The security is based on the difficulty of the Diffie-Hellman-like Conjugacy Problem in B_n. The recommended parameters are as in Ko *et al.*'s exchange-key protocol (see Sec. 6.2.2).

6.4. *Authentication schemes*

Three authentication schemes were introduced by Sibert, Dehornoy and Girault [124], which are based on the Conjugacy Search Problem and Root Extraction Problem. Concerning the cryptanalysis of the Root Extraction Problem, see [68].

We present here their first scheme. This scheme is related to the Diffie-Hellman based exchange-key in its idea of verifying that the secret key computed at the two ends is the same.

Note that any encryption scheme can be transformed into an authentication scheme, by sending to Alice both an encrypted version and a hashed image of the same message m, then requesting her to reply with the deciphered message m (she will do it only if the hashed image of the deciphered message is the same as the one sent by Bob).

Their first scheme is based on the difficulty of the Diffie-Hellman-like Conjugacy Problem. It uses the fact that braids involving non-adjacent families of strands commute. The data consist of a public key, which is a pair of braids, and of Alice's private key, which is also a braid. We assume that n is even, and denote again by LB_n (resp. UB_n) the subgroup of B_n generated by $\sigma_1, \ldots, \sigma_{\frac{n}{2}-1}$, i.e. braids where the $\frac{n}{2}$ lower strands only are braided (resp. in the subgroup generated by $\sigma_{\frac{n}{2}+1}, \ldots, \sigma_{n-1}$). The point is that every element in LB_n commutes with every element in UB_n, and alternative subgroups with this property could be used instead. We assume that H is a fixed collision-free hash function from braids to sequences of 0's and 1's or, possibly, to braids.

- Phase 1. Key generation:

  (1) Choose a public braid $b \in B_n$ such that the Diffie-Hellman-like Conjugacy Problem for b is hard enough;
  (2) Alice chooses a secret braid $s \in LB_n$, her private key; she publishes $b' = sbs^{-1}$; the pair (b, b') is her public key.

- Phase 2. Authentication phase:

  (1) Bob chooses a braid $r \in UB_n$, and sends the challenge $x = rbr^{-1}$ to Alice;
  (2) Alice sends the response $y = H(sxs^{-1})$ to Bob, and Bob checks $y = H(rb'r^{-1})$.

For active attacks, the security is ensured by the hash function H: if H is one-way, these attacks are ineffective.

Two more authentication schemes were suggested by Lal and Chaturvedi [83]. Their cryptanalyses are discussed in [68, 129].

7. Attacks on the Conjugacy Search Problem using Summit Sets

In this section, we explain the algorithms for solving the Conjugacy Decision Problem and the Conjugacy Search Problem (CDP/CSP) in braid groups that are based on Summit sets. These algorithms are given in [44, 45, 47, 54, 56, 58]. We start with the basic idea, and then we continue with its implementations.

We follow here the excellent presentation of Birman, Gebhardt and González-Meneses [14]. For more details, see their paper.

7.1. *The basic idea*

Given an element $x \in B_n$, the algorithm computes a finite subset I_x of the conjugacy class of x which has the following properties:

(1) For every $x \in B_n$, the set I_x is finite, non-empty and only depends on the conjugacy class of x. It means that two elements $x, y \in B_n$ are conjugate if and only if $I_x = I_y$.
(2) For each $x \in B_n$, one can compute efficiently a representative $\tilde{x} \in I_x$ and an element $a \in B_n$ such that $a^{-1}xa = \tilde{x}$.
(3) There is a finite algorithm which can construct the whole set I_x from any representative $\tilde{x} \in I_x$.

Now, for solving the CDP/CSP for given $x, y \in B_n$ we have to perform the following steps:

(a) Find representatives $\tilde{x} \in I_x$ and $\tilde{y} \in I_y$.
(b) Using the algorithm from Property (3), compute further elements of I_x (while keeping track of the conjugating elements), until either:

  (i) $\tilde{y}$ is found as an element of I_x, proving x and y to be conjugate and providing a conjugating element, or
  (ii) the entire set I_x has been constructed without encountering $\tilde{y}$, proving that x and y are not conjugate.

We now survey the different algorithms based on this approach.

In Garside's original algorithm [54], the set I_x is the *Summit Set* of x, denoted $SS(x)$, which is the set of conjugates of x having maximal infimum.

Remark 7.1. All the algorithms presented below for the different types of Summit Sets work also for Garside's groups (defined by Dehornoy and Paris [37]), which are a generalization of the braid groups. In our survey, for simplification, we present them in the language of braid groups. For more details on Garside's groups and the generalized algorithms, see [14].

7.2. *The Super Summit Sets*

The Summit Set is improved by El-Rifai and Morton [44], who consider $I_x = SSS(x)$, the *Super Summit Set* of x, consisting of the conjugates of x having minimal canonical length $\ell(x)$. They also show that $SSS(x)$ is the set of conjugates of x having maximal infimum and minimal supremum at the same time. El-Rifai and Morton [44] show that $SSS(x)$ is finite. In general, $SSS(x)$ is much smaller than $SS(x)$. For example, take the element $x = \Delta_4 \sigma_1 \sigma_1 \in B_4$, $SSS(x) = \{\Delta_4 \cdot \sigma_1 \sigma_3\}$ while

$$SS(x) = \{\Delta_4 \cdot \sigma_1 \sigma_3, \Delta_4 \cdot \sigma_1 \cdot \sigma_1, \Delta_4 \cdot \sigma_3 \cdot \sigma_3\}$$

(the permutation braids in each left normal form are separated by a dot) [14].

Starting with a given element x, one can find an element $\tilde{x} \in SSS(x)$ by a sequence of special conjugations, called *cyclings* and *decyclings*:

Definition 7.2. Let $x = \Delta^p x_1 \cdots x_r \in B_n$ be given in Garside's normal form and assume $r > 0$.

The *cycling* of x, denoted by $\mathbf{c}(x)$, is:

$$\mathbf{c}(x) = \Delta^p x_2 \cdots x_r \tau^{-p}(x_1),$$

where τ is the involution which maps σ_i to σ_{n-i}, for all $1 \le i \le n$.

The *decycling* of x, denoted by $\mathbf{d}(x)$, is:

$$\mathbf{d}(x) = x_r \Delta^p x_1 x_2 \cdots x_{r-1} = \Delta^p \tau^p(x_r) x_1 x_2 \cdots x_{r-1}.$$

If $r = 0$, we have $\mathbf{c}(x) = \mathbf{d}(x) = x$.

Note that $\mathbf{c}(x) = (\tau^{-p}(x_1))^{-1} x (\tau^{-p}(x_1))$ and $\mathbf{d}(x) = x_r x x_r^{-1}$. This means that for an element of positive canonical length, the cycling of x is computed by moving the first permutation braid of x to the end, while the decycling of x is computed by moving the last permutation braid of x to the front. Moreover, for every $x \in B_n$, $\inf(x) \le \inf(\mathbf{c}(x))$ and $\sup(x) \ge \sup(\mathbf{d}(x))$.

Note that the above decompositions of $\mathbf{c}(x)$ and $\mathbf{d}(x)$ are not, in general, Garside's normal forms. Hence, if one wants to perform iterated cyclings or decyclings, one needs to compute the left normal form of the resulting element at each iteration.

Given x, one can use cyclings and decyclings to find an element in $\mathrm{SSS}(x)$ in the following way: Suppose that we have an element $x \in B_n$ such that $\inf(x)$ is not equal to the maximal infimum in the conjugacy class of x. Then, we can increase the infimum by repeated cycling (due to [18, 44]): there exists a positive integer k_1 such that $\inf(\mathbf{c}^{k_1}(x)) > \inf(x)$. Therefore, by repeated cycling, we can conjugate x to another element $\hat{x}$ of maximal infimum. Once $\hat{x}$ is obtained, if the supremum is not minimal in the conjugacy class, we can decrease its supremum by repeated decycling, since, due to [18, 44], there exists a positive integer k_2 such that $\sup(\mathbf{d}^{k_2}(\hat{x})) < \sup(\hat{x})$. Hence, using repeated cycling and decycling a finite number of times, one obtains an element in $\mathrm{SSS}(x)$.

If we denote by m the length of Δ in Artin's generators and r is the canonical length of x, then we have (see [18, 44]):

Proposition 7.3. *A sequence of at most rm cyclings and decyclings applied to x produces a representative $\tilde{x} \in \mathrm{SSS}(x)$.*

Now, we have to explore the whole set $\mathrm{SSS}(x)$. We have the following result (see [44]):

Proposition 7.4. *Let $x \in B_n$ and $V \subset \mathrm{SSS}(x)$ be non-empty. If $V \ne \mathrm{SSS}(x)$, then there exist $y \in V$ and a permutation braid s such that $s^{-1}ys \in \mathrm{SSS}(x) \setminus V$.*

Since $\mathrm{SSS}(x)$ is a finite set, the above proposition allows us to explore all the elements in $\mathrm{SSS}(x)$. More precisely, if one knows a subset $V \subset \mathrm{SSS}(x)$ (we start with: $V = \{\tilde{x}\}$), one conjugates each element in V by all permutation braids ($n!$ elements). If one encounters a new element z with the same canonical length as $\tilde{x}$ (which is a new element in $\mathrm{SSS}(x)$), then add z to V and start again. If no new element is found, this means that $V = \mathrm{SSS}(x)$, and we are done.

One important remark is that this algorithm not only computes the set $\mathrm{SSS}(x)$, but it also provides conjugating elements joining the elements in $\mathrm{SSS}(x)$.

Now the algorithm for checking if x and y are conjugate, is performed as follows: Compute representatives $\tilde{x} \in \mathrm{SSS}(x)$ and $\tilde{y} \in \mathrm{SSS}(y)$. If $\inf(\tilde{x}) \neq \inf(\tilde{y})$ or $\sup(\tilde{x}) \neq \sup(\tilde{y})$, then x and y are not conjugate. Otherwise, start computing $\mathrm{SSS}(x)$ as described above. The elements x and y are conjugate if and only if $\tilde{y} \in \mathrm{SSS}(x)$. Note that if x and y are conjugate, an element conjugating x to y can be found by keeping track of the conjugations during the computations of $\tilde{x}$, $\tilde{y}$ and $\mathrm{SSS}(x)$. Hence, this algorithm solves the Conjugacy Decision Problem and the Conjugacy Search Problem simultaneously.

From the algorithm, we see that the computational cost of computing $\mathrm{SSS}(x)$ depends mainly on two ingredients: the size of $\mathrm{SSS}(x)$ and the number of permutation braids. In B_n, all known upper bounds for the size of $\mathrm{SSS}(x)$ are exponential in n, although it is conjectured that for fixed n, a polynomial bound in the canonical length of x exists [45].

Franco and González-Meneses [47] reduce the size of the set we have to conjugate with, by the following observation:

Proposition 7.5. *Let $x \in B_n$ and $y \in \mathrm{SSS}(x)$. For every positive braid u, there is a unique $\preceq$-minimal element $c_y(u)$ satisfying $u \preceq c_y(u)$ and $(c_y(u))^{-1}y(c_y(u)) \in \mathrm{SSS}(x)$.*

Definition 7.6. Given $x \in B_n$ and $y \in \mathrm{SSS}(x)$, we say that a permutation braid $s \neq \varepsilon$ is *minimal* for y with respect to $\mathrm{SSS}(x)$ if $s^{-1}ys \in \mathrm{SSS}(x)$, and no proper prefix of s satisfies this property.

It is easy to see that the number of minimal permutation braids for y is bounded by the number of Artin's generators.

Now, we have:

Proposition 7.7. *Let $x \in B_n$ and $V \subseteq \mathrm{SSS}(x)$ be non-empty. If $V \neq \mathrm{SSS}(x)$, then there exist $y \in V$ and a generator σ_i such that $c_y(\sigma_i)$ is a minimal permutation braid for y, and $(c_y(\sigma_i))^{-1}y(c_y(\sigma_i)) \in \mathrm{SSS}(x) \setminus V$.*

Using these propositions, the set $\mathrm{SSS}(x)$ can be computed as in [44], but instead of conjugating each element $y \in \mathrm{SSS}(x)$ by all permutation braids ($n!$ elements), it suffices to conjugate y by the minimal permutation braids $c_y(\sigma_i)$ ($1 \leq i \leq n - 1$, $n - 1$ elements).

Figure 15 (taken from [30]) summarizes the solution of the conjugacy problem using the Super Summit Set for an element b.

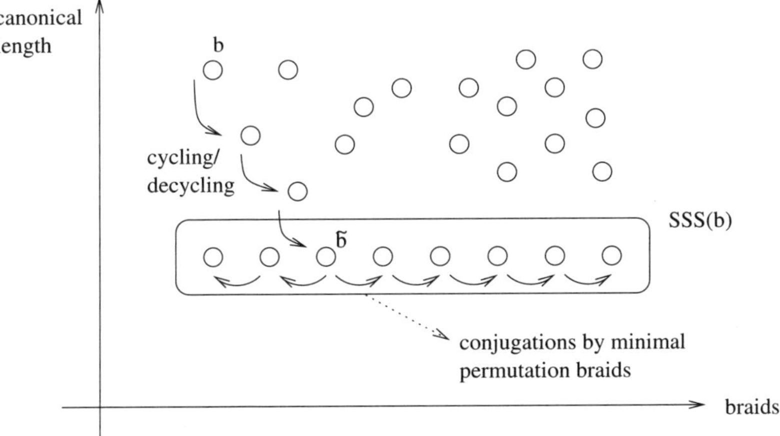

Fig. 15. Solving the conjugacy problem: going to SSS(b) and then exploring it (the points represent the conjugates of b).

Note that the algorithm computes a directed graph whose vertices are the elements in $\mathrm{SSS}(x)$, and whose arrows are defined as follows: for any two elements $y, z \in \mathrm{SSS}(x)$, there is an arrow labeled by the minimal permutation braid p_i starting at y and ending at z if $p_i^{-1} y p_i = z$.

An example for such a graph can be seen in Fig. 16, for the set $\mathrm{SSS}(\sigma_1)$ in B_4 (taken from [14, pp. 10–11]). Note that there are exactly 3 arrows starting at every vertex (the number of Artin's generators of B_4). In general, the number of arrows starting at a given vertex can be smaller or equal, but never larger than the number of generators.

Hence, the size of the set of permutation braids is no longer a problem for the complexity of the algorithm (since we can use the minimal permutation braids instead), but there is still a big problem to handle: The size of $\mathrm{SSS}(x)$ is, in general, very big. The next improvement tries to deal with this.

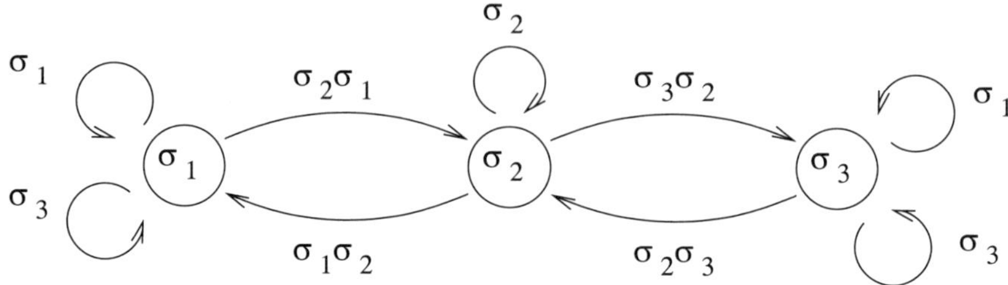

Fig. 16.   The graph of $\mathrm{SSS}(\sigma_1)$ in B_4.

7.3. *The Ultra Summit Sets*

Gebhardt [56] defines a small subset of $\mathrm{SSS}(x)$ satisfying all the good properties described above, so that a similar algorithm can be used to compute it. The definition of this new subset appears after observing that the cycling function maps $\mathrm{SSS}(x)$ to itself. As $\mathrm{SSS}(x)$ is finite, iterated cycling of any representative of $\mathrm{SSS}(x)$ must eventually become periodic. Hence it is natural to define the following:

Definition 7.8. Given $x \in B_n$, the *Ultra Summit Set* of x, $\mathrm{USS}(x)$, is the set of elements $y \in \mathrm{SSS}(x)$ such that $\mathbf{c}^m(y) = y$ for some $m > 0$.

Hence, the Ultra Summit Set $\mathrm{USS}(x)$ consists of a finite set of disjoint orbits, closed under cycling (see some schematic example in Fig. 17, taken from [30]).

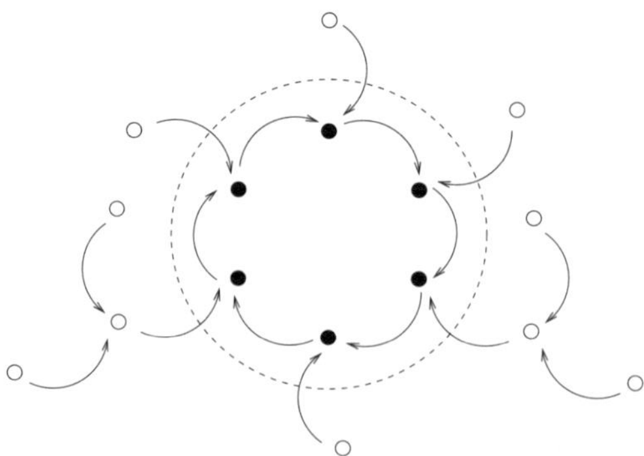

Fig. 17.   Action of cycling inside the Super Summit Set; the elements of the Ultra Summit Set are in black and perform some orbits under cycling.

Example 7.9 ([14]). One has

$$\mathrm{USS}(\sigma_1) = \mathrm{SSS}(\sigma_1) = \mathrm{SS}(\sigma_1) = \{\sigma_1, \ldots, \sigma_{n-1}\},$$

and each element corresponds to an orbit under cycling, since $\mathbf{c}(\sigma_i) = \sigma_i$ for $i = 1, \ldots, n-1$.

A more interesting example is given by the element

$$x = \sigma_1\sigma_3\sigma_2\sigma_1 \cdot \sigma_1\sigma_2 \cdot \sigma_2\sigma_1\sigma_3 \in B_4.$$

In this example, $\mathrm{USS}(x)$ has 6 elements, while $\mathrm{SSS}(x)$ has 22 elements. More precisely, $\mathrm{USS}(x)$ consists of 2 closed orbits under cycling:

$$\mathrm{USS}(x) = O_1 \cup O_2,$$

each one containing 3 elements:

$$O_1 = \left\{ \begin{array}{l} \sigma_1\sigma_3\sigma_2\sigma_1 \cdot \sigma_1\sigma_2 \cdot \sigma_2\sigma_1\sigma_3, \\ \sigma_1\sigma_2 \cdot \sigma_2\sigma_1\sigma_3 \cdot \sigma_1\sigma_3\sigma_2\sigma_1, \\ \sigma_2\sigma_1\sigma_3 \cdot \sigma_1\sigma_3\sigma_2\sigma_1 \cdot \sigma_1\sigma_2 \end{array} \right\},$$

$$O_2 = \left\{ \begin{array}{l} \sigma_3\sigma_1\sigma_2\sigma_3 \cdot \sigma_3\sigma_2 \cdot \sigma_2\sigma_3\sigma_1, \\ \sigma_3\sigma_2 \cdot \sigma_2\sigma_3\sigma_1 \cdot \sigma_3\sigma_1\sigma_2\sigma_3, \\ \sigma_2\sigma_3\sigma_1 \cdot \sigma_3\sigma_1\sigma_2\sigma_3 \cdot \sigma_3\sigma_2 \end{array} \right\}.$$

Notice that $O_2 = \tau(O_1)$.

Note also that the cycling of every element in $\mathrm{USS}(x)$ gives another element which is already in left normal form, hence iterated cycling corresponds to cyclic permutations of the factors in the left normal form. Elements that satisfy this property are called *rigid* (see [14]).

Remark 7.10. The size of the Ultra Summit Set of a *generic* braid of canonical length ℓ is either ℓ or 2ℓ [56]. This means that, in the generic case, Ultra Summit Sets consist of one or two orbits (depending on whether $\tau(O_1) = O_1$ or not), containing rigid braids. But, there are exceptions: for example, the following braid in B_{12}:

$$E = (\sigma_2\sigma_1\sigma_7\sigma_6\sigma_5\sigma_4\sigma_3\sigma_8\sigma_7\sigma_{11}\sigma_{10}) \cdot (\sigma_1\sigma_2\sigma_3\sigma_2\sigma_1\sigma_4\sigma_3\sigma_{10})$$

$$\cdot (\sigma_1\sigma_3\sigma_4\sigma_{10}) \cdot (\sigma_1\sigma_{10}) \cdot (\sigma_1\sigma_{10}\sigma_9\sigma_8\sigma_7\sigma_{11}) \cdot (\sigma_1\sigma_2\sigma_7\sigma_{11})$$

has an Ultra Summit Set of size 264, instead of the expected size 12 (see [15, Example 5.1]).

In the case of braid groups, the size and structure of the Ultra Summit Sets happen to depend very much on the geometrical properties of the braid, more precisely, on its Nielsen-Thurston type: periodic, reducible or pseudo-Anosov (see [14, 15]). A braid α is called *periodic* if there exist integers k, m such that $\alpha^k = \Delta^{2m}$. A braid α is called *reducible* if it preserves a family of curves, called a *reduction system*. A braid is called *pseudo-Anosov* if it is neither periodic nor reducible.

The algorithm given in [56] to solve the CDP/CSP in braid groups (using Ultra Summit Sets) is analogous to the previous ones, but this time one needs to compute $\text{USS}(x)$ instead of $\text{SSS}(x)$. In order to do this, we first have to obtain an element $\hat{x} \in \text{USS}(x)$. We do this as follows: take an element $\tilde{x} \in \text{SSS}(x)$. Now, start cycling it. Due to the facts that cycling an element in $\text{SSS}(x)$ will result in another element in $\text{SSS}(x)$ and that the Super Summit Set of x is finite, we will have two integers m_1, m_2 $(m_1 < m_2)$, which satisfy:

$$\mathbf{c}^{m_1}(\tilde{x}) = \mathbf{c}^{m_2}(\tilde{x}).$$

Given this, the element $\hat{x} = \mathbf{c}^{m_1}(\tilde{x})$ is in $\text{USS}(x)$, since:

$$\mathbf{c}^{m_2-m_1}(\hat{x}) = \hat{x}.$$

After finding a representative $\hat{x} \in \text{USS}(x)$, we have to explore all the set $\text{USS}(x)$. This we do using the following results of Gebhardt [56] (which are similar to the case of the Super Summit Set):

Proposition 7.11. *Let $x \in B_n$ and $y \in \text{USS}(x)$. For every positive braid u, there is a unique $\preceq$-minimal element $c_y(u)$ satisfying $u \preceq c_y(u)$ and $(c_y(u))^{-1}y(c_y(u)) \in \text{USS}(x)$.*

Definition 7.12. Given $x \in B_n$ and $y \in \text{USS}(x)$, we say that a permutation braid $s \neq \varepsilon$ is *minimal* for y with respect to $\text{USS}(x)$ if $s^{-1}ys \in \text{USS}(x)$, and no proper prefix of s satisfies this property.

It is easy to see that the number of minimal permutation braids for y is bounded by the number of Artin's generators.

Now, we have:

Proposition 7.13. *Let $x \in B_n$ and $V \subseteq \text{USS}(x)$ be non-empty. If $V \neq \text{USS}(x)$, then there exist $y \in V$ and a generator σ_i such that $c_y(\sigma_i)$ is a minimal permutation braid for y, and $(c_y(\sigma_i))^{-1}y(c_y(\sigma_i)) \in \text{USS}(x) \setminus V$.*

In [56], it is shown how to compute the minimal permutation braids (they are called there *minimal simple elements* in the Garside group's language) corresponding to a given $y \in \text{USS}(x)$ (a further discussion on the minimal simple elements with some examples can be found in [15]). Hence, one can compute the whole USS(x) starting with a single element $\hat{x} \in$ USS(x), and then we are done.

For a better characterization of the minimal permutation braids, let us introduce some notions related to a braid given in a left normal form (see [15]):

Definition 7.14. Given $x \in B_n$ whose left normal form is $x = \Delta^p x_1 \cdots x_r$ ($r > 0$), we define the *initial factor of* x as $\iota(x) = \tau^{-p}(x_1)$, and the *final factor of* x as $\varphi(x) = x_r$. If $r = 0$ we define $\iota(\Delta^p) = \varepsilon$ and $\varphi(\Delta^p) = \Delta$.

Definition 7.15. Let u, v be permutation braids such that $uv = \Delta$. The *right complement of* u, $\partial(u)$, is defined by $\partial(u) = u^{-1}\Delta = v$.

Note that a cycling of x is actually a conjugation of x by the initial factor $\iota(x)$: $\mathbf{c}(x) = \iota(x)^{-1}x\iota(x)$, and a decycling of x is actually a conjugation of x by the inverse of final factor $\varphi(x)^{-1}$: $\mathbf{d}(x) = \varphi(x)x\varphi(x)^{-1}$.

The notions of Definition 7.14 are closely related (see [14]):

Lemma 7.16. *For every $x \in B_n$, one has $\iota(x^{-1}) = \partial(\varphi(x))$ and $\varphi(x^{-1}) = \partial^{-1}(\iota(x))$.*

The following proposition from [15] characterizes the minimal permutation braids for x as prefixes of x or of x^{-1}:

Proposition 7.17. *Let $x \in$ USS(x) with $\ell(x) > 0$ and let $c_x(\sigma_i)$ be a minimal permutation braid for x. Then $c_x(\sigma_i)$ is a prefix of either $\iota(x)$ or $\iota(x^{-1})$, or both.*

As in the case of the Super Summit Set, the algorithm of Gebhardt [56] not only computes USS(x), but also a graph Γ_x, which determines the conjugating elements. This graph is defined as follows.

Definition 7.18. Given $x \in B_n$, the directed graph Γ_x is defined by the following data:

(1) The set of vertices is USS(x).
(2) For every $y \in$ USS(x) and every minimal permutation braid s for y with respect to USS(x), there is an arrow labeled by s going from y to $s^{-1}ys$.

Example 7.19. Let us give some example for the graph Γ_x. We follow Example 2.10 of [15].

Let $x = \sigma_1\sigma_2\sigma_3\sigma_2 \cdot \sigma_2\sigma_1\sigma_3 \cdot \sigma_1\sigma_3 \in B_4$. This braid x is pseudo-Anosov and rigid. A computation shows that $\mathrm{USS}(x)$ has exactly two cycling orbits, with 3 elements each, namely:

$$
x_1 = \left\{
\begin{array}{l}
x_{1,1} = \sigma_1\sigma_2\sigma_3\sigma_2 \cdot \sigma_2\sigma_1\sigma_3 \cdot \sigma_1\sigma_3, \\
x_{1,2} = \sigma_2\sigma_1\sigma_3 \cdot \sigma_1\sigma_3 \cdot \sigma_1\sigma_2\sigma_3\sigma_2, \\
x_{1,3} = \sigma_1\sigma_3 \cdot \sigma_1\sigma_2\sigma_3\sigma_2 \cdot \sigma_2\sigma_1\sigma_3
\end{array}
\right\} ,
$$

$$
x_2 = \left\{
\begin{array}{l}
x_{2,1} = \sigma_1\sigma_3\sigma_2\sigma_1 \cdot \sigma_2\sigma_1\sigma_3 \cdot \sigma_1\sigma_3, \\
x_{2,2} = \sigma_2\sigma_1\sigma_3 \cdot \sigma_1\sigma_3 \cdot \sigma_1\sigma_3\sigma_2\sigma_1, \\
x_{2,3} = \sigma_1\sigma_3 \cdot \sigma_1\sigma_3\sigma_2\sigma_1 \cdot \sigma_2\sigma_1\sigma_3
\end{array}
\right\} .
$$

The graph Γ_x of $\mathrm{USS}(x)$ is illustrated in Fig. 18. The solid arrows are conjugations by minimal permutation braids which are prefixes of the initial factors, while the dashed arrows are conjugations by minimal permutation braids which are prefixes of the initial factors of the inverse braid. Note that the definitions imply that the cycles x_1 and x_2 of $\mathrm{USS}(x)$ are connected by solid arrows.

Concerning the complexity of this algorithm for solving the Conjugacy Search Problem, the number m_2 of times one needs to apply cycling for finding an element in $\mathrm{USS}(x)$ is not known in general. Nevertheless, in practice, the algorithm based on the Ultra Summit Set is substantially better for braid groups (see [14]). For more information on the Ultra Summit Set and its structure, see [15].

Remark 7.20. One might think that for a given element $x \in B_n$, it is possible that its Ultra Summit Set with respect to Garside's normal form will be different from its Ultra Summit Set with respect to the *right normal form* (see Sec. 3.1). If this happens, it is possible that even though one of the Ultra Summit Sets is large, the other will be small.

Gebhardt and González-Meneses [57] show that at least for *rigid braids*, the size of the above two Ultra Summit Sets is equal, and their associated graphs are isomorphic (a braid w is called *rigid*, if the cycling of w, $\mathbf{c}(w)$, is already given in Garside's normal form, with no need for changing the permutation braids; see also [14, Sec. 3] and Example 7.9 here). They conjecture that this is the situation for any braid.

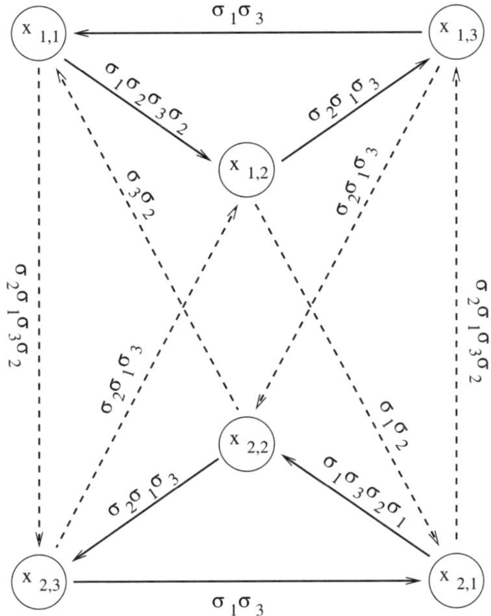

Fig. 18. The graph of $\mathrm{USS}(\sigma_1\sigma_2\sigma_3\sigma_2\sigma_2\sigma_1\sigma_3\sigma_1\sigma_3) \subset B_4$.

7.4. *Some variants of the Ultra Summit Sets*

In this section, we sketch some variants of the Super Summit Sets and the Ultra Summit Sets suggested by several authors.

7.4.1. *Reduced Super Summit Sets*

Lee, in his thesis [88] (2000), suggests a variant of the Super Summit Set, which is actually a subset of the Ultra Summit Set which was defined later (2005) by Gebhardt:

Definition 7.21. The *Reduced Super Summit Set of* x, denoted by $\mathrm{RSSS}(x)$, is:

$$\mathrm{RSSS}(x) = \{y \in C(x) | \mathbf{c}^m(y) = y \text{ and } \mathbf{d}^n(y) = y \text{ for some } m, n \geq 1\}$$

where $C(x)$ is the conjugacy class of x.

Lee's motivation to look on $\mathrm{RSSS}(x)$ comes from the following facts:

- It is still easy to find algorithmically an element in $\mathrm{RSSS}(x)$ for a given x.
- This set is invariant under cyclings and decyclings.
- This set is usually smaller than $\mathrm{SSS}(x)$.

Indeed, it is easy to see (by [44] and [58]) that:

$$\mathrm{RSSS}(x) \subseteq \mathrm{USS}(x) \subseteq \mathrm{SSS}(x).$$

Lee indicates that there is no known algorithm to generate $\mathrm{RSSS}(x)$ without generating $\mathrm{SSS}(x)$ before. Despite this, he has succeeded to compute $\mathrm{RSSS}(x)$ in polynomial time for the case of rigid braids in B_4.

7.4.2. *A general cycling operation and its induced set*

Zheng [133] suggests to generalize the idea of cyclings. He defines:

Definition 7.22. The *cycling operation of order q on x* is the conjugation $\mathbf{c}_q(x) = s^{-1}xs$, where s is the maximal common prefix of x and Δ^q. (Next section, this will be denoted as: $s = x \wedge \Delta^q$.)

The corresponding set is:

$$G_q = \{x \in B_n \mid \mathbf{c}_q^N(x) = x \text{ for some } N > 0\}.$$

The new cycling operations are indeed natural generalizations of the cycling and decycling operation:

$$\mathbf{c}(x) = \tau^{-\inf(x)}\left(\mathbf{c}_{\inf(x)+1}(x)\right), \qquad \mathbf{d}(x) = \mathbf{c}_{\sup(x)-1}(x).$$

Recall that $C(x)$ is the conjugacy class of x. For getting the Super Summit Sets and the Ultra Summit Sets in the language of G_q, we define:

$$\inf_s(x) = \max\{\inf(y) \mid y \in C(x)\}, \quad \sup_s(x) = \min\{\sup(y) \mid y \in C(x)\}.$$

Hence, we get that:

$$\mathrm{SSS}(x) = C(x) \cap \left(\bigcap_{q \in \{\inf_s(x),\sup_s(x)\}} G_q \right),$$

$$\mathrm{USS}(x) = C(x) \cap \left(\bigcap_{q \in \{\inf_s(x),\inf_s(x)+1,\sup_s(x)\}} G_q \right).$$

Zheng [133] defines a new summit set:

$$C^*(x) = C(x) \cap \left(\bigcap_{q \in \mathbb{Z}} G_q \right) = C(x) \cap \left(\bigcap_{\inf_s(x) \leq q \leq \sup_s(x)} G_q \right).$$

It is straightforward that:

$$C^*(x) \subseteq \mathrm{USS}(x) \subseteq \mathrm{SSS}(x).$$

Given an element x, computing an element $\hat{x} \in C^*(x)$ is done by applying iterated general cyclings $\mathbf{c}_q$ until getting repetitions, for $\inf(x) < q < \sup(x)$. A more complicated algorithm is presented for finding all the elements of $C^*(x)$ (see [133, Algorithm 3.8]). Having these ingredients for $C^*(x)$, we can solve the Conjugacy Search Problem based on $C^*(x)$.

Zheng [133, Sec. 6] presents some computational results, and he emphasizes that the new set $C^*(x)$ is important especially for the case of reducible braids, where there are cases that $\mathrm{USS}(x) = \mathrm{SSS}(x)$.

7.4.3. *Stable Super Summit Sets and stable Ultra Summit Sets*

The stable Super Summit Sets and stable Ultra Summit Sets were defined simultaneously by Birman, Gebhardt and González-Meneses [14] and Lee and Lee [85]:

Definition 7.23. Given $x \in B_n$, the *stable Super Summit Set of x* is defined as:

$$\mathrm{SSSS}(x) = \{y \in \mathrm{SSS}(x) \mid y^m \in \mathrm{SSS}(x^m), \forall m \in \mathbb{Z}\}.$$

The *stable Ultra Summit Set of x* is defined as:

$$\mathrm{SU}(x) = \{y \in \mathrm{USS}(x) \mid y^m \in \mathrm{USS}(x^m), \forall m \in \mathbb{Z}\}.$$

Birman, Gebhardt and González-Meneses [14, Proposition 2.23] and Lee and Lee [85, Theorem 6.1(i)] have proved that for every $x \in B_n$ the stable sets $\mathrm{SSSS}(x)$ and $\mathrm{SU}(x)$ are non-empty.

We give here an example from [85], which shows that: (i) the stable Super Summit Set is different from both the Super Summit Set and the Ultra Summit Set; (ii) one cannot obtain an element of the stable Super Summit Set by applying only cyclings and decyclings.

Example 7.24 ([85], p. 11). Consider the positive monoid B_4^+. Let

$$g_1 = \sigma_1\sigma_2\sigma_3, \quad g_2 = \sigma_3\sigma_2\sigma_1, \quad g_3 = \sigma_1\sigma_3\sigma_2, \quad g_4 = \sigma_2\sigma_1\sigma_3.$$

Note that g_i's are permutation braids and conjugate to each other.

It is easy to see that

$$\mathrm{SSS}(g_1) = \mathrm{USS}(g_1) = \{g_1, g_2, g_3, g_4\}.$$

Now, we show that the stable Super Summit Set of g_1 is different from the Super/Ultra Summit Set of g_1. The normal forms of g_i^2 are as follows:

$$g_1^2 = (\sigma_1\sigma_2\sigma_3\sigma_1\sigma_2)\sigma_3; \quad g_2^2 = (\sigma_3\sigma_2\sigma_1\sigma_3\sigma_2)\sigma_1; \quad g_3^2 = \Delta_4; \quad g_4^2 = \Delta_4.$$

Therefore, $\inf(g_1^2) = \inf(g_2^2) = 0$ and $\inf(g_3^2) = \inf(g_4^2) = 1$. Hence,

$$\mathrm{SSSS}(g_1) = \{g_3, g_4\}.$$

Note that $\mathbf{c}^k(g_i) = \mathbf{d}^k(g_i) = g_i$ for $i = 1, \ldots, 4$ and all $k > 1$. In particular, we cannot obtain an element of the stable Super Summit Set by applying only cyclings and decyclings to g_1 or g_2.

A finite-time algorithm for computing the stable Super Summit Sets (i.e. when given $x \in B_n$, first compute an element $\hat{x} \in \mathrm{SSSS}(x)$ and then compute the whole set $\mathrm{SSSS}(x)$) is given by Lee and Lee in [87, Sec. 6].

Birman, Gebhardt and González-Meneses [14] remark that their proof for the non-emptiness of the stable Ultra Summit Set (Proposition 2.23 there) actually yields an algorithm for computing that set.

Zheng [133], as a continuation of his idea of general cyclings, suggests to generalize also the stable sets. He defines:

Definition 7.25. $\mathbf{c}_{p,q}(x) = s^{-1}xs$, where s is the maximal common prefix of x^p and Δ^q (i.e. $s = x^p \wedge \Delta^q$).

The corresponding set is:

$$G_{p,q} = \{x \in B_n \mid \mathbf{c}_{p,q}^N(x) = x \text{ for some } N > 0\}.$$

Note that $\mathbf{c}_q(x^p) = (\mathbf{c}_{p,q}(x))^p$, so applying a $\mathbf{c}_q$ operation on x^p is equivalent to applying a $\mathbf{c}_{p,q}$ operation on x. In particular, $x^p \in G_q$ if and only if $x \in G_{p,q}$.

Similarly, one can define:

$$C^{[m,n],*}(x) = C(x) \cap \left(\bigcap_{m \le p \le n, q \in \mathbb{Z}} G_{p,q} \right).$$

Zheng claims that, with a suitable modification, the algorithms for computing $C^*(x)$ can be used to compute the set $C^{[m,n],*}(x)$.

An even more generalized set is:

$$C^{*,*}(x) = C(x) \cap \left(\bigcap_{p,q \in \mathbb{Z}} G_{p,q} \right),$$

but currently there is no algorithm for computing it, because he does not know how to bound the order p. Nevertheless, Zheng [133, Theorem 7.3] has proved that the set $C^{*,*}(x)$ is non-empty.

The set $C^{*,*}(x)$ is indeed a generalization of the stable sets, since:

$$\mathrm{SSSS}(x) = C(x) \cap \left(\bigcap_{p \geq 1, q \in \{\inf_s(x^p), \sup_s(x^p)\}} G_{p,q} \right),$$

$$\mathrm{SU}(x) = C(x) \cap \left(\bigcap_{p \geq 1, q \in \{\inf_s(x^p), \inf_s(x^p)+1, \sup_s(x^p)\}} G_{p,q} \right).$$

By the non-emptiness result of Zheng, we have an alternative proof that the stable sets are non-empty.

7.5. *Cyclic sliding*

The last step to-date for seeking a polynomial-time solution to the conjugacy search problem has been done by Gebhardt and González-Meneses [58, 59].

Their idea is introducing a new operation, called *cyclic sliding*, and they suggest to replace the usual cycling and decycling operations by this new one, as it is more natural from both the theoretical and computational points of view. Then, the Ultra Summit Set $\mathrm{USS}(x)$ of x, will be replaced by its analogue for cyclic sliding: the set of *sliding circuits*, $\mathrm{SC}(x)$. The sets of sliding circuits and their elements naturally satisfy all the good properties that were already shown for Ultra Summit Sets, and sometimes even better properties: For example, for elements of canonical length 1, cycling and decycling are trivial operations, but cyclic sliding is not.

One more advantage of considering the set $\mathrm{SC}(x)$ is that it yields a simpler algorithm to solve the Conjugacy Decision Problem and the Conjugacy Search Problem in the braid group. The worst case complexity of the algorithm is not better than the previously known ones [56], but it is conceptually simpler and easier to implement. The details of the implementation and the study of complexity are presented in [59].

For any two braids u, v, denote by $u \wedge v$ the largest common prefix of u and v (the notation comes from the corresponding operation on the lattice generated by the partial order $\preceq$ on the elements of B_n, see Sec. 3.1).

                                      D. Garber

The following is an interesting observation:

Observation 7.26. Given two permutation braids u and v, the decomposition $u \cdot v$ is *left-weighted* if $\partial(u) \wedge v = \varepsilon$ or, equivalently, if $uv \wedge \Delta = u$. The condition $\partial(u) \wedge v = \varepsilon$ actually means that if we move any crossing from v to u, then u will not be anymore a permutation braid.

By this observation, it is easy to give a procedure to find the left-weighted factorization of the product of two permutation braids u and v as follows. If the decomposition uv is not left-weighted, this means that there is a nontrivial prefix $s \preceq v$ such that us is still a permutation braid (i.e. $s \preceq \partial(u)$). The maximal element that satisfies this property is $s = \partial(u) \wedge v$. Therefore, for transforming the decomposition uv into a left-weighted one, we have to slide the prefix $s = \partial(u) \wedge v$ from the second factor to the first one. That is, write $v = st$ and then consider the decomposition $uv = (us)t$, with us as the first factor and t as the second one. The decomposition $us \cdot t$ is left-weighted by the maximality of s. This action will be called *local sliding* (see Fig. 19).

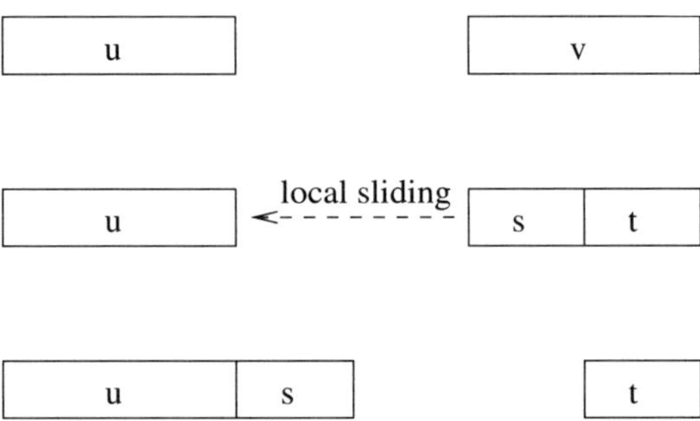

Fig. 19.   An illustration of a local sliding.

Motivated by the idea of local sliding, one wants now to do a cycling in the same manner. Given a braid in a left normal form $x = \Delta^p x_1 \cdots x_r$, we want to slide a part of x_1 to x_r. This will be done by conjugating a prefix of $\tau^{-p}(x_1)$. The appropriate prefix is: $\partial(x_r) \wedge \tau^{-p}(x_1)$, which is equal to: $\iota(x^{-1}) \wedge \iota(x)$. Hence, Gebhardt and González-Meneses [58] define:

Definition 7.27. Given $x \in B_n$, define the *cyclic sliding* $\mathfrak{s}(x)$ *of* x as the conjugate of x by $\mathfrak{p}(x) = \iota(x^{-1}) \wedge \iota(x)$, that is:

$$\mathfrak{s}(x) = \mathfrak{p}(x)^{-1} x \mathfrak{p}(x).$$

By a series of results, Gebhardt and González-Meneses (see [58, Results 3.4–3.10]) show that the cyclic sliding is indeed a generalization of cycling and decycling, and the fact that for every $x \in B_n$, iterated application of cyclic sliding eventually reaches a period, that is, there are integers $N \geq 0$ and $M > 0$ such that $\mathfrak{s}^{M+N}(x) = \mathfrak{s}^N(x)$.

Now, one can define the set of sliding circuits of x:

Definition 7.28. An element $y \in B_n$ *belongs to a sliding circuit* if $\mathfrak{s}^m(y) = y$ for some $m \geq 1$.

Given $x \in B_n$, *the set of sliding circuits of x*, denoted by $\mathrm{SC}(x)$, is the set of all conjugates of x that belong to a sliding circuit.

Note that $\mathrm{SC}(x)$ does not depend on x but only on its conjugacy class. Hence, two elements $x, y \in B_n$ are conjugate if and only if $\mathrm{SC}(x) = \mathrm{SC}(y)$. Therefore, the computation of $\mathrm{SC}(x)$ and of one element of $\mathrm{SC}(y)$ will solve the Conjugacy Decision Problem in B_n.

The set $\mathrm{SC}(x)$ is usually much smaller than $\mathrm{USS}(x)$. For example, for

$$B_{12} \ni x = \sigma_7\sigma_8\sigma_7\sigma_6\sigma_5\sigma_4\sigma_9\sigma_8\sigma_7\sigma_6\sigma_5\sigma_4\sigma_3\sigma_2\sigma_{10}\sigma_9\sigma_8\sigma_7\sigma_6\sigma_5\sigma_4\sigma_3$$

$$\cdot\, \sigma_2\sigma_1\sigma_{11}\sigma_{10}\sigma_9\sigma_8\sigma_7\sigma_6\sigma_5\sigma_4\sigma_3\sigma_2\sigma_1$$

we have that $|\mathrm{SC}(x)| = 6$, but $|\mathrm{SSS}(x)| = |\mathrm{USS}(x)| = 126498$ (see [58, Sec. 5], based on an example from [61]). On the other hand, the size of the set $\mathrm{SC}(x)$ still might be exponential in the length of x (for example, if $\delta = \sigma_{n-1}\cdots\sigma_1 \in B_n$, one has $|\mathrm{SC}(\delta)| = 2^{n-2} - 2$, see [58, Proposition 5.1]).

Gebhardt and González-Meneses have proved [58, Proposition 3.13] that:

$$\mathrm{SC}(x) = \mathrm{RSSS}(x)$$

for x satisfying $\ell_s(x) > 1$ (where $\ell_s(x) = \sup_s(x) - \inf_s(x)$, i.e. the canonical length of elements in $\mathrm{SSS}(x)$), and

$$\mathrm{SC}(x) \subseteq \mathrm{RSSS}(x)$$

for x satisfying $\ell_s(x) = 1$, and in general $\mathrm{SC}(x)$ is a *proper* subset of $\mathrm{RSSS}(x)$ in this case.

They remark that the case $\ell_s(x) = 1$ in which the sets differ is not irrelevant, since, for example, a periodic braid x which is not conjugate to a power of Δ has $\ell_s(x) = 1$, but the conjugacy problem for such braids is far from being easy [16].

As in the previous Summit Sets, the algorithm to solve the CDP/CSP in braid groups (using sliding circuits) starts by obtaining an element $\hat{x} \in \mathrm{SC}(x)$. We do this as follows: take an element x, and apply iterated cyclic sliding on it. Due to the periodic property of the sliding operation, we will have two integers m_1, m_2 ($m_1 < m_2$), which satisfy:

$$\mathfrak{s}^{m_1}(x) = \mathfrak{s}^{m_2}(x).$$

Given this, we have $\hat{x} = \mathfrak{s}^{m_1}(x) \in \mathrm{SC}(x)$, since:

$$\mathfrak{s}^{m_2 - m_1}(\hat{x}) = \hat{x}.$$

After finding a representative $\hat{x} \in \mathrm{SC}(x)$, we have to explore all the set $\mathrm{SC}(x)$. This we do in a similar way to the Ultra Summit Set case: There are $\preceq$-minimal elements that conjugate an element in $\mathrm{SC}(x)$ to another element there. The number of such possible minimal conjugators for a given element in $\mathrm{SC}(x)$ is bounded by the number of Artin's generators. Hence, one can compute the whole set $\mathrm{SC}(x)$ starting with a single element $\hat{x} \in \mathrm{SC}(x)$, and then we are done (for more information, see [58, Sec. 4.1] and [59]).

Again, as in the previous Summit Sets, the algorithm of Gebhardt and González-Meneses [58] not only computes $\mathrm{SC}(x)$, but also a graph $\mathrm{SCG}(x)$, which determines the conjugating elements. This graph is defined as follows.

Definition 7.29. Given $x \in B_n$, the directed graph $\mathrm{SCG}(x)$ is defined by the following data:

(1) The set of vertices is $\mathrm{SC}(x)$.
(2) For every $y \in \mathrm{SC}(x)$ and every minimal permutation braid s for y with respect to $\mathrm{SC}(x)$, there is an arrow labeled by s going from y to $s^{-1}ys$.

More information about these sorts of Summit sets can be found in the series of papers [14, 15, 16] and [84, 85, 87].

7.6. *An updated summary of the theoretical solution for the conjugacy search problem*

In this section, we give an updated summary for the current status of the complexity of the theoretical solution for the Conjugacy Search Problem. We follow here the nice presentation of González-Meneses in his talk at Singapore (2007) [63].

As already mentioned, according to Nielsen-Thurston geometric classification (based on [109] and [127]), there are three types of braids: periodic braids, reducible braids and pseudo-Anosov braids.

For the case of **periodic braids**, Birman, Gebhardt and González-Meneses [16] present a polynomial-time algorithm for solving the conjugacy search problem. Almost at the same time, Lee and Lee [86] suggest another entirely different solution for this case.

For the case of **reducible braids**, there is a result of Gebhardt and González-Meneses [63] that these braids fall into exactly two cases:

(1) The braid α is conjugate to a braid with a *standard* reducing curve, which means that the reducing curves are round circles, and hence the Conjugacy Search Problem can be decomposed into smaller problems (inside the tubes).

There is only one problem here: the conjugate braid (with a standard reducing curve) is in $\mathrm{USS}(\alpha)$, and for reaching it, one has to make an *unknown* number of cycling/decycling (or sliding) steps.

(2) The braid α is rigid (i.e. a cycling of Garside's normal form of α is left-weighted as written, or alternatively, it is a fixed point with respect to cyclic slidings).

For the case of **pseudo-Anosov braids**: Due to a result of Birman, Gebhardt and González-Meneses [14, Corollary 3.24], there exists a small power of a pseudo-Anosov braid which is conjugate to a rigid braid. Another result [62] claims that in the case of pseudo-Anosov braids, the conjugating elements of the pair (x, y) and the pair (x^m, y^m) coincide, and hence instead of solving the Conjugacy Search Problem in the pair (x, y), one can solve it in the pair (x^m, y^m). Therefore, one can restrict himself to the case of rigid braids.

If we summarize all cases, we get that the main challenges in this direction are:

(1) Solve the Conjugacy Search Problem for rigid braids in polynomial time.

(2) Given a braid x, find a polynomial bound for the number of cycling/decycling steps one has to perform for reaching an element in $\mathrm{USS}(x)$.

8. More Attacks on the Conjugacy Search Problem

There are some more ways to attack the Conjugacy Search Problem, apart from solving it theoretically in a complete way. In this section, we present some techniques to attack the Conjugacy Search Problem without actually solving it theoretically.

8.1. *A heuristic algorithm using the Super Summit Sets*

Hofheinz and Steinwandt [70] use a heuristic algorithm for attacking the Conjugacy Search Problem that is the basis of the cryptosystems of Anshel-Anshel-Goldfeld [4] and Ko *et al.* [79].

Their algorithm is based on the idea that it is probable that if we start with two elements in the same conjugacy class, their representatives in the Super Summit Set will not be too far away, i.e. one representative is a conjugate of the other by a permutation braid.

So, given a pair (x, x') of braids, where $x' = s^{-1}xs$, we do the following steps:

(1) By a variant of cycling (adding a multiplication by Δ to the first permutation braid, based on [89, Proposition 1]) and decycling, we find $\tilde{x} \in \mathrm{SSS}(x)$ and $\tilde{x}' \in \mathrm{SSS}(x')$.
(2) Try to find a permutation braid P, such that $\tilde{x}' = P^{-1}\tilde{x}P$.

In case we find such a permutation braid P, since we can combine it with the conjugators in the cycling/decycling process, at the end of the algorithm we will have at hand the needed conjugator for breaking the cryptosystem. Note that we do not really need to find exactly s, since each $\tilde{s}$ that satisfies $x' = \tilde{s}^{-1}x\tilde{s}$ will do the job as well and reveal the shared secret key.

Their experiments show that they succeed to reveal the shared secret key in almost 100% of the cases in the Anshel-Anshel-Goldfeld protocol (where the cryptosystem is based on the Multiple Simultaneous Conjugacy Problem) and in about 80% of the cases in the Diffie-Hellman-type protocol.

Note that their attack is special to cryptosystems that are based on the conjugacy problem, since it depends very much on the fact that x and x' are conjugate.

8.2. *Reduction of the Conjugacy Search Problem*

Maffre [94, 95] presents a deterministic, polynomial algorithm that simplifies the Conjugacy Search Problem in the braid group.

The algorithm is based on the decomposition of braids into products of canonical factors and gives a partial factorization of the secret: a divisor and a multiple. The tests that were performed on different keys of existing protocols showed that many protocols in their current form are vulnerable and that the efficiency of the attack depends on the random generator used to create the key.

8.3. *Length-based attacks*

A different probabilistic attack on the braid group cryptosystems is the *length-based attack*. In this section, we will sketch its basic idea, and different variants of this attack on the braid group cryptosystems. We finish this section with a short discussion about the applicability of the length-based attack to other groups.

8.3.1. *The basic idea*

The basic idea was introduced by Hughes and Tannenbaum [73].

Let ℓ be a length function on the braid group B_n. In the Conjugacy Search Problem, we have an instance of (p, p') where $p' = s^{-1}ps$, and we look for s. The idea of a probabilistic length-based attack to this problem is: if we can write $s = s'\sigma_i$ for a given i, then the length $\ell(\sigma_i s^{-1}ps\sigma_i^{-1})$ should be strictly smaller than the length $\ell(\sigma_j s^{-1}ps\sigma_j^{-1})$ for $j \neq i$.

Thus, for using such an attack, one should choose a good length function on B_n and run it iteratively until he gets the correct conjugator.

8.3.2. *Choosing a length function*

In [52], we suggest some length functions for this purpose. The first option is the *Garside length*, which is the length of Garside's normal form by means of Artin's generators, i.e. if $w = \Delta_n^r P_1 P_2 \cdots P_k$, then

$$\ell_{\mathrm{Gar}}(w) = r|\Delta_n| + |P_1| + |P_2| + \cdots + |P_k|.$$

A better length function is the *Reduced Garside length* (which is called *Mixed Garside length* in [45]). The motivation for this length function is that a part of the negative powers of Δ_n can be canceled with the positive permutation braids. Hence, it is defined as follows: if $w = \Delta_n^{-r} P_1 P_2 \cdots P_k$, then:

$$\ell_{\mathrm{RedGar}}(w) = \ell_{\mathrm{Gar}}(w) - 2 \sum_{i=1}^{\min\{r,k\}} |P_i|.$$

This length function is much more well-behaved, and hence it gives better performance. But even this length function did not give a break of the cryptosystems (by the basic length-based attack).

In [69], Hock and Tsaban checked the corresponding length functions for the Birman-Ko-Lee presentation, and they found that the reduced length function with respect to the Birman-Ko-Lee presentation behaves even better than the reduced Garside length function. We refer the reader to their paper [69] for a detailed discussion on length functions in Garside groups.

For another option for a length function that is based on a normal form, see [48].

8.3.3. *The memory approach*

The main contribution of [51] is new improvements to the length-based attack.

First, it introduces a new approach which uses memory: In the basic length-based attack, we hold each time only the best conjugator so far. The problem with this is that sometimes a prefix of the correct conjugator is not the best conjugator at some iteration and hence is thrown out. In such a situation, we just miss the correct conjugator, and hence the length-based algorithm fails. Moreover, even if we use a "look ahead" approach, which means that instead of adding one generator in each iteration we add several generators in each iteration, we still get total failure for the suggested parameters, and some success for small parameters [52].

In the memory approach, we hold each time a given number (which is the size of the memory) of possible conjugators which are the best among all the other conjugators of this length. In the next step, we add one more generator to all the conjugators in the memory, and we choose again only the best ones among all the possibilities. In this approach, in a successful search, we will often have the correct conjugator in the first place of the memory.

The results of [51] show that the length-based attack with memory is applicable to the cryptosystems of Anshel-Anshel-Goldfeld and Ko *et al.*, and hence their cryptosystems are not secure. Moreover, the experiments show that if we increase the size of the memory, the success rate of the length-based attack with memory becomes higher.

8.3.4. *A different variant of length-based attack by Myasnikov and Ushakov*

Recently, Myasnikov and Ushakov [107] suggest a different variant of the length-based approach.

They start by mentioning the fact that the *geodesic length*, i.e. the length of the shortest path in the corresponding Cayley graph, seems to be the best candidate for a length function in the braid group, but there is no known efficient algorithm for computing it. Moreover, it was shown by Paterson and Razborov [111] that the set of geodesic braids in B_n is co-

NP-complete. On the other hand, many other length functions are bad for the length-based attacks (like the *canonical length*, which is the number of permutation braids in Garside's normal form).

As a length function, they choose some approximation function for the geodesic length: they use Dehornoy's handles reduction and conjugations by Δ (this length function appears in [103, 104]). This length function satisfies $|a^{-1}ba| > |b|$ for almost all a and b.

Next, they identify a type of braid words, which they call *peaks*, which causes problems for length-based attacks:

Definition 8.1. Let G be a group, and let ℓ_G be a length function on G, and $H = \langle w_1, \ldots, w_k \rangle$. A word $w = w_{i_1} \cdots w_{i_n}$ is called an *n-peak in H relative to* ℓ_G if there is no $1 \le j \le n - 1$ such that

$$\ell_G(w_{i_1} \cdots w_{i_n}) \ge \ell_G(w_{i_1} \cdots w_{i_j}).$$

An example of a *commutator-type peak* is given in [107, Example 1]: if $a_1 = \sigma_{39}^{-1}\sigma_{12}\sigma_7\sigma_3^{-1}\sigma_1^{-1}\sigma_{70}\sigma_{25}\sigma_{24}^{-1}$ and $a_2 = \sigma_{42}\sigma_{56}^{-1}\sigma_8\sigma_{18}^{-1}\sigma_{19}\sigma_{73}\sigma_{33}^{-1}\sigma_{22}^{-1}$, then their commutator is a peak: $a_1^{-1}a_2^{-1}a_1a_2 = \sigma_7\sigma_8^{-1}$.

The main idea behind their new variant of the length-based attack is to add elements from the corresponding subgroup to cut the peaks. By an investigation of the types of peaks, one can see that this is done by adding to the vector of elements all the conjugators and commutators of its elements. By this means, the length-based attack will be more powerful. For more information and for an exact implementation, see [107].

8.3.5. *Applicability of the length-based approach*

One interesting point about the length-based approach is that it is applicable not only for the Conjugacy Search Problem, but also for solving equations in groups. Hence, it is a threat also to the Decomposition Problem and for the Shifted Conjugacy Problem which was introduced by Dehornoy (see [31] and Sec. 9.3 below).

Moreover, the length-based approach is applicable in any group which has a reasonable length function, e.g. the Thompson group, as indeed has been done by Ruinskiy, Shamir and Tsaban (see [115] and Sec. 11.1.2 below).

8.4. *Attacks based on linear representations*

A different way to attack these cryptographic schemes is by using linear representations of the braid groups. The basic idea is to map the braid groups into groups of matrices in which the Conjugacy Search Problem is easy. In this way, we might solve the Conjugacy Search Problem of B_n by lifting the conjugator from the group of matrices back to the braid group B_n.

For more information on the linear representations of the braid group, we refer the reader to the surveys of Birman and Brendle [13] and Paris [110].

8.4.1. *The Burau representations*

The best known linear representation of the braid group B_n is the Burau representation [21]. We present it here (we partially follow [89]).

Let $\mathbb{Z}[t^{\pm 1}]$ be the ring of Laurent polynomials

$$f(t) = a_k t^k + a_{k+1} t^{k+1} + \cdots + a_m t^m$$

with integer coefficients (and possibly with negative degree terms). Let $\mathrm{GL}_n(\mathbb{Z}[t^{\pm 1}])$ be the group of $n \times n$ invertible matrices over $\mathbb{Z}[t^{\pm 1}]$. The Burau representation is the homomorphism $B_n \to \mathrm{GL}_n(\mathbb{Z}[t^{\pm 1}])$ that sends a generator $\sigma_i \in B_n$ to the matrix:

$$\begin{pmatrix} 1 & & & & & \\ & \ddots & & & & \\ & & 1-t & t & & \\ & & 1 & 0 & & \\ & & & & \ddots & \\ & & & & & 1 \end{pmatrix} \in \mathrm{GL}_n(\mathbb{Z}[t^{\pm 1}]),$$

where $1 - t$ occurs in row and column i of the matrix.

This representation is reducible, since it can be decomposed into the trivial representation of dimension 1 and an irreducible representation $B_n \to \mathrm{GL}_{n-1}(\mathbb{Z}[t^{\pm 1}])$ of dimension $n - 1$, called *the reduced Burau representation*, which sends a generator $\sigma_i \in B_n$ to the matrix:

$$C_i(t) = \begin{pmatrix} 1 & & & & & \\ & \ddots & & & & \\ & & 1 & & & \\ & & t & -t & 1 & \\ & & & 1 & & \\ & & & & \ddots & \\ & & & & & 1 \end{pmatrix} \in \mathrm{GL}_{n-1}(\mathbb{Z}[t^{\pm 1}]),$$

where t occurs in row i of the matrix. If $i = 1$ or $i = n - 1$, the central block is truncated accordingly (see [89]).

Note that these matrices satisfy the braid group's relations:

$$C_i(t)C_j(t) = C_j(t)C_i(t) \qquad \text{for } |i - j| > 2$$
$$C_i(t)C_{i+1}(t)C_i(t) = C_{i+1}(t)C_i(t)C_{i+1}(t) \quad \text{for } i = 1, \ldots, n - 1.$$

The Burau representation of B_n is faithful for $n = 3$ and is known to be unfaithful for $n \geq 5$ (i.e. the map from B_n to the matrices is not injective) [100, 101, 90, 10]. The case of $n = 4$ remains unknown. In the case of $n \geq 5$, the kernel is very small [130], and the probability that different braids admit the same Burau image is negligible.

Here is a variant of the Burau representation introduced by Morton [102] (we follow the presentation of [89]). The *colored Burau matrix* is a refinement of the Burau matrix by sending σ_i to $C_i(t_{i+1})$, so that the entries of the resulting matrix have several variables. This naive construction does not give a group homomorphism. Thus the induced permutations are considered simultaneously. We label the strands of an n-braid by $t_1, \ldots, t_n$, putting the label t_j on the strand which starts from the jth point on the right.

Now we define:

Definition 8.2. Let $a \in B_n$ be given by a word $\sigma_{i_1}^{e_1} \cdots \sigma_{i_k}^{e_k}$, $e_j = \pm 1$. Let t_{j_r} be the label of the under-crossing strand at the rth crossing. Then *the colored Burau matrix $M_a(t_1, \ldots, t_n)$ of a is defined by*

$$M_a(t_1, \ldots, t_n) = \prod_{r=1}^{k} (C_{i_r}(t_{j_r}))^{e_r}.$$

The permutation group S_n acts on $\mathbb{Z}[t_1^{\pm 1}, \ldots, t_n^{\pm 1}]$ from the left by changing variables: for $\alpha \in S_n$, $\alpha(f(t_1, \ldots, t_n)) = f(t_{\alpha(1)}, \ldots, t_{\alpha(n)})$. Then S_n also acts on the matrix group $GL_{n-1}(\mathbb{Z}[t_1^{\pm 1}, \ldots, t_n^{\pm 1}])$ entry-wise: for $\alpha \in S_n$ and $M = (f_{ij})$, then $\alpha(M) = (\alpha(f_{ij}))$. Then we have

Definition 8.3. The *colored Burau group CB_n* is:

$$S_n \times GL_{n-1}(\mathbb{Z}[t_1^{\pm 1}, \ldots, t_n^{\pm 1}])$$

with multiplication $(\alpha_1, M_1) \cdot (\alpha_2, M_2) = (\alpha_1 \alpha_2, (\alpha_2^{-1} M_1) M_2)$. The *colored Burau representation $C : B_n \to CB_n$* is defined by

$$C(\sigma_i) = ((i, i+1), C_i(t_{i+1})).$$

It is easy to see the following facts:

(1) CB_n is a group, with identity element (e, I_{n-1}) and

$$(\alpha, M)^{-1} = (\alpha^{-1}, \alpha M^{-1}).$$

(2) The $C(\sigma_i)$ satisfy the braid relations and so $C : B_n \to CB_n$ is a group homomorphism.

(3) For $a \in B_n$, $C(a) = (\pi_a, M_a)$, where π_a is the induced permutation and M_a is the colored Burau matrix.

Using the Burau representation, the idea of Hughes [72] to attack the Anshel-Anshel-Goldfeld scheme [4, 5] is as follows: take one or several pairs of conjugate braids (p, p') associated with the same conjugating braids. Now, it is easy to compute their classical Burau image and to solve the Conjugacy Search Problem in the linear group.

In general, this is not enough for solving the Conjugacy Search Problem in B_n, because it is possible that the conjugating matrix that has been found does not belong to the image of the Burau representation, or that one cannot effectively find a possible preimage. However, since the kernel of the classical Burau representation is small [130], there is a non-negligible probability that we will find the correct conjugator and hence we break the cryptosystem.

In a different direction, Lee and Lee [89] indicate a weakness in the Anshel-Anshel-Goldfeld protocol in a different point. The shared key of the Anshel-Anshel-Goldfeld protocol is the colored Burau representation of a commutator element.

The motivation for this attack is that despite the change of variables in the colored Burau matrix by permutations, the matrix in the final output, which is the shared key, is more manageable than braids. They show that the security of the key-exchange protocol is based on the problems of listing all solutions to some Multiple Simultaneous Conjugacy Problems in a permutation group and in a matrix group over a finite field. So if both of the two listing problems are feasible, then we can guess correctly the shared key, without solving the Multiple Simultaneous Conjugacy Problem in braid groups.

Note that the Lee-Lee attack is special to this protocol, since it uses the colored Burau representation of a commutator element, instead of using the element itself. In case we change the representation in the protocol, this attack is useless.

8.4.2. *The Lawrence-Krammer representation*

The Lawrence-Krammer representation is another linear representation of B_n, which is faithful [11, 81]. It associates with every braid in B_n a matrix of order $\binom{n}{2}$ with entries in a 2-variable Laurent polynomial ring $\mathbb{Z}[t^{\pm 1}, q^{\pm 1}]$.

Cheon and Jun [24] develop an attack against the scheme of the Diffie-Hellman-type protocol based on the Lawrence-Krammer representation: as in the case of the Burau representation, it is easy to compute the images of the involved braids in the linear group and to solve the Conjugacy Problem there, but in general, there is no way to lift the solution back to the braid groups.

The task here is easier, since we only have to find a solution to the derived Diffie-Hellman-like Conjugacy Problem:

Problem 8.4. *Given p, sps^{-1} and rpr^{-1}, with $r \in LB_n$ and $s \in UB_n$, find $(rs)p(rs)^{-1}$.*

Taking advantage of the particular form of the Lawrence-Krammer matrices, which contain many 0's, Cheon and Jun obtain a solution with a polynomial complexity and they show that, for the parameters suggested by Ko *et al.* [79], the procedure is doable, and so the cryptosystem is not secure.

9. Newly Suggested Braid Group Cryptosystems, Their Cryptanalyses and Their Future Applications

In this section, we present recent updates on some problems in the braid group, by which one can construct a cryptosystem. We also discuss some newly suggested braid group cryptosystems.

9.1. *Cycling problem as a potential hard problem*

In their fundamental paper, Ko *et al.* [79] suggested some problems which can be considered as hard problems, by which one can construct a cryptosystem. One of the problems is the *Cycling Problem*:

Problem 9.1. *Given a braid y and a positive integer t such that y is in the image of the operator $\mathbf{c}^t$, find a braid x such that $\mathbf{c}^t(x) = y$.*

Maffre, in his thesis [93], shows that the Cycling Problem for $t = 1$ has a very efficient solution. That is, if y is the cycling of some braid, then one can find x such that $\mathbf{c}(x) = y$ very fast.

Following this result, Gebhardt and Gonzáles-Meneses [57] have shown that the general Cycling Problem has a polynomial solution. The reason for that is their following result: The cycling operation is surjective on the braid group [57]. Hence, one can easily find a tth preimage of y under this operation.

Note that the decycling operation and cyclic sliding operation are surjective too (the decycling operation is a composition of surjective maps: $\mathbf{d}(x) = (\tau(\mathbf{c}(x^{-1})))^{-1}$, and the cyclic sliding operation can be written as a composition of a cycling and a decycling, see [58, Lemma 3.8]). Hence, these problems cannot be considered as hard problems, on which one can base a cryptosystem [64].

It will be interesting to find new operations on the braid group whose solution can be considered as an hard problem, from which one can construct a cryptosystem.

9.2. *A cryptosystem based on the shortest braid problem*

A different type of problem consists in finding the shortest words representing a given braid (see Dehornoy [30, Sec. 4.5.2]). This problem depends on a given choice of a distinguished family of generators for B_n, e.g. Artin's generators or the band generators of Birman-Ko-Lee.

We consider this problem in B_∞ which is the group generated by an infinite sequences of generators $\{\sigma_1, \sigma_2, \dots\}$ subject to the usual braid relations.

The *Minimal Length Problem* (or *Shortest Word Problem*) is:

Problem 9.2. *Starting with a word w in the $\sigma_i^{\pm 1}$'s, find the shortest word w' that is equivalent to w, i.e. that satisfies $w' = w$.*

This problem is considered to be hard due to the following result of Paterson and Razborov [111]:

Proposition 9.3. *The Minimal Length Problem (in Artin's presentation) is co-NP-complete.*

This suggests introducing new schemes in which the secret key is a short braid word, and the public key is another longer equivalent braid word. It must be noted that the NP-hardness result holds in B_∞ only, but it is not known in B_n for fixed n.

The advantage of using an NP-complete problem lies in the possibility of proving that some instances are difficult; however, from the point of view of cryptography, the problem is not to prove that some specific instances are difficult (worst-case complexity), but rather to construct relatively large families of provably difficult instances in which the keys may be randomly chosen.

Based on some experiments, Dehornoy [30] suggests that braids of the form $w(\sigma_1^{e_1}, \sigma_2^{e_2}, \ldots, \sigma_n^{e_n})$ with $e_i = \pm 1$, i.e. braids in which, for each i, at least one of σ_i or σ_i^{-1} does not occur, could be relevant.

The possible weakness of this approach is that the shortest word problem in B_n for a fixed n is not so hard. In B_3, there are polynomial-time algorithms for the Shortest Word Problem (see [8] and [131] for the presentation by Artin's generators and [132] for the presentation by band generators). Also, this problem was solved in polynomial time in B_4 for the presentation by the band generators (see [77] and [88, Chap. 5]). For small fixed n, Wiest [131] conjectures an efficient algorithm for finding shortest representatives in B_n. Also, an unpublished work [53] indicates that a heuristic algorithm based on a random walk on the Cayley graph of the braid group might give good results in solving the Shortest Word Problem.

In any case, further research is needed here in several directions:

(1) **Cryptosystem direction:** Can one suggest a cryptosystem based on the Shortest Word Problem in B_∞, for using its hardness due to Paterson-Razborov?

(2) **Cryptanalysis direction:** What is the final status of the Shortest Word Problem in B_n for a fixed n?

(3) **Cryptanalysis direction:** What is the hardness of the Shortest Word Problem in Birman-Ko-Lee's presentation of the braid group?

9.3. *A cryptosystem based on the Shifted Conjugacy Search Problem*

Dehornoy [31] has suggested an authentication scheme that is based on the Shifted Conjugacy Search Problem.

Before we describe the scheme, let us define the Shifted Conjugacy Search Problem. Let $x, y \in B_\infty$. We define:

$$x * y = x \cdot \mathrm{d}y \cdot \sigma_1 \cdot \mathrm{d}x^{-1}$$

where $\mathrm{d}x$ is the *shift* of x in B_∞, i.e. d is the injective function on B_∞ that sends the generator σ_i to the generator σ_{i+1} for each $i \geq 1$. In this context, the Shifted Conjugacy Search Problem is:

Problem 9.4. *Let $s, p \in B_\infty$ and $p' = s * p$. Find a braid $\tilde{s}$ satisfying $p' = \tilde{s} * p$.*

Now, the suggested scheme is based on the Fiat-Shamir authentication scheme: We assume that S is a set and $(F_s)_{s \in S}$ is a family of functions of S to itself that satisfies the following condition:

$$F_r(F_s(p)) = F_{F_r(s)}(F_r(p)), \qquad r, s, p \in S.$$

Alice is the prover who wants to convince Bob that she knows the secret key s. Then the scheme works as follows:

Protocol 9.5.

Public keys: Two elements $p, p' \in S$ such that $p' = F_s(p)$.
Private key: Alice: $s \in S$.

Alice: Chooses a random $r \in S$ and sends Bob $x = F_r(p)$ and $x' = F_r(p')$.
Bob: Chooses a random bit c and sends it to Alice.
Alice: If $c = 0$, sends $y = r$ (then Bob checks: $x = F_y(p)$ and $x' = F_y(p')$);
If $c = 1$, sends $y = F_r(s)$ (then Bob checks: $x' = F_y(x)$).

Dehornoy [31] suggests to implement this scheme on Left-Distributive (LD)-systems. An *LD-system* is a set S with a binary operation that satisfies:

$$r * (s * p) = (r * s) * (r * p).$$

The Fiat-Shamir-type scheme on LD-systems works as follows:

Protocol 9.6.

Public keys: Two elements $p, p' \in S$ such that $p' = s * p$.
Private key: Alice: $s \in S$.

Alice: Chooses a random $r \in S$ and sends Bob $x = r * p$ and $x' = r * p'$.
Bob: Chooses a random bit c and sends it to Alice.
Alice: If $c = 0$, sends $y = r$ (then Bob checks: $x = y * p$ and $x' = y * p'$);
If $c = 1$, sends $y = r * s$ (then Bob checks: $x' = y * x$).

Now, one can use the shifted conjugacy operation as the $*$ operation on B_∞ in order to get a LD-system. So, in this way, one can achieve an authentication scheme on the braid group with a non-trivial operation [31].

Remark 9.7. For attacking the Shifted Conjugacy Search Problem, one cannot use the Summit Sets theory, since it is not a conjugation problem anymore. Nevertheless, one still can apply on it the length-based attack, since it is still an equation with x. Preliminary results indeed show that the length-based attack is applicable to this case, and its success rate is not negligible [49].

Longrigg and Ushakov [91] cryptanalyze the suggestion of Dehornoy, and they show that they can break the scheme (e.g. 24% success rate for keys of length 100 in B_{40}, and the success rate becomes higher as the length of the keys increases). Their idea is that in general cases they can reduce the Shifted Conjugacy Search Problem into the well-studied Conjugacy Search Problem. Based on some simple results, they construct an algorithm for solving the Shifted Conjugacy Search Problem in two steps:

(1) Find a solution $s' \in B_{n+1}$ for the equation $p'\delta_{n+1}^{-1} = s'\mathrm{d}(p)\sigma_1\delta_{n+1}^{-1}(s')^{-1}$ in B_{n+1}. This part can be done using the relevant Ultra Summit Set.
(2) Correct the element $s' \in B_{n+1}$, to obtain a solution $s \in B_n$. This can be done by finding a suitable element $c \in C_{B_{n+1}}(\mathrm{d}(p)\sigma_1\delta_{n+1}^{-1})$ (the centralizer of $\mathrm{d}(p)\sigma_1\delta_{n+1}^{-1}$ in B_{n+1}).

The algorithm for computing centralizers presented in [46] is based on computing the Super Summit Set, which is hard in general (note that actually the Super Summit Set can be replaced by the Ultra Summit Set and the Sliding Circuits set in Franco and González-Meneses' algorithm [64]). Hence, Longrigg and Ushakov use some subgroup of the centralizer which is much easier to work with.

In the last part of their paper, they discuss possibilities for hard instances for Dehornoy's scheme that will resist their attack. Their attack is based on two ingredients:

(1) The Conjugacy Search Problem is easy for the pair
$$(p'\delta_{n+1}^{-1}, \mathrm{d}(p)\sigma_1\delta_{n+1}^{-1})$$
in B_{n+1}.
(2) The centralizer $C_{B_{n+1}}(\mathrm{d}(p)\sigma_1\delta_{n+1}^{-1})$ is "small" (i.e. isomorphic to an abelian group of small rank).

Hence, if one can find keys for which one of the properties above is not satisfied, then the attack probably fails.

Kalka, Liberman and Teicher [75] show that a solution to the Multiple Simultaneous Conjugacy Problem can easily solve the Shifted Conjugacy

Problem. Moreover, they present a complete solution to the Subgroup Conjugacy Problem for B_m in B_n where $m < n$.

With respect to this scheme, it is interesting to check the following directions (see also [31]):

(1) **Cryptanalysis direction:** What is the success rate of a length-based attack on this scheme?

(2) **Cryptanalysis direction:** Can one develop a theory for the Shifted Conjugacy Search Problem that will be parallel to the Summit Sets theory?

(3) **Cryptosystem direction:** Can one suggest an LD-system on the braid group, that will be secure with respect to the length-based attack?

(4) **Cryptosystem direction:** Can one find keys for which the properties above are not satisfied, and for which Longrigg-Ushakov's attack fails?

(5) **Cryptosystem direction:** Can one suggest an LD-system on a different group, that will be secure?

9.4. *Algebraic Eraser*

Recently, Anshel, Anshel, Goldfeld and Lemieux [6] introduce a new scheme for a cryptosystem that is based on combinatorial group theory. We present here the main ideas of the scheme and the potential attacks on it.

9.4.1. *The scheme and the implementation*

We follow the presentation of [76]. Let G be a group acting on a monoid M on the left, that is, to each $g \in G$ and each $a \in M$, we associate a unique element denoted ${}^{g}a \in M$, such that:

$$ {}^{1}a = a; \qquad {}^{gh}a = {}^{g}({}^{h}a); \qquad {}^{g}(ab) = {}^{g}a \cdot {}^{g}b $$

for all $a, b \in M$ and $g, h \in G$. The set $M \times G$, with the operation $(a, g) \circ (b, h) = (a \cdot {}^{g}b, gh)$ is a monoid, which is denoted by $M \rtimes G$.

Let N be a monoid, and $\varphi : M \to N$ a homomorphism. The *Algebraic Eraser* operation is the function $\star : (N \times G) \times (M \rtimes G) \to (N \times G)$ defined by:

$$ (a, g) \star (b, h) = (a\varphi({}^{g}b), gh). $$

The function $\star$ satisfies the following identity:

$$ ((a, g) \star (b, h)) \star (c, r) = (a, g) \star ((b, h) \circ (c, r)) $$

for all $(a, g) \in N \times G$ and $(b, h), (c, r) \in M \rtimes G$.

We say that two submonoids A, B of $M \rtimes G$ are $\star$-*commuting* if

$$(\varphi(a), g) \star (b, h) = (\varphi(b), h) \star (a, g)$$

for all $(a, g) \in A$ and $(b, h) \in B$. In particular, if A, B $\star$-commute, then: $\varphi(a)\varphi(^g b) = \varphi(b)\varphi(^h a)$ for all $(a, g) \in A$ and $(b, h) \in B$.

Based on these settings, Anshel, Anshel, Goldfeld and Lemieux suggest the *Algebraic Eraser Key Agreement Scheme*. It consists of the following public information:

(1) A positive integer m.
(2) $\star$-commuting submonoids A, B of $M \rtimes G$, each given in terms of a generating set of size k.
(3) Elementwise commuting submonoids C, D of N.

Here is the protocol:

Protocol 9.8.

Alice: Chooses $c \in C$ and $(a_1, g_1), \ldots, (a_m, g_m) \in A$, and sends $(p, g) = (c, 1) \star (a_1, g_1) \star \cdots \star (a_m, g_m) \in N \times G$ (where the $\star$-multiplication is carried out from left to right) to Bob.

Bob: Chooses $d \in D$ and $(b_1, h_1), \ldots, (b_m, h_m) \in B$, and sends $(q, h) = (d, 1) \star (b_1, h_1) \star \cdots \star (b_m, h_m) \in N \times G$ to Alice.

Alice and Bob can compute the shared key:

$$(cq, h) \star (a_1, g_1) \star \cdots \star (a_m, g_m) = (dp, g) \star (b_1, h_1) \star \cdots \star (b_m, h_m).$$

For the reason why it is indeed a shared key, see [6] and [76].

Anshel, Anshel, Goldfeld and Lemieux apply their general scheme to a particular case, which they call *Colored Burau Key Agreement Protocol* (CBKAP): Fix positive integers n and r, and a prime number p. Let $G = S_n$, the symmetric group on the n symbols $\{1, \ldots, n\}$. The group $G = S_n$ acts on $GL_n(\mathbb{F}_p(t_1, \ldots, t_n))$ by permuting the variables $\{t_1, \ldots, t_n\}$ (note that in this case the monoid M is in fact a group, and hence, the semi-direct product $M \rtimes G$ also forms a group, with inversion $(a, g)^{-1} = (^{g^{-1}} a^{-1}, g^{-1})$ for all $(a, g) \in M \rtimes G$).

Let $N = GL_n(\mathbb{F}_p)$. The group $M \rtimes S_n$ is the subgroup of $GL_n(\mathbb{F}_p(t_1, \ldots, t_n)) \rtimes S_n$, generated by $(x_1, s_1), \ldots, (x_{n-1}, s_{n-1})$, where $s_i = (i, i+1)$, and $x_i = C_i(t_i)$ (see Sec. 8.4.1 above), for $i = 1, \ldots, n-1$. Recall that the colored Burau group $M \rtimes G$ is a representation of Artin's braid group B_n, determined by mapping each Artin's generator σ_i to (x_i, s_i), $i = 1, \ldots, n - 1$.

$\varphi : M \to GL_n(\mathbb{F}_p)$ is the evaluation map sending each variable t_i to a fixed element $\tau \in \mathbb{F}_p$. Let $C = D = \mathbb{F}_p(\kappa)$ be the group of matrices of the form:

$$\ell_1 \kappa^{j_1} + \cdots + \ell_r \kappa^{j_r},$$

with $\kappa \in GL_n(\mathbb{F}_p)$ of order $p^n - 1$, $\ell_1, \ldots, \ell_r \in \mathbb{F}_p$, and $j_1, \ldots, j_r \in \mathbb{Z}$.

Commuting subgroups of $M \rtimes G$ are chosen in a similar way to LB_n and UB_n in Sec. 6.2.2. This part is done by a Trusted Third Party (TTP), before the key-exchange protocol starts.

Fix $I_1, I_2 \subseteq \{1, \ldots, n-1\}$ such that for all $i \in I_1$ and $j \in I_2$, $|i - j| > 2$, and $|I_1|$ and $|I_2|$ are both $\leq n/2$. Then, define $L = \langle \sigma_i : i \in I_1 \rangle$ and $U = \langle \sigma_j : j \in I_2 \rangle$, subgroups of B_n generated by Artin's generators. From the construction of I_1 and I_2, L and U commute elementwise. Adjoin to both groups the central element Δ^2 of B_n.

Now, choose a random secret $z \in B_n$. Next, choose

$$w_1 = z w_1' z^{-1}, \ldots, w_k = z w_k' z^{-1} \in zLz^{-1}$$

and $v_1 = z v_1' z^{-1}, \ldots, v_k = z v_k' z^{-1} \in zUz^{-1}$, each a product of t generators. Transform them into Garside's normal form, and remove all even powers of Δ. Reuse the names $w_1, \ldots, w_k; v_1, \ldots, v_k$ for the resulting braids. These braids are made public.

Anshel, Anshel, Goldfeld and Lemieux have cryptanalyzed their scheme and the TTP protocol, and conclude that if the conjugating element z is known, there is a successful linear algebraic attack on CBKAP (see [6, Sec. 6]). On the other hand, if z is not known, this attack cannot be implemented. Moreover, they claim that the length-based attack is ineffective against CBKAP because w_i and v_i are not known, as well as for other reasons.

9.4.2. *The attacks*

There are several attacks on this cryptosystem. Kalka, Teicher and Tsaban [76] attack the general scheme and then show that the attack can be applied to CBKAP, the specific implementation of the scheme.

For the general scheme, they show that the secret part of the shared key can be computed (under some assumptions, which also include the assumption that the keys are chosen with standard distributions). They do it in two steps: First they compute d and $\varphi(b)$ up to a scalar, and using that they can compute the secret part of the shared key. They remark that if the keys are chosen by a distribution different from the standard, it is possible that this attack is useless (see [76, Sec. 8] for a discussion on this point).

In the next part, they show that the assumptions are indeed satisfied for the specific implementation of the scheme. The first two assumptions (that it is possible to generate an element $(\alpha, 1) \in A$ with $\alpha \neq 1$, and that N is a subgroup of $GL_n(\mathbb{F})$ for some field $\mathbb{F}$ and some n) can be easily checked. The third assumption (that given an element $g \in \langle s_1, \ldots, s_k \rangle$, where $(a_1, s_1), \ldots, (a_k, s_k) \in M \rtimes G$ are the given generators of A, then g can be explicitly expressed as a product of elements of $\{s_1^\pm, \ldots, s_k^\pm\}$), can be reformulated as the *Membership Search Problem in generic permutation groups*:

Problem 9.9. *Given random $s_1, \ldots, s_k \in S_n$ and $s \in \langle s_1, \ldots, s_k \rangle$, express s as a short (i.e. of polynomial length) product of elements from $\{s_1^\pm, \ldots, s_k^\pm\}$.*

They provide a simple and very efficient heuristic algorithm for solving this problem in generic permutation groups. The algorithm gives expressions of length $O(n^2 \log(n))$, in time $O(n^4 \log(n))$ and space $O(n^2 \log(n))$, and is the first practical one for $n \geq 256$. Hence, the third assumption is satisfied too. So the attack can be applied to the CBKAP implementation.

Myasnikov and Ushakov [108] attack the scheme of Anshel, Anshel, Goldfeld and Lemieux from a different direction. Anshel, Anshel, Goldfeld and Lemieux [6] discuss the security of their scheme and indicate that if the conjugator z generated randomly by the TTP algorithm is known, then one can attack their scheme by an efficient linear attack, which can reveal the shared key of the parties. The problem of recovering the exact z seems like a very difficult mathematical problem since it reduces to solving the following system of equations:

$$
\begin{cases}
w_1 = \Delta^{2p_1} z w_1' z^{-1} \\
\quad \vdots \\
w_k = \Delta^{2p_k} z w_k' z^{-1} \\
v_1 = \Delta^{2r_1} z v_1' z^{-1} \\
\quad \vdots \\
v_k = \Delta^{2r_k} z v_k' z^{-1}
\end{cases} ,
$$

which has too many unknowns, since only the left hand sides are known. Hence, it might be difficult to find the original z.

The attack of Myasnikov and Ushakov is a variant of the length-based attack. It is based on the observation that actually any solution z' for the system of equations above can be used in a linear attack on the scheme. Hence, they start by recovering the powers of Δ that were inserted, so one can peel the Δ^{2p} part. In the next step, they succeed in revealing the conjugator z (or any equivalent solution z').

Experimental results with instances of the TTP protocol generated using $|z| = 50$ (which is almost three times greater than the suggested value) showed 100% success rate. They indicate that the attack may fail when the length of z is large relative to the length of Δ^2 (for more details, see [108, Sec. 3.4]).

Chowdhury [28] shows that the suggested implementation of the Algebraic Eraser scheme to the braid group (the TTP protocol) is actually based on the Multiple Simultaneous Conjugacy Search Problem, and then it can be cracked. He gives some algorithms for attacking the implementation.

It will be interesting to continue the research on the Algebraic Eraser key-agreement scheme in several directions:

(1) **Cryptosystem direction:** Can one suggest a different distribution for the choice of keys, so the cryptosystem can resist the attack of Kalka-Teicher-Tsaban?

(2) **Cryptosystem direction:** Can one suggest a different implementation (different groups, etc.) for the Algebraic Eraser scheme which can resist the attack of Kalka-Teicher-Tsaban?

(3) **Cryptanalysis direction:** Can the *usual* length-based approach [51] be applied to attack the TTP protocol?

(4) **General:** One should perform a rigorous analysis of the algorithm of Kalka-Teicher-Tsaban for the Membership Search Problem in generic permutation groups (see [76, Sec. 8]).

9.5. *Cryptosystems based on the decomposition problem and the triple decomposition problem*

This section deals with two cryptosystems which are based on different variants of the decomposition problem: Given a and $b = xay \in G$, find x, y.

Shpilrain and Ushakov [120] suggest the following protocol, which is based on the decomposition problem:

Protocol 9.10.

Public key: $w \in G$.

Alice: Chooses an element $a_1 \in G$ of length ℓ, chooses a subgroup of the centralizer $C_G(a_1)$, and publicizes its generators $A = \{\alpha_1, \ldots, \alpha_k\}$.

Bob: Chooses an element $b_2 \in G$ of length ℓ, chooses a subgroup of $C_G(b_2)$, and publicizes its generators $B = \{\beta_1, \ldots, \beta_m\}$.

Alice: Chooses a random element $a_2 \in \langle B \rangle$ and sends publicly the normal form $P_A = N(a_1 w a_2)$ to Bob.

Bob: Chooses a random element $b_1 \in \langle A \rangle$ and sends publicly the normal form $P_B = N(b_1 w b_2)$ to Alice.

Shared secret key: $K_A = a_1 P_B a_2 = b_1 P_A b_2 = K_B$.

Since $a_1 b_1 = b_1 a_1$ and $a_2 b_2 = b_2 a_2$, we indeed have $K = K_A = K_B$, the shared secret key. Alice can compute K_A and Bob can compute K_B.

They suggest the following values of parameters for the protocol: $G = B_{64}$, $\ell = 1024$. For computing the centralizers, Alice and Bob should use the algorithm from [46], but actually they have to compute only some elements from them and not the whole sets.

Two key-exchange protocols which are based on a variant of the decomposition problem have been suggested by Kurt [82]. We describe here the second protocol which is an extension of the protocol of Shpilrain and Ushakov to the *triple decomposition problem*:

Problem 9.11. *Given $v = x_1^{-1} a_2 x_2$, find $x_1 \in H$, $a_2 \in A$ and $x_2 \in H'$ where $H = C_G(g_1, \ldots, g_{k_1})$, $H' = C_G(g'_1, \ldots, g'_{k_2})$, and A is a subgroup of G given by its generators.*

Here is Kurt's second protocol (his first protocol is similar): Let G be a non-commutative monoid with a large number of invertible elements.

Protocol 9.12.

Alice: Picks two invertible elements $x_1, x_2 \in G$, chooses subsets $S_{x_1} \subseteq C_G(x_1)$ and $S_{x_2} \subseteq C_G(x_2)$, and publicizes S_{x_1} and S_{x_2}.

Bob: Picks two invertible elements $y_1, y_2 \in G$, chooses subsets $S_{y_1} \subseteq C_G(y_1)$ and $S_{y_2} \subseteq C_G(y_2)$, and publicizes S_{y_1} and S_{y_2}.

Alice: Chooses random elements $a_1 \in G$, $a_2 \in S_{y_1}$ and $a_3 \in S_{y_2}$ as her private keys. She sends Bob publicly (u, v, w) where $u = a_1 x_1$, $v = x_1^{-1} a_2 x_2$, $w = x_2^{-1} a_3$.

Bob: Chooses random elements $b_1 \in S_{x_1}$, $b_2 \in S_{x_2}$ and $b_3 \in G$ as his private keys. He sends Alice publicly (p, q, r) where $p = b_1 y_1$, $q = y_1^{-1} b_2 y_2$, $r = y_2^{-1} b_3$.

Shared secret key: $K = a_1 b_1 a_2 b_2 a_3 b_3$.

Indeed, K is a shared key, since Alice can compute

$$a_1 p a_2 q a_3 r = a_1 b_1 a_2 b_2 a_3 b_3$$

and Bob can compute $u b_1 v b_2 w b_3 = a_1 b_1 a_2 b_2 a_3 b_3$.

As parameters, Kurt suggests to use $G = B_{100}$ and each secret key should be of length 300 Artin's generators.

Chowdhury [27] attacks the two protocols of Kurt, by observing that by some manipulations one can gather the secret information by solving only the Multiple Simultaneous Conjugacy Search Problem. Hence, the security of Kurt's protocols is based on the solution of the Multiple Simultaneous Conjugacy Search Problem. Since the Multiple Simultaneous Conjugacy Search Problem can be attacked by several methods, Chowdhury has actually shown that Kurt's protocols are not secure.

Although Shpilrain and Ushakov indicate that their key-exchange scheme resists length-based attack, it will be interesting to check if this indeed is the situation (see Sec. 8.3). Also, it is interesting to check whether one can change the secrets of Kurt's protocols in such a way that it cannot be revealed merely by solving the Simultaneous Conjugacy Search Problem. If such a change exists, one should check whether the new scheme resists length-based attacks.

10. Future Directions I: Alternate Distributions

In this section and the next, we discuss some more future directions of research in this and related areas. This section deals with the interesting option of changing the distribution of the generators. In this way, one can increase the security of cryptosystems that are vulnerable when assuming a standard distribution. In the next section, we deal with some suggestions of cryptosystems that are based on other non-commutative groups, apart from the braid group.

For overcoming some of the attacks, one can try to change the distribution of the generators. For example, one can require that if the generator σ_i appears, then in the next place we give more probability for the appearance of $\sigma_{i\pm1}$. In general, such a situation is called a *Markov walk*, i.e. the distribution of the choice of the next generator depends on the choice of the current chosen generator.

A work in this direction is the paper of Maffre [95]. After suggesting a deterministic polynomial algorithm that reduces the Conjugacy Search Problem in braid group (by a partial factorization of the secret), he proposes a new random generator of keys that are secure against his attack and the one of Hofheinz and Steinwandt [70].

This situation appears also in the Algebraic Eraser scheme (Sec. 9.4). The attack of Kalka, Teicher and Tsaban [76] assumes that the distribution of the generators is standard. They indicate that if the distribution is not standard, it is possible that the attack fails.

Recently, Gilman, Myasnikov, Myasnikov and Ushakov [60] announce that a proper choice of subgroups for the Anshel-Anshel-Goldfeld protocol yields a key-exchange scheme which is resistant to all its known attacks.

11. Future Directions II: Cryptosystems Based on Different Non-Commutative Groups

The protocols presented here for the braid groups can be applied to other non-commutative groups, so the natural question here is:

Problem 11.1. *Can one suggest a different non-commutative group where the existing protocols on the braid group can be applied, and the cryptosystem will be secure?*

We survey here some suggestions.

11.1. *The Thompson group*

When some of the cryptosystems on the braid groups were attacked, it was natural to look for other groups, with a hope that a similar cryptosystem on a different group will be more secure and more successful. The Thompson group is a natural candidate for such a group: there is a normal form which can be computed efficiently, but the decomposition problem seems difficult. On this base, Shpilrain and Ushakov [119] suggest a cryptosystem.

In this section, we will define the Thompson group, the Shpilrain-Ushakov cryptosystem, and we discuss its cryptanalysis.

11.1.1. *Definitions and the Shpilrain-Ushakov cryptosystem*

Thompson's group F is the infinite non-commutative group defined by the following generators and relations:

$$F = \langle x_0, x_1, x_2, \cdots \mid x_i^{-1} x_k x_i = x_{k+1} (k > i) \rangle .$$

Each $w \in F$ admits a unique *normal form* [22]:

$$w = x_{i_1} \cdots x_{i_r} x_{j_t}^{-1} \cdots x_{j_1}^{-1},$$

where $i_1 \leq \cdots \leq i_r$, $j_1 \leq \cdots \leq j_t$, and if x_i and x_i^{-1} both occur in this form, then either x_{i+1} or x_{i+1}^{-1} occurs as well. The transformation of an element of F into its normal form is very efficient [119].

We define here a natural length function on the Thompson group:

Definition 11.2. *The normal form length of an element $w \in F$, $\mathrm{LNF}(w)$, is the number of generators in its normal form: If* $w = x_{i_1} \cdots x_{i_r} x_{j_t}^{-1} \cdots x_{j_1}^{-1}$ *is in normal form, then* $\mathrm{LNF}(w) = r + t$.

Shpilrain and Ushakov [119] suggest the following key-exchange protocol based on the Thompson group:

Protocol 11.3.

> *Public subgroups:* A, B, W of F, where $ab = ba$ for all $a \in A$, $b \in B$.
> *Public key:* A braid $w \in W$.
> *Private keys:* Alice: $a_1 \in A$, $b_1 \in B$; Bob: $a_2 \in A, b_2 \in B$.
>
> *Alice:* Sends Bob $u_1 = a_1 w b_1$.
> *Bob:* Sends Alice $u_2 = b_2 w a_2$.
>
> *Shared secret key:* $K = a_1 b_2 w a_2 b_1$.

K is a shared key since Alice can compute $K = a_1 u_2 b_1$ and Bob can compute $K = b_2 u_1 a_2$, and both are equal to K since a_1, a_2 commute with b_1, b_2.

Here is a suggestion for implementing the cryptosystem [119]: Fix a natural number $s \geq 2$. Let $S_A = \{x_0 x_1^{-1}, \ldots, x_0 x_s^{-1}\}$, $S_B = \{x_{s+1}, \ldots, x_{2s}\}$ and $S_W = \{x_0, \ldots, x_{s+2}\}$. Denote by A, B and W the subgroups of F generated by S_A, S_B and S_W, respectively. A and B commute elementwise, as required.

The keys $a_1, a_2 \in A$, $b_1, b_2 \in B$ and $w \in W$ are all chosen of normal form length L, where L is a fixed integer, as follows: Let X be A, B or W. Start with the unit word, and multiply it on the right by a (uniformly) randomly selected generator, inverted with probability $\frac{1}{2}$, from the set S_X. Continue this procedure until the normal form of the word has length L.

For practical implementation of the protocol, it is suggested in [119] to use $s \in \{3, 4, \ldots, 8\}$ and $L \in \{256, 258, \ldots, 320\}$.

11.1.2. *Length-based attack*

We present some attacks on the Ushakov-Shpilrain cryptosystem.

As mentioned before, the length-based attack is applicable for any group with a reasonable length function. Ruinskiy, Shamir and Tsaban [115] applied this attack to the Thompson group.

As before, the basic length-based attack without memory always fails for the suggested parameters. If we add the memory approach, there is some improvement: for a memory of size 1024, there is 11% success. But if the memory is small (up to 64), even the memory approach always fails. They suggest that the reason for this phenomenon (in contrast to a significant success for the length-based attack with memory on the braid group) is that the braid group is much closer to the free group than the Thompson group, which is relatively close to an abelian group.

Their improvement is trying to avoid repetitions. The problem is that many elements return over and over again, and hence the algorithm goes into loops which make its way to the solution more difficult. The solution of this is holding a list of the already-checked conjugators, and when we generate a new conjugator, we check in the list if it has already appeared (this part is implemented by a hash table). In case of appearance, we just ignore it. This improvement increases significantly the success rate of the algorithm: instead of 11% for a memory of size 1024, we now have 49.8%, and instead of 0% for a memory of size 64, we now have 24%.

In the same paper [115], they suggest some more improvements for the length-based algorithm. One of their reasons for continuing with the improvements is the following interesting fact which was pointed out by Shpilrain [118]: there is a very simple fix for key-agreement protocols that are broken in probability less than p: Agree on k independent keys in parallel, and use the exclusive-or command XOR on them all to obtain the shared key. The probability of breaking the shared key is at most p^k, which is much smaller.

In a different paper, Ruinskiy, Shamir and Tsaban [123] attack the key-agreement protocols based on non-commutative groups from a different direction: by using functions that estimate the distance of a group element to a given subgroup. It is known that in general the Membership Problem is hard, but one can use some heuristic approaches for determining the distance of an element to a given subgroup, e.g. to count the number of generators that are not in the subgroup.

They test it against the Shpilrain-Ushakov protocol, which is based on Thompson's group F, and show that it can break about half the keys within a few seconds.

11.1.3. *Special attack by Matucci*

Some interesting special attacks for the Ushakov-Shpilrain cryptosystem can be found in Matucci and Kassabov [98] and Matucci [97].

11.2. *Polycyclic groups*

Eick and Kahrobaei [42] suggest to use polycyclic groups as the basis of a cryptosystem. These groups are a natural generalization of cyclic groups, but they are much more complex in their structure than cyclic groups. Hence, their algorithmic theory is more difficult, and thus it seems promising to investigate classes of polycyclic groups as candidates to have a more substantial platform, perhaps more secure.

Here is one presentation for polycyclic groups:

$$\langle a_1, \ldots, a_n \mid a_i^{-1} a_j a_i = w_{ij}, \ a_i a_j a_i^{-1} = v_{ij}, \ a_k^{r_k} = u_{kk},$$
$$\text{for } 1 \le i < j \le n, \ k \in I \rangle$$

where $I \subseteq \{1, \ldots, n\}$ and $r_i \in \mathbb{N}$ if $i \in I$ and the right hand sides w_{ij}, v_{ij}, u_{jj} of the relations are words in the generators $a_{j+1}, \ldots, a_n$. Using induction, it is straightforward to show that every element in the group defined by this presentation can be written in the form $a_1^{e_1} \cdots a_n^{e_n}$ with $e_i \in \mathbb{Z}$ and $0 \le e_i < r_i$ if $i \in I$ (see [71] and [126] for more information).

Eick and Kahrobaei introduce a Diffie-Hellman-type key-exchange which is based on the polycyclic group. As in the braid groups' case, the cryptosystem is based on the fact that the word problem can be solved effectively in polycyclic groups, while the known solutions to the conjugacy problem are far less efficient. For more information, see [42].

In a different direction, Kahrobaei and Khan [74] introduce a non-commutative key-exchange scheme which generalizes the classical El-Gamal Cipher [43] to polycyclic groups.

11.3. *Miller groups*

Mahalanobis [96] suggested some Diffie-Hellman-type key-exchange on Miller groups [99], which are groups with an abelian automorphism group.

11.4. *Grigorchuk group*

Garzon and Zalcstein [55] suggest a cryptosystem which is based on the word problem of the Grigorchuk group [67]. Both Petrides [112] and González-Vasco, Hofheinz, Martinez and Steinwandt [66] cryptanalyze this cryptosystem.

The Conjugacy Decision Problem in this group is also polynomial [92], so this problem cannot serve as a base for a cryptosystem.

11.5. *Twisted conjugacy problem in the semigroup of* 2×2 *matrices over polynomials*

Shpilrain and Ushakov [121] suggest an authentication scheme that is based on the *twisted conjugacy search problem*:

Problem 11.4. *Given a pair of endomorphisms (i.e. homomorphisms into itself) φ, ψ of a group G and a pair of elements w, $t \in G$, find an element $s \in G$ such that $t = \psi(s^{-1})w\varphi(s)$ provided at least one such s exists.*

Their suggested platform semigroup G is the semigroup of all 2×2 matrices over truncated one-variable polynomials over $\mathbb{F}_2$, the field of two elements. For more details, see their paper.

Gonzáles-Meneses and Ventura [65] show that the twisted conjugacy problem is solvable in braid groups, and hence one cannot use it as a base for a cryptosystem.

Acknowledgments

First, I wish to thank the chairmen of the PRIMA school and conference on Braids which took place at Singapore in June 2007, Jon Berrick and Fred Cohen, for giving me the opportunity to give tutorial talks and a conference talk on the fascinating topic of braid group cryptography. Also, I wish to thank the Institute for Mathematical Sciences at the National University of Singapore for hosting my stay.

Second, I wish to thank Patrick Dehornoy who has let me use his survey on braid group cryptography [30]. I have followed his presentation in many places.

I also want to thank Joan Birman, Rainer Steinwandt and Bert Wiest for useful communications. I owe special thanks to Jon Berrick, Juan González-Meneses and Boaz Tsaban who have helped me in some stages of the preparation of these lecture notes. I want to thank Ben Chacham for some useful corrections.

References

1. S. I. Adjan, *Fragments of the word Delta in a braid group*, Mat. Zam. Acad. Sci. SSSR **36-1** (1984), 25–34; translated: Math. Notes of the Acad. Sci. USSR **36-1** (1984), 505–510.

2. L. M. Adleman and J. DeMarrais, *A subexponential algorithm for discrete logarithms over all finite fields*, Proceedings of the 13th Annual International Cryptology Conference on Advances in Cryptology (1994), 147–158.

3. D. Aharonov, *Quantum computation*, Annual Review of Computational Physics, World Scientific, Volume VI (Dietrich Stauffer, ed.) (1998), 259–346.

4. I. Anshel, M. Anshel, B. Fisher and D. Goldfeld, *New key agreement protocols in braid group cryptography*, CT-RSA 2001 (San Francisco), Springer Lect. Notes in Comp. Sci. **2020** (2001), 1–15.

5. I. Anshel, M. Anshel and D. Goldfeld, *An algebraic method for public-key cryptography*, Math. Research Letters **6** (1999), 287–291.

6. I. Anshel, M. Anshel and D. Goldfeld, *Key agreement, the Algebraic Eraser^{TM}, and lightweight cryptography*, Contemp. Math. **418** (2006), 1–34.

7. E. Artin, *Theory of braids*, Ann. Math. **48** (1947), 101–126.

8. M. A. Berger, *Minimum crossing numbers for 3-braids*, J. Phys. A: Math. Gen. **27** (1994), 6205–6213.

9. D. Bessis, F. Digne and J. Michel, *Springer theory in braid groups and the Birman-Ko-Lee monoid*, Pacific J. Math. **205** (2002), 287–310.

10. S. Bigelow, *The Burau representation is not faithful for $n = 5$*, Geometry and Topology **3** (1999), 397–404.

11. S. Bigelow, *Braid groups are linear*, J. Amer. Math. Soc. **14** (2001), 471–486.

12. J. Birman, *Braids, Links, and Mapping Class Groups*, Annals of Math. Studies **82**, Princeton Univ. Press (1975).

13. J. S. Birman and T. E. Brendle, *Braids: A survey*, in: Handbook of Knot Theory, Elsevier, B. V., Amsterdam (2005), pp. 19–103.

14. J. S. Birman, V. Gebhardt and J. González-Meneses, *Conjugacy in Garside Groups I: Cyclings, powers, and rigidity*, Groups, Geometry and Dynamics **1**(3) (2007), 221–279.

15. J. S. Birman, V. Gebhardt and J. González-Meneses, *Conjugacy in Garside Groups II: Structure of the Ultra Summit Set*, Groups, Geometry and Dynamics **2**(1) (2008), 16–31.

16. J. S. Birman, V. Gebhardt and J. González-Meneses, *Conjugacy in Garside Groups III: Periodic braids*, J. Algebra **316**(2) (2007), 746–776.

17. J. Birman, K. Ko and S. Lee, *A new approach to the word problem in the braid groups*, Adv. Math. **139** (1998), 322–353.

18. J. Birman, K. Ko and S. Lee, *The infimum, supremum, and geodesic length of a braid conjugacy class*, Adv. Math. **164** (2001), 41–56.

19. T. Brady, *A partial order on the symmetric group and new $K(\pi, 1)$'s for the braid groups*, Adv. Math. **161** (2001), 20–40.

20. X. Bressaud, *A normal form for braids*, Journal of Knot Theory and its Ramifications **17**(6) (2008), 697–732.

21. W. Burau, *Über Zopfgruppen und gleichsinnig verdrilte Verkettungen*, Abh. Math. Sem. Hanischen Univ. **11** (1936), 171–178.

22. J. W. Cannon, W. J. Floyd and W. R. Parry, *Introductory notes to Richard Thompson's groups*, L'Enseignement Mathematique **42** (1996), 215–256.

23. J. Cha, J. Cheon, J. Han, K. Ko and S. Lee, *An efficient implementation of braid groups*, AsiaCrypt 2001, Springer Lect. Notes in Comp. Sci. **2048** (2001), 144–156.

24. J. Cheon and B. Jun, *A polynomial time algorithm for the braid Diffie-Hellman conjugacy problem*, CRYPTO 2003, Springer Lect. Notes in Comp. Sci. **2729** (2003), 212–225.

25. M. Cho, D. Choi, K. Ko and J. Lee, *New signature scheme using conjugacy problem*, preprint (2002).
(online: http://eprint.iacr.org/2002/168/).

26. W.-L. Chow, *On the algebraical braid group*, Ann. Math. **49** (1948), 654–658.

27. M. M. Chowdhury, *On the security of new key exchange protocols based on the triple decomposition problem*, preprint (2007).
(online: http://www.arxiv.org/abs/cs.CR/0611065).

28. M. M. Chowdhury, *On the AAGL Protocol*, preprint (2007).
(online: http://www.arxiv.org/abs/cs.CR/0708.2397).

29. P. Dehornoy, *A fast method for comparing braids*, Adv. Math. **125** (1997), 200–235.

30. P. Dehornoy, *Braid-based cryptography*, Contemp. Math. **360** (2004), 5–33.

31. P. Dehornoy, *Using shifted conjugacy in braid-based cryptography*, Contemp. Math. **418** (2006), 65–73.

32. P. Dehornoy, *Alternating normal forms in braid monoids and locally Garside monoids*, J. Pure Appl. Alg. **212**(11) (2008), 2416–2439.

33. P. Dehornoy, *Efficient solutions to the braid isotopy problem*, Disc. Appl. Math. **156** (2008), 3094–3112.

34. P. Dehornoy, *Convergence of handle reduction of braids*, preprint.
(online: http://www.math.unicaen.fr/~dehornoy/Surveys/Dhn.pdf).

35. P. Dehornoy, I. Dynnikov, D. Rolfsen and B. Wiest, *Why are braids orderable?*, Panoramas & Synthèses, Vol. **14**, Soc. Math. France (2002).

36. P. Dehornoy, M. Girault and H. Sibert, *Entity authentication schemes using braid word reduction*, Proc. Internat. Workshop on Coding and Cryptography, 153–164, Versailles (2003).

37. P. Dehornoy and L. Paris, *Gaussian groups and Garside groups, two generalizations of Artin groups*, Proc. London Math. Soc. **79**(3) (1999), 569–604.

38. P. Deligne, *Les immeubles des groupes de tresses généralisés*, Invent. Math. **17** (1972), 273–302.

39. W. Diffie and M. Hellman, *New directions in cryptography*, IEEE Trans. on Inf. Theory **22** (1976), 644–654.

40. I. Dynnikov, *On a Yang-Baxter mapping and the Dehornoy ordering*, Uspekhi Mat. Nauk **57**(3) (2002), 151–152; English translation: Russian Math. Surveys, 57-3 (2002).

41. I. Dynnikov and B. Wiest, *On the complexity of braids*, preprint.
(online: http://hal.archives-ouvertes.fr/hal-00001267/en/).

42. B. Eick and D. Kahrobaei, *Polycyclic groups: A new platform for cryptology?*, preprint (2004).
(online: http://www.arxiv.org/abs/math.GR/0411077).

43. T. El-Gamal, *A public-key cryptosystem and a signature scheme based on discrete logarithms*, IEEE Transactions on Information Theory **IT-31**(4) (1985), 469–472.

44. E. A. El-Rifai and H. R. Morton, *Algorithms for positive braids*, Quart. J. Math. Oxford **45**(2) (1994), 479–497.

45. D. Epstein, J. Cannon, D. Holt, S. Levy, M. Paterson and W. Thurston, *Word Processing in Groups*, Jones & Bartlett Publ. (1992).

46. N. Franco and J. González-Meneses, *Computation of centralizers in Braid groups and Garside groups*, Rev. Mat. Iberoamericana **19**(2) (2003), 367–384.

47. N. Franco and J. Gonzáles-Meneses, *Conjugacy problem for braid groups and Garside groups*, J. Algebra **266** (2003), 112–132.

48. J. Fromentin, *Every braid admits a short sigma-definite representative*, preprint (2008).
(online: http://www.arxiv.org/abs/math.GR/0811.3902).

49. D. Garber and D. H. Gootvilig, *Length-based attack on the shifted conjugacy search problem*, in preparation. (slides online: http://www.algebraforum.org/conference/slides/Length_based_shifted_2009.pdf).

50. D. Garber, S. Kaplan and M. Teicher, *A new algorithm for solving the word problem in braid groups*, Adv. Math. **167**(1) (2002), 142–159.

51. D. Garber, S. Kaplan, M. Teicher, B. Tsaban and U. Vishne, *Probabilistic solutions of equations in the braid group*, Adv. Appl. Math. **35** (2005), 323–334.

52. D. Garber, S. Kaplan, M. Teicher, B. Tsaban and U. Vishne, *Length-based conjugacy search in the braid group*, Contemp. Math. **418** (2006), 75–87.

53. D. Garber, S. Kaplan and B. Tsaban, *A heuristic approach to the shortest word problem*, unpublished.

54. F. A. Garside, *The braid group and other groups*, Quart. J. Math. Oxford **20-78** (1969), 235–254.

55. M. Garzon and Y. Zalcstein, *The complexity of Grigorchuk groups with application to cryptography*, Theoret. Comput. Sci. **88** (1991), 83–98.

56. V. Gebhardt, *A new approach to the conjugacy problem in Garside groups*, J. Algebra **292**(1) (2005), 282–302.

57. V. Gebhardt and J. Gonzáles-Meneses, *On the cycling operation in braid groups*, Disc. Appl. Math., to appear.
(online: http://www.arxiv.org/abs/math.GT/0704.2600).

58. V. Gebhardt and J. Gonzáles-Meneses, *The cyclic sliding operation in Garside groups*, preprint (2008).
(online: http://www.arxiv.org/abs/math.GR/0808.1430).

59. V. Gebhardt and J. González-Meneses, *Solving the conjugacy problem in Garside groups by cyclic sliding*, preprint (2008).
(online: http://www.arxiv.org/abs/math.GR/0809.0948).

60. R. Gilman, A. D. Myasnikov, A. G. Myasnikov and A. Ushakov, *New developments in commutator key exchange*, Proc. First Int. Conf. Symb. Comp. and Crypto., Beijing (2008).
(online: http://www.math.stevens.edu/~rgilman/rhg/china.pdf).

61. P. Gonzalez-Manchon, *There exist conjugate simple braids whose associated permutations are not strongly conjugate*, Math. Proc. Cambridge Phil. Soc. **143** (2007), 663–667.

62. J. González-Meneses, *The n-th root of a braid is unique up to conjugacy*, Alg. Geom. Topo. **3** (2003), 1103–1118.

63. J. González-Meneses, *Towards a polynomial solution to the conjugacy problem in braids groups*, Talk's slides, Singapore, June 2007.
(online: http://www.ims.nus.edu.sg/Programs/braids/files/juan_conf.pdf).

64. J. González-Meneses, Private communications (2008).

65. J. González-Meneses and E. Ventura, *Twisted conjugacy in braid groups*, Talk's slides, Paris, September 2008.
(online: http://www.tresses08.institut.math.jussieu.fr/talks/
GonzalezMeneses/2008_Paris_02.pdf).

66. M. I. González-Vasco, D. Hofheinz, C. Martinez and R. Steinwandt, *On the security of two public key cryptosystems using non-abelian groups*, Designs, Codes and Cryptography **32** (2004), 207–216.

67. R. I. Grigorchuk, *Burnside's problem on periodic groups*, Funct. Anal. Appl. **14** (1980), 41–43.

68. A. Groch, D. Hofheinz and R. Steinwandt, *A practical attack on the root problem in braid groups*, Contemp. Math. **418** (2006), 121–131.

69. M. Hock and B. Tsaban, *Solving random equations in Garside groups using length functions*, submitted (2008).
(online: http://www.arxiv.org/abs/math/0611918).

70. D. Hofheinz and R. Steinwandt, *A practical attack on some braid group based cryptographic primitives*, PKC 2003; Springer Lect. Notes in Comp. Sci. **2567** (2002), 187–198.

71. D. F. Holt, B. Eick and E. A. O'Brien, *Handbook of computational group theory*, Chapman & Hall/CRC (2005).

72. J. Hughes, *A linear algebraic attack on the AAFG1 braid group cryptosystem*, ACISP 2002; Springer Lect. Notes in Comp. Sci. **2384** (2002), 176–189.

73. J. Hughes and A. Tannenbaum, *Length-based attacks for certain group based encryption rewriting systems*, Inst. for Math. and its Applic. (2000).
(online: http://www.ima.umn.edu/preprints/apr2000/1696.pdf).

74. D. Kahrobaei and B. Khan, *A non-commutative generalization of El-Gamal key exchange using polycyclic groups*, Proc. IEEE (2006), 1–5.

75. A. Kalka, E. Liberman and M. Teicher, *A note on the shifted conjugacy problem in braid groups*, Groups, Complexity and Cryptology, to appear.

76. A. Kalka, M. Teicher and B. Tsaban, *Cryptanalysis of the Algebraic Eraser and short expressions of permutations as products*, submitted (2008).
(online: http://arxiv.org/abs/math.GR/0804.0629).

77. E. S. Kang, K. H. Ko and S. J. Lee, *Band-generator presentation for the 4-braids*, Topology Appl. **78** (1997), 39–60.

78. C. Kassel and V. Turaev, *Braid Groups*, Springer (2007).

79. K. H. Ko, S. J. Lee, J. H. Cheon, J. W. Han, J. S. Kang and C. Park, *New public-key cryptosystem using braid groups*; Crypto 2000; Springer Lect. Notes in Comp. Sci. **1880** (2000), 166–184.

80. N. Koblitz and A. Menezes, *A survey of public-key cryptosystems*, SIAM Review **46** (2004), 599–634.

81. D. Krammer, *Braid groups are linear*, Ann. Math. **151** (2002), 131–156.

82. Y. Kurt, *A new key exchange primitive based on the triple decomposition problem*, preprint (2006).
(online: http://eprint.iacr.org/2006/378).

83. S. Lal and A. Chaturvedi, *Authentication schemes using braid groups*, preprint (2005).
(online: http://arxiv.org/pdf/cs.CR/0507066).

84. E. K. Lee and S. J. Lee, *Translation numbers in a Garside group are rational with uniformly bounded denominators*, J. Pure Appl. Alg. **211**(3) (2007), 732–743.

85. E. K. Lee and S. J. Lee, *Abelian subgroups of Garside groups*, Comm. Alg. **36**(3) (2008), 1121–1139.

86. E. K. Lee and S. J. Lee, *Conjugacy classes of periodic braids*, preprint (2007).
(online: http://www.arxiv.org/abs/math.GT/0702349).

87. E. K. Lee and S. J. Lee, *Some power of an element in a Garside group is conjugate to a periodically geodesic element*, Bull. London Math. Soc. **40** (2008), 593–603.

88. S. J. Lee, *Algorithmic solutions to decision problems in the braid groups*, Ph.D. thesis, Korea Advanced Institute of Science and Technology, 2000.

89. S. J. Lee and E. K. Lee, *Potential weaknesses of the commutator key agreement protocol based on braid groups*, Springer Lect. Notes in Comp. Sci. **2332** (2002), 14–28.

90. D. D. Long and M. Paton, *The Burau representation is not faithful for $n \geq 6$*, Topology **32**(2) (1993), 439–447.

91. J. Longrigg and A. Ushakov, *Cryptanalysis of shifted conjugacy authentication protocol*, preprint (2007).
(online: http://www.arxiv.org/abs/math.GR/0708.1768).

92. I. Lysenok, A. Myasnikov and A. Ushakov, *The conjugacy problem in the Grigorchuk group is polynomial time decidable*, preprint (2008).
(online: http://www.arxiv.org/abs/math.GR/0808.2502).

93. S. Maffre, *Conjugaison et cyclage dans les groupes de Garside, applications cryptographiques*, Ph.D. Lab. LACO, 2005.
(online: http://www.unilim.fr/theses/2005/sciences/
2005limo0028/maffre_s.pdf).

94. S. Maffre, *Reduction of conjugacy problem in braid groups, using two Garside structures*, WCC (2005), 214–224.

95. S. Maffre, *A weak key test for braid-based cryptography*, Designs, Codes and Cryptography **39** (2006), 347–373.

96. A. Mahalanobis, *Diffie-Hellman key exchange protocol and non-abelian nilpotent groups*, Israel J. Math. **165** (2008), 161–187.

97. F. Matucci, *The Shpilrain-Ushakov protocol for Thompson's Group F is always breakable*, J. Crypto. **21**(3) (2008), 458–468.

98. F. Matucci and M. Kassabov, *The simultaneous conjugacy problem in Thompson's group F*, preprint (2006).
(online: https://www.arxiv.org/math/0607167).

99. G. A. Miller, *A non-abelian group whose group of isomorphisms is abelian*, Messenger Math. **43** (1913), 124–125.

100. J. A. Moody, *The Burau representation of the braid group is unfaithful for large n*, Bull. Amer. Math. Soc. New Ser. **25**(2) (1991), 379–384.

101. J. Moody, *The faithfulness question for the Burau representation*, Proc. Amer. Math. Soc. **119**(2) (1993), 671–679.

102. H. R. Morton, *The multivariable Alexander polynomial for a closed braid*, Low dimensional topology (Funchal, 1998), 167–172, Contemp. Math. **233**, Amer. Math. Soc., Providence, RI (1999).

103. A. G. Myasnikov, V. Shpilrain and A. Ushakov, *A practical attack on some braid group based cryptographic protocols*, Crypto 2005, Springer Lect. Notes in Comp. Sci. **3621** (2005), 86–96.

104. A. G. Myasnikov, V. Shpilrain and A. Ushakov, *Random subgroups of braid groups: An approach to cryptanalysis of a braid group based cryptographic protocol*, in: PKC 2006, Springer Lect. Notes in Comp. Sci. **3958** (2006), 302–314.

105. A. G. Myasnikov, V. Shpilrain and A. Ushakov, *Group-based cryptography*, Birkhauser (2008).

106. A. D. Myasnikov, *Generic case complexity and one-way functions*, Groups, Complexity and Cryptology **1** (2009), 13–31.

107. A. D. Myasnikov and A. Ushakov, *Length based attack and braid groups: Cryptanalysis of Anshel-Anshel-Goldfeld key exchange protocol*, in: Public Key Cryptography (PKC, Beijing, 2007), T. Okamoto *et al.* (ed.), Springer Lect. Notes in Comp. Sci. **4450** (2007), 76–88.

108. A. D. Myasnikov and A. Ushakov, *Cryptanalysis of the Anshel-Anshel-Goldfeld-Lemieux key agreement protocol*, Groups, Complexity and Cryptology **1** (2009), 63–75.

109. J. Nielsen, *Collected Mathematical Papers of Jakob Nielsen*, edited by V. Hansen, Birkhauser (1986), [N-18], [N-20] and [N-21].

110. L. Paris, *Braid groups and Artin groups*, in: Handbook of Teichmuller Theory (A. Papadopoulus, ed.), Vol. 2, EMS Publishing House, Zurich (2008).

111. M. S. Paterson and A. A. Razborov, *The set of minimal braids is co-NP-complete*, J. Algorithms **12**(3) (1991), 393–408.

112. G. Petrides, *Cryptanalysis of the public key cryptosystem based on the word problem on the Grigorchuk groups*, 9th IMA International Conference on Cryptography and Coding, Springer Lect. Notes in Comp. Sci. **2898** (2003), 234–244.

113. R. L. Rivest, A. Shamir and L. Adleman, *On digital signatures and public key cryptosystems*, Commun. Ass. Comp. Mach. **21** (1978), 120–126.

114. D. Rolfsen, *Minicourse on the braid groups*, preprint (2006). (online: http://www.math.ubc.ca/~rolfsen/reprints.html); See also a chapter in this volume.

115. D. Ruinskiy, A. Shamir and B. Tsaban, *Length-based cryptanalysis: The case of Thompson's group*, J. Math. Crypt. **1** (2007), 359–372.

116. D. Ruinskiy, A. Shamir and B. Tsaban, *Cryptanalysis of group-based key agreement protocols using subgroup distance functions*, PKC07, Springer Lect. Notes in Comp. Sci. **4450** (2007), 61–75.

117. P. W. Shor, *Polynomial-time algorithm for prime factorization and discrete logarithms on a quantum computer*, SIAM J. Comp. **26**(5) (1997), 1484–1509.

118. V. Shpilrain, *Assessing security of some group based cryptosystems*, Contemp. Math. **360** (2004), 167–177.

119. V. Shpilrain and A. Ushakov, *Thompson's group and public key cryptography*, Springer Lect. Notes in Comp. Sci. **3531** (2005), 151–164.

120. V. Shpilrain and A. Ushakov, *A new key exchange protocol based on the decomposition problem*, Contemp. Math. **418** (2006), 161–167.

121. V. Shpilrain and A. Ushakov, *An authentication scheme based on the twisted conjugacy problem*, ACNS 2008, Springer Lect. Notes in Comp. Sci. **5037** (2008), 366–372.

122. V. Shpilrain and G. Zapata, *Combinatorial group theory and public key cryptography*, Applicable Algebra in Engineering, Communication and Computing **17** (2006), 291–302.

123. V. Shpilrain and G. Zapata, *Using the subgroup membership search problem in public key cryptography*, Contemp. Math. **418** (2006), 169–179.

124. H. Sibert, P. Dehornoy and M. Girault, *Entity authentication schemes using braid word reduction*, Discrete Appl. Math. **154**(2) (2006), 420–436.

125. V. M. Sidelnikov, M. A. Cherepnev and V. Y. Yashcenko, *Systems of open distribution of keys on the basis of noncommutative semigroups*, Russ. Acad. Nauk Dokl. **332-5** (1993); English translation: Russian Acad. Sci. Dokl. Math. **48-2** (1994), 384–386.

126. C. C. Sims, *Computation with finitely presented groups*, Encyclopedia of Mathematics and its Applications, **48**, Cambridge University Press (1994).

127. W. Thurston, *On the topology and geometry of diffeomorphisms of surfaces*, Bull. Amer. Math. Soc. **19** (1988), 109–140.

128. W. Thurston, *Finite state algorithms for the braid group*, Circulated notes (1988), 23 pages. (see http://www.archivum.info/sci.math/2005-11/msg06699.html for more details on these notes).

129. B. Tsaban, *On an authentication scheme based on the root problem in the braid group*, preprint (2005). (online: http://arxiv.org/ps/cs.CR/0509059).

130. V. Turaev, *Faithful linear representations of the braid groups*, Séminaire Bourbaki, Vol. 1999/2000. Astérisque No. **276** (2002), 389–409.

131. B. Wiest, *An algorithm for the word problem in braid groups*, preprint (2002). (online: http://arxiv.org/abs/math.GT/0211169).

132. P. Xu, *The genus of closed 3-braids*, Journal of Knot Theory and its Ramifications, **1**(3) (1992), 303–326.

133. H. Zheng, *General cycling operations in Garside groups*, preprint (2006). (online: http://arxiv.org/abs/math.GT/0605741).